NO SHADOW OF A DOUBT

No Shadow of a Doubt

The 1919 Eclipse That Confirmed Einstein's Theory of Relativity

Daniel Kennefick

PRINCETON UNIVERSITY PRESS

PRINCETON AND OXFORD

Copyright © 2019 by Princeton University Press

Published by Princeton University Press
41 William Street, Princeton, New Jersey 08540
6 Oxford Street, Woodstock, Oxfordshire OX20 1TR
press.princeton.edu

All Rights Reserved

LCCN 2019930002

ISBN 978-0-691-18386-2

British Library Cataloging-in-Publication Data is available

Editorial: Eric Crahan, Kristin Zodrow, and Pamela Weidman
Jacket Image: 1919 solar eclipse © Royal Astronomical Society
Production: Erin Suydam
Publicity: Sara Henning-Stout (US), Katie Lewis (UK)
Copyeditor: Wendy Lawrence

This book has been composed in Adobe Text Pro and Gotham

Printed on acid-free paper ∞

Printed in the United States of America

10 9 8 7 6 5 4 3 2 1

To my wife, Julia Kennefick

CONTENTS

NO SHADOW OF A DOUBT

Prologue

MAY 29, 1919

The night of May 29, 1919, was a busy one for Arthur Stanley Eddington. Busy nights are part of an astronomer's routine, and in 1919 such nights were often followed by days in the developing room, preparing photographic plates of the night sky. That particular May, the process was reversed for Eddington. The plate developing took place at night. Eddington, on the tropical island of Principe, hoped for cooler temperatures after sunset. Principe lies just off the western coast of Equatorial Africa, far from Eddington's usual haunts as the director of the Cambridge Observatory in England. His nights were free from observing duties because the plates had been exposed during the daytime. Eddington was not a solar astronomer used to taking images of the Sun. The images he had taken on the afternoon of May 29 were of a field of stars. Normally, such images cannot be taken during daylight hours because sunlight scattered across the sky overwhelms the relatively puny light of the stars. But at 2:13 p.m. (Greenwich time) that day, there had been an eclipse of the Sun, visible in totality on Principe. To observe stars near the Sun during totality, Eddington had traveled all the way from England.

A decade previously, Albert Einstein had predicted that the presence of the Sun close to the stars would alter their positions. To test this prediction, Eddington had made this arduous journey. Perhaps the day after the eclipse, and certainly as soon as he could, he began to study his first good plate, the only one he had developed with enough stars visible to make his measurements. He used an eyepiece with an attached micrometer screw to measure the positions on the plate of the five visible stars. These stars were well-known members of the Hyades star cluster, whose positions in the sky were also known. Eddington was trying to determine if these positions had changed slightly during the eclipse. He had brought with him plates, taken before his departure, showing the same stars by night in England. These comparison plates permitted him to measure the normal positions of these stars against those he had imaged during the few minutes of the day when the totality made them visible. He knew that other scientists were waiting to hear how Einstein's theory had fared. He quickly sent a telegram back to his collaborator, England's Astronomer Royal, Sir Frank Watson Dyson. It read, "Through cloud, hopeful." Dyson and Eddington had together planned the expedition amid great difficulty during the final months of World War I. Eddington had entertained hopes of sending his results home within days, but instead the difficult process of measurement took months. He was handicapped by the small number of stars on his plates, the result of clouds obscuring the area around the Sun during the eclipse. Only the brightest stars had been visible on Principe and only at the very end of totality. Fortunately, a team from Dyson's own observatory at Greenwich, near London, had gone to Brazil to observe the eclipse and had better luck with the weather. But they encountered technical difficulties that complicated their data analysis, conducted at Greenwich under Dyson's direction after the expedition's return from Brazil.

Einstein's theory of general relativity predicts that light is affected by gravity and that starlight passing close to the Sun falls toward it, no longer following a straight path through space. This deviation causes an apparent shift in the position of the stars, as seen from Earth. It was this shift that the expedition sought to measure, in spite of its tiny size. As Eddington put it, the purpose of the expedition was to weigh light. He and Dyson hoped to discover if light has mass and is therefore affected by the force of gravity. Most physicists of the day thought that light, as a wave phenomenon, was massless and not subject to gravity's pull. Today, we have no doubts that Einstein was right. Yet, looking back, today's scientists and historians are often very critical of Eddington and Dyson's work. Astronomers and physicists of the last few decades who have tried to improve on the measurements made by the English team of 1919 have found the experiment very difficult, even with more modern equipment. Members of an expedition from the University of Texas, who also went to Africa during an eclipse in 1973, expressed doubts that Eddington and Dyson could have achieved the accuracy they claimed with the instruments available in 1919. Then historians and philosophers of science weighed in, claiming that Eddington was biased in favor of Einstein's theory and questioning certain decisions taken during the difficult data analysis process. Thus, measurements conducted a century ago are still causing controversy today.

Faced with such controversy, it is important to look closely to see what really happened. This means looking at all the people involved, beginning with Dyson and Eddington, who are introduced in chapter 1. It means paying attention to their institutions, their observatories, their learned societies and their committees (chapters 1 and 2); the places where they did their work; and the instruments they used (chapter 7). It means studying the people who made those instruments (chapter 7), as well as the assistants, "computers," and others who aided the famous astronomers whose names are well remembered (see chapter 8). Of course, we must pay attention to the theoretical background, to Einstein and his prediction (in chapter 3), and to the turbulent state of the scientists' contemporary war-torn world in 1919 (chapters 4 and 5). Not all scientists were favorably disposed to Einstein's ideas, and their views are discussed in chapter 6. Only then can we follow the protagonists on their journey (chapter 9) and through the eclipse itself (chapter 10) to the difficult data analysis that followed (chapter 11) and the enthusiastic worldwide reception of their results (chapter 12). Of course, the story does not end there, and we will discuss the work of astronomers at subsequent eclipses, which confirmed the results of 1919 (chapter 13).

Remarkably, even Dyson, the main organizer of the expeditions, has been unfairly overlooked in some recent histories. By focusing on the expertise of Dyson and his assistants, we will learn something of the key techniques and methods used in 1919 and appreciate more fully the experimental art that went into the science of the day. How this experimental art evolves over time is discussed in the book's final chapters, 14 and 15. The story that results from such careful study is not only more detailed but richer, more exciting, and more interesting than the story of the great scientists told as if they acted alone. The historian of science Steven Shapin famously showed how the chemist Robert Boyle's research depended on many technicians and assistants who, though unnamed and almost invisible, made important contributions to his work (Shapin 1989). Most of the people involved in 1919 were applauded in their day, but the passage of time has caused us to lose sight of their crucial roles. This book aims to correct this oversight, which could cripple serious attempts to understand how the science of the eclipse was accomplished. Only with this complete history can we reach firm conclusions about the reliability of this famous experiment.

And a famous experiment it was. May 29, 1919, was indeed the date of probably the most important eclipse in history. Eclipses have been credited with ending wars and with preserving the rule of colonizing explorers like Christopher Columbus. Yet for all the wonder and admiration they excite, the observation of this one eclipse by a few astronomers from Britain and Ireland has outdone them all. Their work may well have been the most important scientific experiment of the entire twentieth century. What they achieved is responsible for the worldwide fame of Albert Einstein. They established his theory of general relativity as the most celebrated theory in physics. They deposed Newton's theory of gravity from its perch as the greatest achievement of the human intellect. And they excited the whole world as no scientific event had before. Yet they were almost foiled by cloud at one site and by the intensity of the Sun's heat at the other. By good fortune they still managed to pull it off. Yet many have charged over the years that something was suspect about their work and the results they claimed to have achieved. In this book we will meet those men, follow their journey, learn what they did, and argue that they did it well, or as well as possible in very difficult circumstances.

Astronomers are anxious weather watchers at the best of times. But while clouds, rain, or wind can spoil a night's observing once in a while, an object that is missed one night will be available for viewing the next. Not so with eclipses. Though total eclipses of the Sun occur regularly, every year and a half on average, they are visible only along narrow tracks in different parts of the world each time and only for a few minutes at any given point along those tracks. Anyone who wishes to observe one will usually be obliged to travel, since it has been calculated that, on average, several centuries pass by between total eclipses at any given spot on the Earth's surface. But you must be prepared to travel in vain, since cloud cover of only a few minute's duration will be enough to entirely block the view. Of course, another eclipse will occur in a year or two, but it will involve more travel. Such travel is arduous because of the increasingly large and sophisticated instruments that astronomers began to use from the mid-nineteenth century onward. Fragile optical equipment is not easily transported. Given the expense and the difficulty and the likelihood of failure, one might wonder why astronomers bother with eclipses at all. The reason is that although eclipse expeditions are apt to end in disappointment, every once in a while they can change the world.

1

The Experiment That Weighed Light

Although Eddington and Dyson collaborated closely in organizing the expeditions of 1919, the two expeditions remained quite separate. Two different English observatories were involved, Dyson being director of one and Eddington of the other. Conscious of the vagaries of the weather, they chose different locations and hoped that the results from each site would confirm each other. Eddington, who was then one of England's most famous astrophysicists, personally led one expedition. He took with him a Northamptonshire clockmaker named Edwin Turner Cottingham to keep their instruments in working order while Eddington himself conducted the experiment. The Royal Obervatory at Greenwich mounted the second expedition. As director of that observatory, Frank Dyson sent two of his assistants, Andrew Claude de la Cherois Crommelin and Charles Rundle Davidson. They would each operate a different instrument, for some further redundancy. Dyson would oversee their data analysis after their return to England. Dyson and Eddington, working together, had hatched the plan and raised the funds, and together they would reveal the results later that year at a packed meeting of two of Britain's leading scientific societies. If May 29 is an important date in the history of science, then November 6, 1919, is also famous. The atmosphere that day was like that of a Greek drama, in the words of one participant. As one might expect in a Greek drama, some people greeted the news of the eclipse results with excitement and others with despair. The arguments as to the interpretation of the data still continue.

Just as the expeditions were mounted by two different observatories working closely in concert, two different societies collaborated in sponsoring them. The Royal Society of London, founded in 1660 at Gresham College in London, is England's premier scientific society. For centuries it has brought scientists of all fields together to discuss their work. In the nineteenth century, science diversified, and the number of scientists greatly increased. As different scientific disciplines grew, the scientists formed various professional societies. One of the oldest is the Royal Astronomical Society (usually known as the RAS), which was founded as the Astronomical Society of London in 1820 and took its current name in 1831. Both societies are still very active today, publishing scholarly journals and meeting regularly to hear talks by distinguished scientists.

Throughout their history, these societies have played an important role in sponsoring scientific research, especially where considerable expense is involved. Expeditions for the purpose of scientific discovery are a classic example of where such sponsorship is necessary. The late nineteenth century saw a revolution in astronomy that gave birth to the field of astrophysics. The experimental techniques of physics, including the use of spectroscopy and photography, began to be applied to astronomy, transforming how astronomers did their work. These new techniques increased interest in eclipses of the Sun, and in 1884 the RAS founded the Permanent Eclipse Committee to oversee planning for expeditions to observe solar eclipses.[1]

It was found helpful for the RAS committee to collaborate with a similar eclipse committee organized by the Royal Society to plan an expedition to West Africa in 1893 (Pang 1993). In the wake of that expedition, and so the two societies could collaborate in the future, the Permanent Eclipse Committee became, in 1894, the Joint

Permanent Eclipse Committee, or JPEC. By 1919 JPEC therefore already had a quarter century of experience in organizing eclipse expeditions. Its chair at that time was Dyson, and it was he who played the leading role in mobilizing resources on behalf of the two expeditions that year. The famous joint meeting of the Royal Society and the RAS was held specifically to hear JPEC's report on the expeditions of 1919.

The astronomers of 1919 thus were blessed to control an institutional framework with the power to mount such expeditions. They would need all the help they could get. They were beset with troubles, most importantly those of war. World War I ended just barely in time (in November 1918) for the expeditions to travel at all. The measurements to be made were much more challenging than those typically attempted during an eclipse. They would attempt to measure the shifts in position of the stars close to the Sun due to the Sun's gravitational pull on the starlight as it passed close by the Sun on its way to Earth. This was what Einstein had predicted. But the amount of the deflection of starlight he had predicted was such that the shift of a star's position, as measured on a photographic plate, would be less than the width of the star's image on the plate. Thus, the effect was quite tiny and would only be measurable with the most sensitive handling of the instruments during the eclipse and the most careful measurements afterward. Even then it would be possible to distinguish the true deflection of starlight from effects caused by changes in optical magnification of the telescope only with long and tedious calculations. These calculations would be performed at Greenwich by *computers*, a word that at that time referred to humans whose job was to crunch numbers using pen and paper only, without the aid of electronic calculating machines.

Fortunately, the science of astrometry, as the measurement of stellar positions is called, had greatly advanced in the decades before 1919. During this era, the distances to a large number of nearby stars were measured for the first time. This was done by very carefully determining tiny shifts in the positions of stars (their *parallax*) between different seasons of the year, due to the Earth's motion. Both Dyson and Eddington had carried out such work when they began their careers in astronomy. They had the experience, but they had previously always performed such exacting measurements in optimal conditions. They had used telescopes permanently and appropriately mounted in an observatory, with the equipment needed for data analysis all readily on hand and on-site. Most important, if there was a problem with any of their work, they could simply take a new image on a suitable night and begin again. In 1919, they would have no second opportunities to get things right. They would have no chance to profit by any mistakes they would make—only to regret them.

Some events in astronomy can be missed in the blink of an eye. A transit of Venus occurs only twice in every century or so. But anyone on one whole hemisphere of the globe can observe such a transit, so even if one observer is clouded out or suffers an equipment malfunction, others will witness it. The implacability of scientific advance is enabled by repeatability. Scientists tinker and modify with each new trial, and the expectation is that over time the precision of measurement will relentlessly improve. Take away the ability of scientists to demonstrate their persistence and they become more mortal. They are prone to the vagaries of fate that bedevil most human endeavors. The caprice of weather or of human fallibility may ruin any amount of careful preparation.

Given everything they had to contend with and the limitations imposed by scarce equipment and limited preparation time, the men of 1919 seem to have been very fortunate to get the results they did. Whether they really achieved the measurement precision they claimed has often been doubted. Whether they really overthrew Newton and vindicated Einstein has been questioned. One even sometimes hears the word *fraud* spoken in connection with their work. This is partly because their experiment was repeated many times over the succeeding decades without the level of precision ever noticeably improving. Surely, this suggests something suspicious. Can it be that nothing was learned from one expedition to another? We have to remember that few people were fortunate enough to repeat the experiment. When it comes to eclipse experiments, persistence is frequently rewarded with disappointment. Einstein's closest colleague in astronomy, Erwin Finlay Freundlich, went on at least six expeditions to test his mentor's theory and was able to observe totality only once. There is plenty of reason to call Dyson and Eddington lucky, but why should they also be called frauds? The answer is that Eddington was a renowned champion of Einstein's theory. He was quite frank in admitting his expectation, or at least his hope, that the theory would be confirmed. Many people have accused him of bias and have even claimed that he had ulterior motives that went beyond the science of the case. Could it be that this most famous of experiments was decided by someone who had made up his mind beforehand? What does this tell us about the way science is conducted? Are the expense and painstaking care of complex experimentation simply wasted efforts in which scientists merely contrive to confirm their expectations? The expeditions of 1919, it turns out, have something to tell us not only about the history of physics but also about the way science itself works.

Two Astronomers: Eddington and Dyson

Arthur Stanley Eddington was born in 1888 into a Quaker family in a scenic part of the North of England, the Lake District. His father was a school headmaster who died when Eddington was still an infant. His mother moved to the West of England, to Weston-super-Mare, and brought up Eddington and his sister in genteel poverty. Eddington remained very close to his mother and his sister, to whom he was known as Stanley, throughout his life. He was a brilliant student and availed himself of a series of scholarships to attend Owens College (now the University of Manchester) and then Trinity College, Cambridge. In 1904 he became the first student to become Senior Wrangler in only his second year at Cambridge. The title of Senior Wrangler is awarded annually to the student achieving the best mark in the mathematics degree exams, which is typically of three years' duration.

After briefly working at the famous Cavendish laboratory in Cambridge, Eddington went to the Greenwich observatory as chief assistant to the director, who is known as England's Astronomer Royal. His work there resulted in a paper that received Cambridge's Smith's

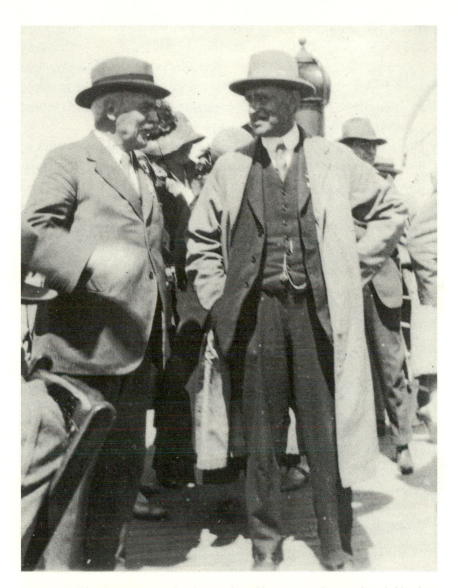

FIGURE 1. Frank Watson Dyson and Arthur Stanley Eddington, in a photograph probably taken at an International Astronomical Union meeting in Cambridge, Massachusetts, in 1932. Dyson and Eddington had much in common, including their religious and educational backgrounds, career paths, and passion for astronomy. They cooperated brilliantly in the planning, execution, and subsequent presentation of the eclipse expedition and its results. This photograph appears to have been taken just after a more formal one by the same photographer. Here, Dyson has turned animatedly toward Eddington, who smiles in response.
(Courtesy of the Meggers collection of the Emilio-Segrè Visual Archive.)

prize in 1907, and this led to him being elected a fellow of his old Cambridge college, Trinity. In 1913 he succeeded Charles Darwin's son George as the Plumian Professor of Astronomy and the next year was made director of the Cambridge Observatory. His rapid ascent to the heights of British science is very reminiscent of the path followed more than a decade previously by his colleague Dyson.

Frank Dyson's background and career were very similar to Eddington's. It is not hard to see why the two men should have gotten along well. Like Eddington, Dyson was religiously nonconformist. His father was a Baptist minister. Although born in Leicestershire in 1868, he mostly grew up in Yorkshire, from where the Dysons had originally hailed. Like Eddington, he was a Trinity student at Cambridge. He was Second Wrangler in 1889, compared to Eddington's achievement of Senior Wrangler fifteen years later. Both men won the Smith's prize (Barrow-Green 1999) and were awarded fellowships at Trinity as a result. Both began their astronomy careers in the important position of chief assistant at the Royal Observatory at Greenwich. In fact, Eddington succeeded Dyson in this position when the latter went off to Edinburgh as Astronomer Royal for Scotland. Of course, the odd similarity of his résumé with Eddington's mostly reflects the well-trodden path of the best and brightest coming up through the Cambridge system. Here were two talented and highly intelligent men who went to the forefront of their profession at each step on the ladder.

Because of the difficulties posed by wartime conditions, there is a mildly improvised air about the 1919 eclipse expeditions. Compromises were made in equipment and personnel since many instruments and people were simply not available because of the war. But in its two leaders, the expedition was blessed with men who were superbly qualified for the task at hand. If they were self-selected, it was because they were among the few who clearly saw the need for such a test at the time. As we shall see, it would have been hard to find two people with more experience of differential astrometry to undertake the experiment, especially since this was a very new field in which Dyson and Eddington were pioneers.

Although their backgrounds were similar, their careers diverged in one sense. Dyson reached the heights of astronomy as England's Astronomer Royal and immersed himself in observational work and in the organizational tasks of the leader of England's astronomical community. Eddington, though also an observatory director, was a Cambridge professor continuing the Cambridge tradition of theoretical physics. Even if he applied that physics training primarily to astronomical problems, he remained at the forefront of European theoretical physics, so he brought a unique double perspective to bear on the problem of testing Einstein's theory of general relativity. On the one hand, he understood the theory as well as anyone did; on the other hand, he had the observational skills to go out and test it, since this could only be done through high-precision astronomy.

Eddington's fame was based upon the study of light and gravity. It was he who showed that the radiation pressure of sunlight trying to escape the interior of the Sun is what keeps the Sun from collapsing under its own gravitational force. This remarkable insight defies ordinary intuition, since the most incredible weight in the solar system, that of the Sun, is held up by literally the "lightest" support imaginable, light itself. So intense is the power of sunlight emerging from the core of the Sun that it can accomplish this Herculean task of supporting the weight of the Sun on its shoulders. This aspect of Eddington's work certainly prepared him for the eclipse experiment of 1919. Einstein had argued that if gravity was a universal force, it ought to affect light. Eddington was well prepared to think of light as a thing having weight and heft. Indeed, it was he himself who presented the eclipse test as an attempt to weigh light. Physicists of the nineteenth century had naively presumed that light was weightless. Now Eddington and Dyson would prove them wrong.

Eddington's ideas about the interior structure of the Sun ultimately led to the modern study of gravitational collapse and the discovery of collapsed stars like neutron stars and black holes. Ironically, Eddington rejected the idea that such ultradense objects could exist, disappointing a young student, Subramanian Chandrasekhar, known as Chandra, who imagined that he was building upon his mentor's groundbreaking work. In a similar way, the inventors of quantum mechanics were shocked that Einstein rejected their ideas. Instead, both men engaged in the quest for a unified field theory (nowadays, sometimes called a "theory of everything") in their later years. Neither received any kudos for this later work. It was viewed as eccentric and outside the mainstream of physics. Eddington was obsessed with numbers and tried to calculate how many fundamental particles are in the universe based on what most physicists regarded as a kind of numerology. Later, the philosopher Bertrand Russell recalled "enjoy[ing] asking him questions to which nobody else would have given a definite answer," such as "How many electrons are there in the Universe? . . . He would give me an answer, not in round numbers, but exact to the last digit." Russell also recalled Eddington's satisfied response to the discovery of the expansion of the universe: "He told me once, with evident pleasure, that the expanding universe would shortly become too large for a dictator, since messages sent with the velocity of light would never reach its more distant portions."[2]

Two Observatories: Greenwich and Cambridge

The eclipse expeditions of 1919 were organized on behalf of two learned societies by the Joint Permanent Eclipse Committee, but they were carried out by personnel from two different English observatories. Typically, this was how such expeditions were mounted at the time. The English system of coordination through the learned societies was used primarily to facilitate government funding and arrange for the sharing of equipment between observatories. Crucially, the data analysis was carried out separately at the two observatories, and the expeditions took their own data quite independently of each other at two different sites. This is important to keep in mind because Eddington's subsequent fame has overshadowed everyone else who participated, and some modern commentators talk almost as if Eddington was solely responsible for every decision taken. At the time, no one in England would have made this mistake. Dyson was a well-respected and widely known public figure. As Eddington himself took pains to point out, it was Dyson's expertise and influence that made the whole enterprise possible.

The Principe team was led by Eddington, in his capacity as director of the Cambridge Observatory. Eddington's role was largely personal. He understood the theory being tested and was experienced in the type of astrometry required to measure the predicted effect. But his observatory was not known for its expertise in eclipses, and the equipment he used was largely borrowed from the Oxford Observatory. He was accompanied not by one of his own staff but by Cottingham, who was familiar with the equipment to be used, as he counted both the Oxford and Cambridge Observatories among his clients. Cottingham's role was largely to maintain the equipment in working order at the site. It is probable that Eddington alone handled the data analysis of the Principe expedition. We cannot determine how this was done because none of the data analysis sheets or photographic plates have survived. We do know that Eddington began the data analysis on Principe by himself, so it is almost certain that he continued on his own when back in England.

The expedition to Sobral in Brazil was under the direction of Dyson, the Astronomer Royal. He sent two members of his own staff, and some of the equipment was from his own observatory. Although he did not go himself, he directed the data analysis after the expedition returned to England. Key points in the data analysis sheets are in his handwriting, and he cowrote the report on the expeditions with Eddington. He took the lead in publicizing the results among the astronomy community and was coequal with Eddington in communicating them to the general public. They were both important public scientific figures who had a gift for popularization.[3]

The Royal Observatory at Greenwich opened in 1676, having been commissioned by King Charles II the previous year. He created the position of Astronomer Royal at the same time. Until recently, the Astronomer Royal also served as the director of the observatory. In the eighteenth century, the observatory played a leading role in solving the problem of longitude. As a result, the prime meridian, the reference longitude for most of the world, runs through the old observatory at Greenwich to this day. Since the best-known method

FIGURE 2. A view of the Royal Observatory at Greenwich in the 1920s. The dome in the foreground at right is where the astrographic telescope was housed. Frank Dyson and Charles Davidson spent many years working on this instrument, and its lens accompanied Davidson to Sobral in 1919. In 1894 high winds blew the shutter off the dome, and the headpiece fell into the room below, narrowly missing Davidson as he and Dyson were at work. The Royal Observatory is now a museum. (Image courtesy of Graham Dolan.)

of finding longitude at sea involved the use of precision timekeeping, the observatory was placed in charge of timekeeping for the British navy. This involved, among other things, dropping a ball down a spire at the top of the observatory at 1:00 p.m. each day by which ships in the Thames below could set their timepieces. By the early twentieth century, timekeeping had become important in civilian life, following the growth of the railways. Previously, towns had kept their own local time, but now it became desirable for everyone in England to follow Greenwich time. In order to facilitate this, Dyson developed the pips system with the BBC (British Broadcasting Corporation). This involved sending out a radio signal broadcasting the sound of six pips marking the seconds leading up to each hour. This permitted everyone with a radio set capable of receiving the BBC to set their clocks against master clocks at Greenwich, which controlled the time signal.

The early twentieth century was a golden age for astronomy. The subject had great prestige after the dramatic discoveries of the previous 150 years, including the addition of two new planets to the solar system. As new universities were founded in the late nineteenth century, many of them built observatories. Even the observatories of the older universities are not as old as one might think. Eddington's observatory in Cambridge was built in 1823, for instance. In the nineteenth century, British eclipse expeditions were dominated by London observatories—not only Greenwich but also newer facilities like Kew Observatory and the Solar Physics Observatory at the Royal College of Science in Kensington (now part of Imperial College, London). The Kew Observatory had been built for King George III in the eighteenth century (it is often known as the King's Observatory) to allow him to observe the transit of Venus in 1769. In the nineteenth century, the British Association for the Advancement of Science took over the building. Its director in the mid-nineteenth century, Warren de la Rue, was a pioneer of the use of photography in astronomy and focused particularly on solar physics. Because of his interest in astronomical photography, he donated the astrographic telescope to the Oxford Observatory for use in the *Carte du Ciel* project.[4] The lens from this telescope would be taken by Eddington to Principe. The Solar Physics Observatory in Kensington was built for Sir Norman Lockyer, a key member of JPEC for decades and the founder of the journal *Nature*.

By 1919 the Kew Observatory had become the home of the Met Office, Britain's national weather forecasting service, and Lockyer's Kensington observatory had been closed down. Lockyer had moved to Sidmouth in southwestern England, where he had a new observatory called the Hill Observatory (now known as the Norman Lockyer Observatory). Most of the equipment from his London observatory went to Cambridge to the new Solar Physics Observatory, a near neighbor to Eddington at the Cambridge Observatory. As an illustration of how common and influential amateur observatories were at this time, Sidmouth played host to not one but two observatories in 1917. The other was the personal observatory of a wealthy

German engineer, Adolph Friedrich Lindemann. Lindemann had been involved in the laying of one of the first successful transatlantic cables for telegraphic communication between Europe and America. He and his son, a promising young physicist named Frederick Alexander Lindemann, were interested in astronomy and in Einstein's theory. Since they both spoke German and spent time in Germany, they were more familiar with his work than most astronomers in Britain and were interested in joining the effort to test the theory. Not being eclipse specialists, they wondered if the light deflection experiment might be accomplished during the daytime, using filters to try to pick out the light of particularly bright stars against the glare of the sky. After some experiments carried out at Sidmouth, they wrote a paper proposing that a bigger observatory should try the observation at the conjunction of the Sun with Regulus on August 21, 1917. Regulus, also known as Alpha Leonis, since it is the brightest star in the constellation Leo, is the brightest star close to the ecliptic and thus the brightest star that the Sun comes close to in the sky. John Evershed, at the Kodaikanal observatory in India, a solar astronomer already used to testing Einstein's theory, took up the challenge and tried the observation on that date. We can argue that this was the first attempt to test Einstein's full theory of general relativity (before 1915 he had a different prediction for light deflection, as we shall see). As such it had a very fitting centenary because on August 21, 2017, Regulus was once again close to the Sun, but on this occasion the Sun was in total eclipse across a swath of land through the middle of the United States. This made the 2017 eclipse easily accessible by professional and amateur astronomers who cared to attempt the Einstein test.

But back in 1917, the attempt failed. Regulus could not be imaged in full daylight near the Sun. So it seemed that if the measurement was to be accomplished at all, it would have to be at an eclipse. In the wake of World War I, with astronomers of many countries still on war duty or struggling to survive amid revolution and upheaval, the field was wide open for Cambridge and Greenwich to make the running in 1919.

2

Eclipses

Eclipses occur when the Moon, in its motion around Earth, passes through a node of its orbit at a time when it is either new (the dark of the Moon) or full. The Moon's orbit is tilted with respect to the ecliptic, which is the plane of the Earth's orbit about the Sun. The nodes of the Moon's orbit are the points at which these two orbits intersect. Briefly, the Moon is passing through the plane of the sky on which the Sun moves. If it does this when the Sun is directly in line with the Moon, as viewed from the Earth, then an eclipse of some kind occurs. When the Moon is between the Earth and the Sun while this takes place, there will be a solar eclipse, as all or part of the Sun may be obscured. When the Earth is between the Moon and the Sun, a lunar eclipse occurs as the Earth blocks the light of the Sun, which permits us to see the Moon by reflection. Obviously, solar eclipses can occur only at a new Moon, since the bright side of the Moon (which always faces the Sun) will be turned away from the Earth. Contrarily, lunar eclipses always occur at the full Moon, since the bright side of the Moon is directly facing the Earth.

A lunar eclipse can be seen over the whole side of the Earth that happens to be facing the Moon because the shadow cast by the Earth is much bigger than the Moon itself. However, the shadow cast by the Moon during a solar eclipse is much smaller than the Earth, so at best, a total solar eclipse is visible only over a small part of the Earth's surface—a track along which the shadow runs eastward and sometimes quite a bit to the northeast or southeast. A solar eclipse comes in several forms. The Moon in its orbit is sometimes nearer, sometimes farther away, from the Earth. When at its farthest, it is too small in the sky to completely obscure the Sun, and an annular solar eclipse occurs, in which the outer rim of the Sun is still visible. More commonly, there is a total solar eclipse for those within the path of totality. Over a much greater area, a partial solar eclipse will take place. A partial eclipse is of little use to science and may even go largely ignored by people on the ground, since even a part of the Sun is still so bright that it is difficult, and dangerous, to observe at this time. A total solar eclipse is a rare event in the solar system. No other planet is privileged to witness such a dramatic event, since it is a coincidence that the Moon happens to be at just the right distance from Earth so as to appear to be the same size as the Sun. Since the Moon is moving away from us due to tidal friction, we should enjoy these events while we can. In less than a billion years, they will probably cease to occur because the Moon will appear too small in our sky. After that, only annular solar eclipses will be observed.

Both Eddington and Dyson had been on eclipse expeditions before 1919. Indeed, as the chair of the Joint Permanent Eclipse Committee (JPEC), Dyson was principally responsible for the planning of such expeditions. However, no British expeditions had been mounted during World War I. The British had sent expeditions to the eclipse of August 1914, which took place just after the outbreak of war. So sudden was the onset of hostilities that the German expedition, which had intended to perform the light deflection experiment, found themselves interned as enemy aliens by the Russian government and unable to observe the eclipse. The British had stations in Sweden and Russia in 1914 but did not attempt to test Einstein's theory during that eclipse. It was only during the war, after the publication of the final version of general relativity, that Eddington and Dyson became so interested in testing the theory. Nevertheless, the eclipse of 1914 had a major impact on planning for 1919 because the war had led various governments, including the Russians, to commandeer civilian vessels for war service. As such, transporting home the bulky equipment taken to observe the eclipse had been impossible, and when planning commenced in 1918, these instruments were still stranded in Russia, which was by then in a condition of postrevolutionary turmoil.

A number of things are necessary to successfully observe a solar eclipse. First, you must be able to predict where and when it will happen. Ancient astronomers were able to predict eclipses to useful accuracy, and the dates of forthcoming eclipses would be noted in medieval and early modern almanacs and ephemerides. An ephemeris is calculated from published astronomical tables, such as those of Copernicus or Kepler, and gives positions of planets and the Moon and Sun a few years ahead in a form permitting the production of a yearly almanac of the seasons. Christopher Columbus famously used an ephemeris to predict a lunar eclipse in 1504 to awe the people of Jamaica, of whom he made himself a colonial overlord. However, predicting whether a given location will experience a total solar eclipse is a much dicier business. Since the path of totality is usually only a hundred kilometers or so across, relatively small errors can still defeat efforts to witness it. A good example is the eclipse of 1715, the time of which astronomer Edmond Halley predicted to an accuracy of four minutes. He even drew a path of totality across a map of England to help people observe the eclipse. However, this track turned out to be off by thirty kilometers or so, and he had to correct it after the fact. By the nineteenth century, calculations had improved greatly, largely because of the need to solve the problem of longitude, which had obliged scientists to improve their understanding of the motion of the Moon. In addition, mapmaking had become much more accurate and reliable. In Halley's time, few countries other than England had been mapped well, but by the nineteenth century, better maps were available for many countries that European powers had colonized.

Before 1919, Eddington's previous solar eclipse expedition was to observe an eclipse in Brazil in 1912. There were, unusually, two total solar eclipses in 1912. The first took place early in the year, on

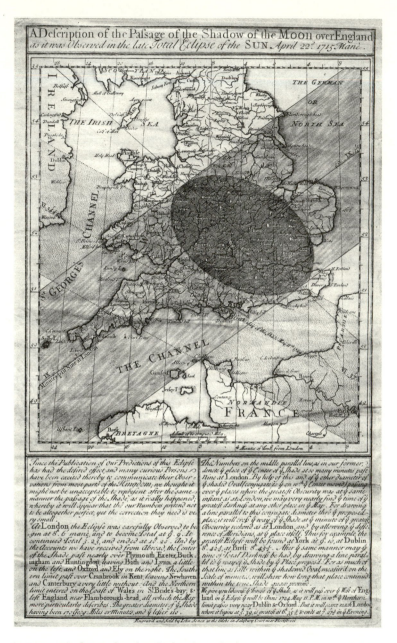

FIGURE 3. The first predicted track of totality to be drawn across a map can probably be traced to English astronomer Edmond Halley. This eclipse, which took place on May 3, 1715, is still known as Halley's eclipse. This map is not the predicted one, which was about twenty miles off the observed eclipse path, likely due to inaccuracies in the state of knowledge of lunar motion. After the eclipse, Halley drew the corrected path that is depicted here and added in a second track, the one trending southwestward from Ireland and then across Britain toward France, of another total solar eclipse visible in 1724.

(Courtesy of the University of Cambridge, Institute of Astronomy Library.)

April 17. It was an annular eclipse along most of its track across the Atlantic and into Spain and Portugal. Such an annular eclipse is of no use for observing stars since the uncovered part of the Sun will still outshine them. The particular new Moon that created that eclipse in 1912 was unlucky, as two nights before, with the Moon close to its darkest on a still night, the RMS *Titanic* had struck an iceberg that her lookout had been unable to spot in the darkness until it was too late.

Astronomical expeditions played a major role in European science in the eighteenth century. One of the best known of all scientifically inspired expeditions, the first voyage of Capt. James Cook, was to observe the transit of Venus in 1769. This important astronomical event is similar to an eclipse, since it occurs when Venus passes in front of the Sun. Although this hardly makes for spectacular viewing, it is a very rare event. It happens only every century or so, occurring twice a few years apart when the conditions are right. In the seventeenth century, Edmond Halley pointed out that observing a transit of Venus from widely separated places on the surface of the Earth would permit the distance between Earth and the Sun to be measured by parallax. The distance from the Earth to the Sun plays such an important role in astronomy that it is known as the *astronomical unit*, and it had never been measured accurately before the eighteenth century. So the expense of the first Cook voyage was amply justified from a scientific point of view. Unfortunately, eclipses were not important enough to command the level of logistical support from the navy necessary to visit far-flung places like Tahiti, from where Cook's expedition observed the transit of Venus.

What really made eclipse expeditions practical was the development of steam-powered transportation in the nineteenth century. With the advent of steamships and railroads, travel—especially travel with expensive equipment—became much more feasible. Until the 1860s, expeditions had tended to prefer European destinations, but after that European colonial governments had improved transportation across Asia and Africa, and transatlantic steamship services were available to both North and South America. Expeditions could now reach most eclipse locations on land. In essence, astronomers could go wherever tourists could go, and the 1840s also marked the beginnings of modern tourism. As one modern historian of eclipses has put it, eclipse expeditions have less in common with James Cook and more in common with Thomas Cook, the originator of the package tour (Pang 1993).

The first steps in eclipse planning involved an astronomer performing the laborious calculations necessary to predict the track of the eclipse. Then, the best available maps of the relevant countries would be selected, and the track would be drawn across these maps. This was possible only because by the late-nineteenth-century maps accurate in latitude and longitude were available for much of the world. Once this step had been completed, the map was examined to find railway stops that lay within the path of totality. The ideal situation involved a field station near such a railway stop, or even better, a port served by steamship. The only considerations competing with these logistical ones were those connected with the weather. Local knowledge was essential here. Meteorological information about arbitrary places on the Earth's surface was not readily available, so a European settler might be consulted about the risks of cloud cover at a certain location at the relevant time of year. It was for this reason that Eddington went to Principe in 1919. The coast of West Africa had also been considered, but they were warned off by accounts of possible bad weather. Principe was considered a better possibility, but when Eddington arrived, he found that locals were surprised that he expected clear skies in May.

The next consideration was money. Passage by steamship or rail, especially with freight, was not cheap and was generally the largest expense involved in such an expedition. This is where the learned societies, such as the Royal Society, came in, and the principal reason why JPEC came to dominate eclipse planning. They had the inside route to funding not only from the societies themselves but also from the British government, since the Royal Society controlled the relevant government grant committee. Indeed, by the late nineteenth century, British eclipse expeditions were becoming very professional in their planning and in their composition. Earlier expeditions often consisted of a group of scientifically interested individuals traveling together and then conducting their own personal observations of the eclipse, typically by recording them and drawing them by hand (Pang 1993). The contrast with the expedition of 1919 is, as one might say, total. Expeditions organized by JPEC discouraged bringing anyone but professional astronomers along. Those astronomers were often engaged upon a disciplined program of observations, recorded by photography. Eddington confessed to his mother and sister that he barely saw the eclipse. He was too busy attending to his equipment and changing photographic plates to do more than glance up to determine when the time of totality began. In 1919 no one from Britain came along, to either observing site, who was not fully engaged with the light deflection experiment.

The centralized planning characteristic of the British expeditions of 1919 was unique to Britain. Few other countries had the resources or the nationally organized scientific bodies for such elaborate planning. Most European countries organized eclipse expeditions on a more ad hoc basis. Even the French, who had been prominent in eclipse observing in the nineteenth century, were not involved in 1919, partly because of a lack of interest in Einstein's theory and partly because of the great disruption caused by the war. In Germany, Einstein himself had to pitch in with fundraising efforts to enable his theories to be tested. In any case, as a defeated country Germany could not even think about sending expeditions in 1919. The United States obviously had the resources for expeditions of this kind but lacked the strong centralized planning. Efforts were largely left up to individual observatories, and the Lick Observatory in California played the leading role here. Its expeditions were often quite elaborate and retained a certain amount of the informality that had vanished from the British teams. The director of the observatory, William Wallace Campbell, was often assisted by his wife, Elizabeth, who would sometimes organize the expedition and even accompany it. Women rarely accompanied British expeditions in the twentieth century, although a number of women were playing important roles in astronomy at that time.

JPEC owed its origins to the increasingly technical demands of eclipses after 1870. The advent of spectroscopy and photography encouraged bringing along more and more equipment, and this increased the expense of expeditions and demanded that elaborate precautions be taken to ensure the safety of delicate instruments. After 1870 the British organized eclipse-planning committees that included representatives from several different observatories, often in the London area (Pang 1993). Many of those involved were gentleman amateurs, though still serious ones, such as Sir Norman Lockyer and William Huggins, who were both pioneers in spectroscopy. As planning expeditions came to be dominated by JPEC, men such as Lockyer and Huggins would often serve on its committee.

Over the course of the nineteenth century, the scientific goals of eclipse expeditions also changed. To begin with, from the 1840s to the 1860s, the focus was on phenomena near the visible surface of the Sun. Because the Moon is so close in apparent size to the Sun, it can cover the disk of the Sun and still leave the uppermost layer of the solar atmosphere visible to observers. This permits one to see phenomena such as prominences, or solar flares. The turbulent upper atmosphere of the Sun frequently erupts to create plumes of superheated gas, which form a sort of bridge or arch-like structure. Eddington photographed a particularly impressive prominence on Principe that is the most famous image from the 1919 eclipse.

As knowledge of the solar spectrum improved, a striking feature of eclipses drew the attention of astronomers. The solar spectrum as usually viewed is an absorption line spectrum. It consists of all colors of light, as Newton demonstrated, with very fine dark lines only visible with a very sensitive, high-resolution spectrometer. Below the layers that produce these colors are layers where the gas is even hotter. There is a thin layer on the surface of the Sun that produces an emission spectrum. This is a spectrum of bright lines at the same wavelengths as the dark lines in an absorption spectrum, with only darkness between. The layer of the Sun that produces this emission spectrum is just below the visible surface but becomes observable during an eclipse. When the Sun's surface is obscured, this *reversing layer* becomes visible at the limb of the Sun, viewed from the side. This layer was termed the *chromosphere* (color sphere) by Norman Lockyer, who, along with Jules Janssen, led the study of it in the

1860s. In 1868 these two men realized that the spectral lines from the reversing layer were bright enough to observe in the limb of the Sun even without an eclipse. In the following few years, their work contributed to the discovery of the element helium, which was discovered not on Earth but through the presence of its emission lines in the solar spectrum (Nath 2013). Lockyer named the element after the Greek word for the Sun, *Helios*.

After this success there was no longer as much reason to travel to an eclipse to study this part of the Sun's atmosphere. However, there was still great interest in the corona, the uppermost part of the solar atmosphere. From 1870 onward, attention focused on this phenomenon. Gradually, the method of investigation changed. To begin with, most observers drew the corona, sketching quickly during the eclipse and then filling in details from memory afterward. As photographic techniques improved, scientists began to prefer to use cameras to study the corona. This encouraged JPEC to permit only professionals to accompany the expeditions. Amateurs, who might have previously included observers with special drawing or artistic skills, were no longer necessary (Pang 1993). No doubt, a serious atmosphere was preferable when the scientists present had to focus their attention not on the marvelous work of nature unfolding above them but on their instruments. I myself once accompanied an astronomer (my wife) on an observing run and noticed the contrast between the beauty of nature and the demands of science. We were at the sixty-inch telescope on Mount Palomar in California. In order to keep light from entering the dome, the room in which the astronomers sat, filled with computers and other equipment used to control the instrument, was windowless. With so much electronic equipment, the room became quite warm, and it was occasionally necessary (for me—the astronomers present had no difficulty staying awake) to step outside the dome to take in some air. Outside, it was a beautifully clear night, and the stars blazed with extraordinary brightness on the mountaintop. Yet none of this could be seen from inside the control room, where I repeatedly failed to keep myself awake in the stifling heat.

It is an odd fact that Dyson and Eddington had each been on hand to observe the only two eclipses before 1913 that played a role in the testing of general relativity. These two eclipses were those of May 28, 1900, and October 10, 1912. The first of these obviously took place before anyone on any of the expeditions had even heard of Albert Einstein. Its role in our story is that the 1919 eclipse took place on what was almost the first eclipse's anniversary (May 29, 1919). It follows, therefore, that the star fields were very similar. This naturally suggests that the data from 1900 might have been useful in 1919. The 1919 eclipse was regarded as auspicious for testing Einstein's theory because the Sun was in the Hyades, the star cluster closest to Earth, and there were therefore several bright stars close to the Sun. This is important because the Sun's corona, which is not obscured during an eclipse, is quite bright (similar to the full Moon), possibly obscuring dimmer stars and making them difficult or impossible to observe. Obviously, the 1900 eclipse was near the same star cluster, and the plates taken then were, as we shall see, later used to attempt to test Einstein's theory. It so happens that the 1900 eclipse was the first of six witnessed by Frank Dyson, who was then a young man and the chief assistant at Greenwich. Even more importantly, one of the lenses used in 1919 was also first employed during the 1900 eclipse.

Some of the plates from the 1900 eclipse were of interest to those wishing to test general relativity because they showed more than just the Sun itself. In traditional eclipse drawings and photography, the item of interest is the Sun, as well as its corona. A wider field of view only serves to reduce the size of the Sun's image on the plate. As such, most existing eclipse photographs were useless for testing the light deflection prediction. But in the late nineteenth century, the search for new planets in our solar system was at its height. The discoveries of Uranus in the eighteenth century and Neptune in the nineteenth had whetted the appetite for one of the great adventures of science, the discovery of new and previously unknown worlds. While the search for a large object beyond the orbit of Neptune has never, to this day, succeeded, one other possibility existed. Mercury is the inner solar system planet that few people have seen. It is too close to the Sun to be easily visible because it is rarely in the sky at nighttime—and then just briefly before rising or setting immediately after or before the Sun. Since it is small and not particularly bright, it is difficult to observe when it is close to a bright horizon. How much more difficult would it be to observe a planet even closer to the Sun? Such a planet could therefore easily exist, with no one being the wiser. A good time to search for such a planet, given the name Vulcan by the astronomers who thought they had discovered it in 1859, would be during an eclipse. And the plates taken at that time would focus on the area, potentially rich with stars, close to but not part of the Sun, especially in an eclipse like that of 1900.

The 1912 eclipse was the first at which anyone set out to test Einstein's theory. By then, Einstein had not only become known for his special theory of relativity but had also published accounts of his quest to find a general theory based upon the principle of equivalence. He had proposed the eclipse test to astronomers as a way of determining whether this principle was as widely applicable as he believed. The Argentine National Observatory, led by Charles Perrine, made the attempt, and it was unsuccessful due to clouds and rain. This eclipse was the first observed by Arthur Stanley Eddington, who had succeeded Dyson as chief assistant at Greenwich in 1906. Eddington's station was only thirty miles from the one occupied by Perrine, and it has been speculated that it was at this eclipse that Eddington first learned of Einstein's theory of gravity and this observational test of it (Stachel 1986; Warwick 2014). But the truth is that neither Eddington nor Perrine, who both subsequently wrote about their expeditions, ever mentioned meeting each other at this time. Although we cannot say anything definite about Eddington's knowledge of Perrine's experiment, we do have a letter sent home by the then twenty-three-year-old man to his mother that at least mentions the existence of Perrine's expedition. Other letters home were written from on board the *Arlanza*, a steamship running between England and Brazil.[1] He would write home to the same relations seven years later while on the 1919 expedition.

In reading Eddington's 1912 letter home, I could not help but be excited by the name *Arlanza*, since I grew up spending a great deal of time at a house in Cork city in Ireland named after that ship. The house then belonged to my aunt but had originally been built for a family friend, Millicent Mary Anderson, a spinster known to our family as Nant. In 1915, three years after Eddington's voyage, the *Arlanza* was converted to service as an armed merchant cruiser, tasked with enforcing the naval blockade of Germany in the Norwegian Sea. According to family tradition, Nant's brother was then an officer on the ship. In October 1915 the *Arlanza* sailed to Archangelsk in Northern Russia with a cargo of platinum bullion and sailed back with a delegation of Russian statesmen bound for a conference with their wartime allies, Britain and France. After the departure of her convoy of minesweepers in the White Sea, she struck a mine and was holed below the waterline near the bow.[2] Her bulkhead doors closed automatically to prevent her from sinking, but two sailors were trapped beneath them as they attempted to escape from the forward section of the ship. The trapped sailors were rescued, and the doors closed, but the ship was left with her bow well down in the water and her propellers raised up behind her. In fact, one of the ships coming to her aid was holed and sank upon striking the *Arlanza*'s raised propeller. In the confusion, the Russian dignitaries were treated to an icy dip in frigid waters when their lifeboat capsized. But in the end, everyone was rescued and returned to the *Arlanza* when it became apparent that she was not sinking (Nicol 2001, 114).

According to family tradition, rearranging her cargo induced her to list and raised the gaping hole out of the water (or perhaps, more importantly, got the propellers back into the water), and she made it to a bay on the Kola peninsula in northern Russia. From there, the crew obtained supplies from the nearest settlement by reindeer sled. After overwintering in her isolated bay, she was able to return for repairs to Belfast, where she had been built, at the Harland and Wolff shipyards. Inspired by this adventure, Nant named her house in Cork after the ship, whose travails passed into family lore. For Eddington, however, it was a much more tranquil voyage through far warmer waters in 1912. The horrors of the Murmansk convoy were unimaginable at that time. Yet the dangers and difficulties of transport to and from Russia in wartime were to play a decisive role in the story of Eddington's next eclipse expedition in 1919.

3

Two Pacifists, Einstein and Eddington

Albert Einstein was born into a Jewish family in the city of Ulm in southern Germany in 1879. Although his family moved to Munich when he was young, he always referred to the Swabian German dialect he spoke at home as setting him apart. His father ran an electricity-generating company in Munich, but the company folded when Einstein was a teenager, and the family moved to Italy. Einstein was initially left behind to finish high school, but he soon fled his school, which he hated, to rejoin his family in Italy. He spoke quite good Italian as a result. He then moved to Switzerland, where he attended the Swiss Federal Polytechnic Institute and eventually obtained his PhD. His academic star did not rise in the way Eddington's did, and he failed to win a position at a university, ending up as a clerk in the Swiss patent office in Bern. While employed there, he made some of his greatest scientific discoveries, publishing five famous papers in just one year in 1905.

Although Einstein's early discoveries are now considered far more important than Eddington's work as an assistant at Greenwich, Eddington's steady ascent to the heights of British science stands in contrast to Einstein's early obscurity. It was not until several years after 1905 that Einstein received his first academic position. Nevertheless, by 1914, when Einstein moved to Berlin, both men were distinguished professors at leading centers of European science. At that point World War I began, and a trait the two men shared came to the fore. They were both passionate proponents of pacifism and internationalism in politics. As such, they were both staunch opponents of the war, willing to risk the wrath of both the government and the public in standing up for their beliefs.

It is possible that the two men's outsider status played a role in their unusually public opposition to the war. As a German Jew, Einstein struggled against anti-Semitism throughout his life. It probably played a role in his early difficulties obtaining an academic job. In later years he became a symbol of everything the Far Right, especially the Nazi Party, hated in German life. A few scientific colleagues attacked him for what they termed "Jewish physics." Eddington was a Quaker, and although the barriers to a dissenter (an Englishman who does not conform to the Church of England) attending or holding tenure at Oxbridge universities had disappeared by his time, he depended on his academic brilliance for advancement, especially because of his relatively impoverished background. Both Einstein and Eddington were leading examples of something new in science, the professional theoretical physicist. Until then, science had often been the preserve of gentlemen amateurs who could afford ample time for thinking. Men such as Eddington and Einstein took advantage of the increasing number of universities in the late nineteenth century to forge scientific careers that would have been impossible without a regular job that permitted time for research.

Traditionally, most physicists performed their own experiments. While both Eddington and Einstein were adept experimenters, both earned reputations for their brilliance in mathematical calculation and theoretical conceptualization. The role of the theorist was relatively new in science, and with their self-confident willingness to put forward bold ideas, Einstein and Eddington came to typify the breed. Even more impressively, they were prepared to let others test their predictions in full view of the scientific public. If traditional scientists, such as Newton, could mold their theoretical views in the privacy of their own laboratories until theoretical prediction and experimental measurement dovetailed nicely, theorists such as Eddington and Einstein had to make their predictions and then abide by whether others found them satisfactory.

In character, the two men differed. Einstein, very much a ladies' man, married twice and had several affairs. Eddington seems to have been completely uninterested in women from a romantic standpoint. It has been widely speculated that he was homosexual, but there is no evidence that he had romantic relationships with men, either. Little of his private correspondence has survived. It is possible that, after his death, friends or family disposed of papers providing evidence of sexual relations that would have been considered criminal at that time. It is also quite possible that he remained celibate throughout his life. He was, at any rate, a very private man, while Einstein became, after 1919, a celebrity willing to employ his fame in the service of various political causes. At the same time, Eddington was the more gifted science popularizer. Both men had a lively sense of humor and were engaging company among their friends.

In 1919 Einstein was best known for the special and general theories of relativity. These theories, the first put forward in 1905, represented a masterly synthesis of the fundamentals of physics, uniting the theories of kinematics (the study of motion), dynamics (forces and acceleration), electromagnetism, and gravity. In doing so, he revolutionized our understanding of such basic concepts as space and time, inaugurating the era of spacetime, in which time is viewed as the fourth dimension. Perhaps even more revolutionary, he showed that light and other forms of energy should be viewed as quantum in nature (again, a project begun in 1905). He argued that energy comes in little packets, called *quanta* by the German physicist Max Planck, which cannot be further subdivided. This view of light as composed of particles, which we now call *photons*, was controversial at the time because nineteenth-century physicists had concluded that light was a wave phenomenon. Einstein's tendency to view light as a particle undoubtedly played a role in his insight that light quanta (photons) should fall as they pass close to a heavy body like the Sun.

For a century, physicists had believed that light was strictly a wave-like entity, leading them to regard it as immune to the effects of gravity. In returning to the earlier notion that light was composed of particles, it naturally made sense to Einstein that they would be affected by gravity and could fall toward the Sun. Of course, particles of light move so quickly that they would not have time to fall very far, but just possibly, the change in direction would be enough to be noticeable to a careful observer. In applying the particle concept to the study of optics, Einstein reintroduced an idea of Newton's, who, in the first query to his famous book *Optics*, had asked, "Do not Bodies act upon Light at a distance [presumably by their gravity], and by their action bend its Rays; and is not this action (*caeteris paribus*) strongest at the least distance?"

An odd aspect of Einstein's and Eddington's sudden ascent to fame in 1919 was not the public's appreciation of the two men's scientific brilliance but rather their shared opposition to the Great War. Both men were pacifists—and outspoken ones, at that. A remarkable feature of the eclipse story is that the public responded so well to these two oddballs, because oddballs they certainly were. Now, a century later, we are used to every war having its associated antiwar movement, but life was very different in 1914. An antiwar movement existed, and it was associated, as now, primarily with the radical Left. But the nationalism of the early twentieth century made no concession to the idea of a principled rejection of militarism. The antiwar movement was viewed as being frankly treacherous to the nation, and socialist radicals embraced this badge of treason with fervor. In countries such as Russia and Germany, with no meaningful democracy, a revolutionary overthrow of the state was arguably the only way to overthrow militarism. By 1919 this radical wing of the antiwar movement had begun to adopt for itself the term *Communist*, inspired by Lenin, one of its most prominent figures. The Communists, while opposed to militarism, were not pacifists, and by 1919 they were in violent conflict with the militarists. Out of this militarist movement emerged what would become fascism. So in one sense, the collision between militarism and its opponents in the Great War set the course of the mid-twentieth century.

However, the conflict between Communists and fascists had little relevance to men like Einstein and Eddington. They were not Communists, though Einstein held many views in common with democratic socialists. They were thoroughgoing pacifists who rejected violence in all cases (though admittedly, like the Communists, they ended up making an exception for fascism). For them, antimilitarism was not merely an aspect of opposition to monarchism but a political statement in itself. Pacifists of their sort existed in 1914, but they were rare birds. Eddington was, like his mother and sister, a member of the Quakers, a small English religious movement. Today, the Quakers are recognized by many countries as the prime example of a group entitled to faith-based exemption from conscription into the military, termed conscientious objector status. During World War I, no such accommodation had been established in most countries, and in Britain many Quakers went to jail for refusing to be conscripted. There is every reason to believe that Eddington was willing to join them. Einstein admired this quiet courage, later saying, "If I were not a Jew, I would be a Quaker" (Douglas 1956, 101).

However, some scientists were unwilling to see a man as distinguished as Eddington go to jail. His employer, Cambridge University, intervened to obtain an exemption for Eddington based upon the importance of his work to science. But this was the era of the white feather, when, allegedly, women would walk up to young men on the street and present them with a token of cowardice, shaming them for their refusal to volunteer. As the war dragged on, with all sides sustaining unprecedented casualty rates, conscription boards looked less and less kindly on exemptions. Finally, Eddington prepared to go to jail. He made out a statement for one of the hearings discussing his case, stating his refusal to accept conscription. At this point Dyson intervened, coming up with the idea that the eclipse expedition itself would represent Eddington's war work. The conscription board may have resented the notion that Eddington would sit out the war in the comforts of home. An onerous task such as being shipped off to Africa in the cause of science at least suggested that Eddington was not just a skiver.

This story was first related by Chandra, who had been a student of Eddington's and knew him well. In later life he rose above any bitterness he might have felt at Eddington's rejection of his ideas and became an important biographer of Eddington. He has had a major influence on subsequent writing about the eclipse expeditions. His notion that Eddington undertook the expedition as a kind of war work for peacenik scientists has played into a bigger narrative that the whole enterprise consisted of a kind of pacifist conspiracy against science. Eddington wanted to avoid being conscripted. He also wished, like other Quakers, to "heal the wounds of . . . war" (Stanley 2003, 68) and avoid the postwar dismemberment of international science, since allied scientists shunned German scientists after the war. What better way to illustrate the internationalism of science than by leading an English expedition that would prove a German theorist correct? How perfect was it that they would thereby overthrow and replace the crowning glory of English science, Newton's theory of gravitation? In considering the truth of this narrative, let us at least remember from the beginning how isolated and despised pacifists like Eddington had been during the war. It would have taken a considerable leap of faith for Eddington to imagine how popular healing the wounds of war would prove to be.

It was not just Eddington who was isolated in his pacifism. Einstein also was shocked by the war and found it to be a depressing and difficult time in his life. He had fled Germany as a high school student, repelled by the militaristic ethos of the German education system and anxious to avoid military service. To do this he was obliged to renounce his German citizenship and later became a Swiss citizen. He returned to Germany in 1914 to take up a very prestigious academic position in Berlin. He enjoyed his new life in the company of the leading figures of European science, such as Max Planck, Walther Nernst, and Fritz Haber. He was also close to his well-connected cousin Elsa, who was his extramarital romantic interest at this time. Doubtless, he felt that the sophisticated circles in which he was moving represented a new reformed Germany. He probably imagined that it was no longer the same country as the one he grew up in. Imagine his horror when, within months of his arrival, this new Germany succumbed to a rabidly nationalistic war fever, and his clever scientific friends joined in the rush to demand that Germany go to war to defeat its enemies.

That he rose above this, as did Eddington, to produce his greatest achievements is a testament to his ability to focus on his work and exclude other parts of his life. Nevertheless, as we shall see, he did not refuse to become involved politically as the world went mad around him. Although he was in no danger of conscription, he nevertheless stood up to be counted as a pacifist in time of war. This was as dangerous for him as Eddington's stand against conscription because Germany was a military dictatorship for much of the war, and civil liberties were in no way guaranteed. After the war, Einstein's scientific stature and his political principles combined to make him a very symbolic figure. He was the good German scientist who had advocated for peace while his colleagues had improved and tended to the engines of war. The Great War, which had witnessed the aerial bombing of cities and the use of poison gas, taught Europe that science was a force for destruction as well as progress. Einstein came to stand for a pure science, divorced from the worst impulses of mankind. Eddington would dramatize and highlight the virtues of this noble and disinterested science in a way that would catch the imagination of a world shocked by the unrestrained violence of modern war. This was a key difference between them: the shyer Eddington was the one with the gift for popularization. He evangelized Einstein's theory to the English-speaking world. Einstein even spoke a little disparagingly of this achievement of Eddington's, according to another Eddington biographer: "He [Einstein] spoke of the literary value, the beauty and brilliance of Eddington's writing in those books aimed at giving to the intelligent lay reader at least some understanding, some insight into the significance of the new scientific ideas—but with a smile he added that a scientist is mistaken if he thinks he is making the layman understand; a scientist should not attempt to popularize his theories, if he does 'he is a fakir—it is the duty of a scientist to remain obscure'" (Douglas 1956, 99–100). Eddington would utterly frustrate Einstein's desire to remain in obscurity.

FIGURE 4. Albert Einstein and Arthur Stanley Eddington together in the garden of the Cambridge Observatory in 1930. The two men had much in common, including a mastery of theoretical physics, firm pacifist convictions, and an idealistic commitment to the cause of internationalism, especially in science. The photograph was taken by Eddington's sister, Winifred, according to Vibert Douglas' biography of Eddington. (Reproduced from Douglas 1957, plate 11, 115.)

Two Theories of Relativity: Special and General

Einstein disliked the name *theory of relativity*. His ideas about what we now call relativity theory were based upon what he referred to as the principle of relativity. Simply put, this principle states that you cannot determine the state of motion of your vehicle without looking outside unless the vehicle is accelerating. Looking out the window permits you to judge your speed in relation to other vehicles. No experiment one can perform in a closed, windowless room inside the vehicle will tell you anything about your motion past objects outside the vehicle. In answer to the question, "What about the vehicle's motion through empty space itself?," the young Einstein replies that empty space does not exist. It cannot be felt or seen. One is aware of it only as a means of describing the distances between the external objects themselves. For Einstein, only those visible objects have reality. Space itself can be observed only to the extent that it is defined by measurements made of the positions of the objects. Take away the objects, and space no longer has meaning. Possibly, it even ceases to exist. This relationalist view of space stands in contrast to the substantivalist view expressed by Newton and his followers, for whom space was the vessel within which material objects existed. From this viewpoint, objects cannot exist without a space to place them in.

The principle of relativity guided Einstein in writing his papers of 1905, which introduced what is now called the special theory of relativity to the world. He used it as a conceptual tool to demonstrate that measurements of length and time using yardsticks and clocks produce results that vary depending on who does the measuring. Two different observers who are in motion with respect to each other will not agree on the measured length of an object or the elapsed time between two events. This is because it takes time for information about measurements to travel from one place to another, even if the information travels at the speed of light. If you are in motion, it takes extra time for light rays bearing news of a measurement to reach you, and thus the measurement you record is affected by how rapidly you are moving with respect to the speed of light.

Now, the principle of relativity claims that any nonaccelerating, or inertial, observers are entitled to think of themselves as at rest. So what happens if two inertial observers compare the measurements they make? Will the measurements not depend on which one of them is "really" at rest with respect to the light rays they use to exchange information? This is the question that many physicists of Einstein's time tried to answer. For years the young Einstein also tried to answer this question. One viewpoint was that light travels through the luminiferous ether, so the real question is whether an observer is at rest relative to the ether. The famous Michelson-Morley experiment was an attempt to measure the Earth's velocity though the ether. Einstein regarded this experiment, which was performed in a basement in Ohio, as a wrongheaded attempt to measure one's own velocity (that of our planet) without looking out the window. Einstein himself thought that light ought to move at a speed that depended on the speed of the object that emitted it. But this also seems not to be true, based on astronomers' observations of binary stars. So in the end, he came up with a second principle, to go along with the principle of relativity, which said that every inertial (nonaccelerating) observer would measure the speed of light as the same regardless of who emitted the light beam or how it traveled.

So this means that every inertial observer measures the same speed of light in vacuum, denoted c (from the Latin word for "speed," as in *celerity*). If two observers are in motion with respect to each other, they can each claim that the other is moving with a measured speed v. Einstein was able to show that differences in measurements of length and duration made by these observers will depend on ratios of v/c. Obviously, a slowly moving observer, for whom v/c is very small, will hardly notice that light is not traveling infinitely quickly. The speed of 186,000 miles per second is actually quite close to infinitely quickly compared to most everyday speeds. But if a human could move very quickly, so that v/c approached 1, then, obviously, discrepancies with the measurements of other humans would arise. These discrepancies would be remarkable. An object such as a yardstick might appear shrunken to only a few inches long as it passed by. A clock moving by at close to the speed of light would seem to have hands that were hardly moving. What of a person holding the rapidly moving clock? Einstein insisted that the reality was in the measurements. If the clock ran slow, so would the person's internal clock. Such a fast-moving person would actually age more slowly! These effects, length contraction and time dilation, posed a serious danger to physics. A cornerstone of physics is the idea of reproducibility. If you perform an experiment, I should be able to repeat it and expect to get the same result. But Einstein said that if we perform the experiments on different planets, moving at different speeds, we should not expect our measurements to agree!

Happily, Einstein did not just pose a terrible dilemma for physics. He also solved that dilemma. Since it turned out that elementary quantities such as length and time varied between observers, Einstein showed how to construct new quantities that did not vary between inertial observers. He called these quantities *invariants*. He would have preferred to call his theory the *theory of invariants* to focus attention on his solution and not on the problem he had identified. To this day, people often talk about relativity as if Einstein showed that there is no way of knowing what is really true. This is exactly why he disapproved of the name. But since he was too modest to name his theory, he lost his chance to win the propaganda war over what it meant. However, Einstein rightly felt that he had demonstrated not how different observers might disagree about the world around them but rather how they could discover that they were in agreement. He showed that while separate measurements of space and time would vary between observers, there existed spacetime quantities, consisting of length and duration combined, that were invariant. These invariant quantities are the basis of relativistic physics (invariant physics). This is the reason Einstein brought spacetime into being. It is not just a cute conflation of space and time; it is essential to the whole relativity project. Only by treating time as a fourth dimension of spacetime can we make physics once again an invariant discipline that is safe for reproducible calculations and experiments.

How exactly does one join space and time? Einstein's idea was, within a few years, put into an elegant mathematical form by one of his teachers, Hermann Minkowski. This relied upon the mathematical concept of the metric. The metric is nothing more than a generalization of one of the oldest theorems in mathematics, the Pythagorean theorem. Recall that this theorem tells us that the square of the hypotenuse of a right triangle is equal to the sum of the squares of the other two sides. If you divide space into dimensions, this theorem shows you how to work out the straightest distance to a given point if you know how far west and north it is. The metric generalizes the theorem in two ways. First, it allows you to write it in a coordinate independent way. You should get the same straight-line distance to a point whether you use Cartesian coordinates or polar coordinates or some other system of your own devising. To do this you need to put in the right coefficients for each term in the sum of squares. The metric is a handy shorthand method of writing down the different coefficients in a compact, matrix-like diagram. It is not just a matrix, though; it is a tensor, which is a geometrical construct that itself conforms to certain strict rules of transformation from one observer to another. It is another example of an invariant quantity. Tensors are generalizations of vectors, which are also invariants. Second, the metric allows you to calculate the Pythagorean theorem in curved spacetime, which is vital for general relativity. Curved space is fairly simple to understand. Just imagine the difference between calculating the combination of three miles west and five miles north on a flat map versus on a spherical globe. The answers are not the same, not to mind the calculation. But the metric encodes the rules of geometry for the two different surfaces. Unfortunately, the metric of general relativity is four-dimensional, not two-dimensional, which makes the metric much more complex, but the principle is still the same. It was this more complex mathematics that underpinned the general theory of relativity when Einstein finally completed it, after years of exhausting effort, in late 1915. Only a few years later, the eclipse expeditions of 1919 would proclaim him to the world as the man who had overthrown Newton's famous theory of gravity.

Three Observational Tests

In 1907 Einstein was looking out the window of the patent office in Bern when he had the happiest thought of his life, as he later described it. He asked himself if a man falling off a roof would feel his own weight. He realized that the answer was, surprisingly, no. No one had ever experienced the weightlessness of free fall up to that time (at least no one who lived to tell of it), yet Einstein was able to intuit its existence. He did this, as he did with his relativity principle, by building on the work of Galileo. The great Italian physicist had shown that all objects fall at the same rate of acceleration regardless of their size or composition. Other physicists had confirmed the result with very careful experiments. Einstein realized that the law of falling meant that a man suspended from a roof in a painter's scaffold will, when the rope snaps, fall to the street while maintaining the same velocity as the scaffold itself. Everything in the scaffold, buckets of paint and all, will be undisturbed. Nothing will move relative to anything else, since everything falls equally under gravity. At first, he may not even know he is falling. But he will then realize that the scaffold is no longer pressing up against his feet to stop him from falling. His own weight has disappeared! This is what happens when an astronaut orbits the Earth inside a spaceship. Gravity still exists and the Earth still pulls the astronaut downward, but since the spaceship is falling at the exact same rate, the astronaut floats freely inside (or alongside) it, with no force between them. This is the weightlessness of free fall.

Einstein took the idea further. Suppose a man is in a closed room inside a rocket in deep space. If the rocket accelerated forward, the floor of the room would start pressing upward on the man's feet. Would this not feel like gravity? How would the man know, if the rocket accelerated at 9.8 meters per second squared, that he was not simply back on the Earth? The answer: he would not know, unless he looked out the window to check. Here, Einstein realized, he had a principle quite like the relativity principle. But whereas the relativity principle allowed him to compare inertial observers but not accelerated observers, this new principle would allow him to compare accelerated observers to observers experiencing gravity. He called this principle the *principle of equivalence* and planned to use it to complete his theory of relativity. The theory of 1905 became the special theory of relativity (or restricted theory of relativity) because it could only deal with inertial observers. The new theory would be the general theory of relativity, and it would be a theory of gravity, as well as accelerated motion.

Armed with the equivalence principle, Einstein was able to make a couple of immediate and very important predictions about his new theory of gravity, without having discovered its equations at all. The first of these followed directly from the equivalence principle. Einstein's work on special relativity predicted that a rapidly moving object appeared to stationary observers to have its onboard clocks slowed in time. An atom is a sort of clock, and the spectral lines it emits have frequencies that tell us the rate at which the clock is running. Accordingly, spectral lines from a speeding atom should have reduced frequencies and a reddish appearance. This effect, predicted by Einstein, is sometimes called the *transverse Doppler shift* because it is similar to the more familiar Doppler effect. This redshift was an important prediction of the special theory. The prediction held not only for an inertial body but also for a body moving under acceleration. But such a body, by the equivalence principle, would behave the same way if it was in a gravitational field. Thus, an atom situated at the surface of the Sun, where gravity is stronger than on Earth, should emit light that is slightly redder than the same atom would on Earth. This solar redshift is the first of Einstein's three classical tests of general relativity.

The second test is the one that concerns us. If we think back to Einstein's thought experiment of a scientist conducting experiments in a room on board an accelerating spaceship, we can ask ourselves what the scientist would report if she were to shoot a beam of light across the room. At the time of emission, the light would partake of the motion of the spaceship, and so it would initially keep pace with it, shooting horizontally across the space between the walls. But as soon as it left one wall, it would no longer be pushed by the spaceship. Since the spaceship accelerates, it would follow that the light beam would now be moving forward more slowly than the spaceship. The scientist, who would still be accelerating with the spaceship, would see the light beam fall toward the floor. The equivalence principle demands that the same thing would happen if the spaceship were stationary on Earth. Could this be correct? Would light really fall in a gravitational field? Most scientists of the time assumed that it would not. Light had no mass; it was merely a disturbance in the ether. There seemed no reason to expect it to interact gravitationally.

Einstein thought otherwise. In 1905 he had an ingenious idea right after publishing his first paper on special relativity. It occurred to him that the theory had one more remarkable prediction. Later, he proposed a clever thought experiment to show that light must have mass. He realized that light could push a spaceship because Maxwell had shown that it carried momentum and therefore could exert pressure. This is the same radiation pressure that Eddington would later put to good use in preventing the Sun from collapsing. This force, which light exerts, could even move a spaceship from the inside, at least briefly, if one simply fired a beam from one side of the ship to the other. This method of moving a spaceship from the inside would work unless the light itself had some mass that would shift the spaceship's center of gravity to counterbalance the motion. In other words, Maxwell had shown that light would push the spacecraft to the left when it moved across the inside of the spacecraft from left to right. Since one cannot normally move a spaceship without a propellant ejected to the outside, Einstein deduced that the light must carry some mass with it to the right so that the center of mass of the spaceship would not move. This follows the principle expounded in Newton's third law of motion that you cannot pick yourself up by your own bootstraps. This argument led to Einstein's famous equation $E = mc^2$ that all energy, in particular light energy, has mass. And if light has mass, it should fall in a gravitational field. This idea came naturally to Einstein, who thought of light as a particle.

How could these two tests actually be carried out? Einstein knew that solar astronomers had already observed a redshift in spectral lines in the Sun. Perhaps this would turn out to be the right size for his theory. Since the Sun has the strongest gravitational field in the solar system, it was also possible that starlight, passing close by it, would be sufficiently deflected from its path to affect the apparent positions of the relevant stars. Stars cannot be observed in daylight when they are close to the Sun, but remarkably, our planet is one where occasionally our Moon completely obscures the Sun. Perhaps during a total solar eclipse, the deflection of starlight could be measured. A few years later, Einstein would begin trying to interest astronomers in testing his ideas.

By 1907 he already had a third test in mind that he hoped his theory might pass. That test involved the most celebrated anomaly in Newtonian gravitational theory at that time, the anomalous perihelion advance of Mercury. Astronomers had noticed a minuscule aberration in the orbit of Mercury during the nineteenth century, and it was the chief reason for their interest in seeking the planet Vulcan. The gravitational pull of such an "intra-Mercurial" planet would explain why Mercury's orbit did not agree with the Newtonian prediction. This anomaly appealed to Einstein because it was the only way then known in which Newtonian theory might actually be wrong in its predictions of the motions of the solar system. But the equivalence principle did not suggest why his theory would predict anything different from Newton's for the motion of Mercury. But there was hope because Mercury, the planet closest to the Sun, experiences the strongest gravitational field of any planet in the solar system. Intuition suggested that strong gravitational fields would be where Newtonian gravity would be most likely to be wrong. Nevertheless, Einstein said nothing publicly about this test in 1907, choosing not to propose a possible test that his future theory might or might not pass. We know only of his hopes in this area from a letter to a close friend, Conrad Habicht, in which Einstein says, "At the moment I am working on a relativistic analysis of the law of gravitation by means of which I hope to explain the still unexplained secular changes in the perihelion of Mercury" (*CPAE*, vol. 5, doc. 69, December 24, 1907). A manuscript from 1913 in which he and his closest friend Michele Besso calculated the amount of perihelion advance predicted for Mercury by his Entwurf theory (a theory he had worked out with his other friend Marcel Grossmann) shows that this "private" third test was a key one for Einstein. It is likely that he would have persevered with the Entwurf theory had it passed this test. As it happened, it was only when he had finished his final theory in 1915 that it all fell into place. Indeed, he wrote to colleagues after this success, giving the Mercury perihelion result as one reason for his rejection of the Entwurf theory (Einstein 1995, 344–59). By then he had realized that his theory required spacetime to be curved, and this realization would not only affect the Mercury perihelion test but also change his prediction of the light deflection test, doubling it in size compared to his original prediction.

Spacetime Curvature and the Three Tests

As we shall see, Einstein had a difficult few years beginning in 1914. Yet he accomplished one of the great achievements of scientific history in those years. He founded the field equations of gravity, called the *Einstein equations*, that we still employ today and used those equations to establish the modern fields of gravitational wave theory and cosmology. Although cosmology existed as a subject before 1916, Einstein completely transformed it. He showed how scientists could meaningfully talk about the shape or the geometry of the universe and how the objects in the universe determined this structure of spacetime. All the matter and radiation we see has mass, which produces a gravitational field, which determines the geometric properties of the universe as a whole. In traditional physics the geometry of space was assumed to be a given, ordained from above, that determined the fates of the objects within it, including ourselves. In modern cosmology the things in the universe determine its geometry, which then determines their fate. In the words of one modern relativist, "Spacetime tells matter how to move; matter tells spacetime how to curve."

As with all of his thinking about gravity, Einstein came to the realization that spacetime might be curved via the equivalence principle. One of the central discoveries of special relativity is the phenomenon of length contraction—the idea that when we measure objects in motion, we find that they appear shorter, in the direction of motion, than they do when they are at rest. This idea originated before Einstein, in the work of the Irish physicist George Fitzgerald and, more fully, the Dutch physicist Hendrik Antoon Lorentz. In their minds, length contraction was a physical shortening of objects that occurred due to forces acting upon them as they attempted to move through the electromagnetic field. The original conception was that objects at rest in the ether (the medium of the field, according to nineteenth-century physicists) would feel no such force. Length contraction, among other things, provided an explanation of why it is impossible to measure your own velocity through the ether using experiments like the famous Michelson-Morley experiment. This experiment was designed to detect the ether wind by measuring the time it took light to travel along a tube going upwind against the ether. The headwind would increase the time it took for the light to travel along the tube. However, the motion against the ether would decrease the length of the tube, canceling out the headwind effect.

Einstein, in typical fashion, took the idea of length contraction and turned it on its head. If length contraction explained why we cannot detect the ether, then his response was to drop the ether. Why spend time worrying about things that you cannot detect? But to Lorentz, the ether explained the contraction by the force it presumably exerted on moving objects. To Einstein, the contraction was not due to a physical force at all. It was a measurement effect created by the finite speed of light. The ether did not create the effect by objecting to things being in motion with respect to it. Rather, when we as observers try to measure things in motion with respect to ourselves, we find that they have moved by the time information about their positions reaches us. This affects our measurement. The key idea is that no inertial observers are entitled to say that they are the ones at rest. If two observers are in constant but different states of motion, then each will claim the other has shrunk. They are not experiencing a force that crushes them until they are shorter. Each feels normal. It is only when they look at each other that they notice anything unusual, and then neither can agree which is "really" moving because they each think the other has shrunk!

Now this raises the question of whether all such relativistic changes of length are more apparent than real. In the years after 1905, people such as Einstein's friend Paul Ehrenfest considered objects that were not in constant inertial motion but were instead accelerating, such as a rotating disk. If you sit on a merry-go-around and I watch from a park bench, we will readily agree that I am at rest, and you are in motion. You are no longer an inertial observer—you can feel the centrifugal force associated with your rotation. So in this case, I think you have shrunk, and you agree that I am now fatter than you, as I sit there on the bench. Does this mean that you are really being physically shrunk? Will you feel discomfort? Could you possibly be killed by having your organs distorted if the merry-go-round starts whipping you around at near light speed?

The problem of the rotating disk was debated quite a bit in the years between 1905 and 1915. Physicists concluded that the disk cannot hold itself together against the forces trying to pull it apart because its own internal forces can make themselves felt across the disk only at the speed of sound. If we assume that sound cannot travel faster than light, then we realize that no object can be perfectly rigid. Therefore, no material can hold the disk together against the kind of forces exerted upon it if it is set in motion too quickly. By introducing acceleration, the length contraction, which is only a disagreement about measurement in special relativity, becomes something that really can rip objects apart. From this realization, Einstein deduced two important points. The field of force induced upon an object by accelerating it can be understood as a change in the geometry of the spacetime in which the object resides. Furthermore, the equivalence principle demanded that the same must be true of a gravitational field. To show that the rotating disk has undergone a change in its geometry, consider the following thought experiment.

Let's have you, sitting on the merry-go-round, measure its circumference using a meterstick and then measure its diameter. Now divide one by the other to calculate pi. What do I see as I watch this from the bench and compare it with measurements I made earlier, before you set the merry-go-round in motion? When you lay your meterstick along the outer edge of the disk to measure its circumference, I will of course see it as shrunk because it is in motion. Thus, you will have to use more metersticks to measure the circumference than I required before it was spinning. The circle of spacetime in which the disk resides has a greater circumference than it did before. But when you place the metersticks along the diameter, though they will appear thinner, they are still metersticks because they are not moving in the direction of their length. So the diameter is unchanged. This means that the ratio of the circumference and the diameter is no longer pi! Something changed about the geometry of the disk when we set it into rotation. This is what we mean when we say that the disk has a different geometry when it is rotating.[1]

For Einstein, if rotational motion, which is a form of acceleration, can alter geometry, then it follows that gravity can also. This is one aspect of the equivalence principle. What applies to acceleration also applies to gravity. Although Einstein took several years to think through all of these arguments, by 1912, when he returned to Zurich from Prague to take up an appointment at his alma mater, the Swiss Federal Polytechnic, he had decided that what he needed for his theory was a branch of mathematics that dealt with non-Euclidean geometries and could do so in four dimensions, the number of dimensions of spacetime. But did such a geometry even exist? Einstein went to his old college friend Marcel Grossmann, by then also a professor in Zurich. "Grossmann, you must help me or else I'll go crazy" (Pais 1982, 212). Luckily, Grossmann was able to tell him that such a geometry, developed in the nineteenth century, did exist. It was known as Riemannian geometry.

It is often said that Einstein, in his student days, depended on Grossmann's dependable note taking. Einstein was, it is said, prone to skipping some of his math classes, probably in order to spend more time on physics. Grossmann, whose focus was on math, was the more diligent student who took good notes, which was useful at exam time. Einstein had paid enough attention to his mathematics to know there was such a thing as non-Euclidean geometry, but the Gaussian geometry that he was taught dealt only with two dimensions. Grossmann was able to reassure him that mathematicians, with their usual thoroughness, had generalized the idea to higher dimensions. He began to help Einstein learn the mathematics he needed to finish his theory, and they collaborated on the Entwurf theory. Einstein had always neglected sophisticated mathematical methods. He suddenly found that building the theory his physical insight was guiding him toward required him to become very sophisticated indeed. It has been argued that this permanently shaped, or even warped, his research style. Some historians even blame the lack of success of his later research on this change in style.

What is non-Euclidean geometry? Euclid systemized (and extended) the vast geometric knowledge of classical Greece. Einstein borrowed from Euclid's methods in that he founded his relativity theories upon certain principles or axioms, from which the rest of the theory could be logically deduced. Famously, *Euclid's Elements*, his book on geometry, begins with five postulates, or axioms, that state certain basic truths about geometry. These axioms cannot themselves be proven, but from them one can prove all the other truths of geometry. Euclid's fifth postulate has always attracted attention. A modern version of this postulate states that two straight lines that intersect one another cannot both be parallel to the same straight line. A group of lines that all intersect each other at the same point are called *congruent* lines. This postulate says that only one of these congruent lines can be parallel to a given line that is not part of the congruence. It follows that none of the congruent lines can be parallel to each other. In other words, parallel lines never intersect each other.

This idea seems so self-evident that over the more than two millennia separating Euclid from Einstein, many mathematicians became obsessed by the idea that the fifth postulate was unnecessary. They tried to find ways to dispense with it, such as proving it from the other four postulates. Euclid's style is elegant and concise. Nothing superfluous is included in his *Elements*, so the five postulates should be the minimum number of axioms needed to prove everything contained in the book. Imagine the fame awaiting the mathematician who could improve upon Euclid by reducing the number of postulates from five to four. One approach assumed the postulate was not true and tried to show that the results were so absurd as to be impossible. An example of this approach is the work of the great medieval Persian mathematician Omar Khayyam, who argued that there are three possible cases for parallel lines. Either they tend to converge (in which case they are probably congruent as well as parallel), or they diverge, or they do neither. The last possibility gives us Euclid's fifth postulate. Khayyam argued that the other two possibilities produce nonsense; therefore, there is really no need to include the fifth postulate, since there can be no non-Euclidean geometry. Eventually, but only in the nineteenth century, some mathematicians began to wonder if these absurd non-Euclidean geometries were really so absurd after all. The first to boldly say so in public were Nikolai Lobachevsky and János Bolyai, though Carl Friedrich Gauss claimed to have previously worked out a non-Euclidean geometry that was published after their work appeared. It was Gauss' work that Einstein had previously encountered when he made his appeal to Grossmann.

What Gauss and others showed is that Euclid's geometry works only on a plane or a flat two-dimensional surface. We can imagine a non-Euclidean geometry quite easily, since we live on a two-dimensional surface that is not flat but curved. Living on a sphere, we understand that lines can be both parallel and congruent. Lines of longitude are parallel lines in spherical geometry, but they all meet at the poles. Why is this so? Consider a person standing on the equator. The lines of longitude point north and south from there and are all perpendicular to the equator itself, which runs east and west. Thus, the lines of longitude are all parallel to each other, according to the definition of Omar Khayyam, which depends upon the idea of right angles or perpendiculars. In plane geometry, the lines of longitude never get any closer than they do at the equator and thus can never intersect. But on a sphere, they get closer to each other as they go farther north because the definition of northward brings them all toward the same point, the North Pole. Paradoxically, they also get closer together as we move to the south! Thus, a group of lines that meet the definition of parallel at the equator also meet the definition of congruence at the poles. They are, in some sense, both parallel and congruent. This is a non-Euclidean geometry because the fifth postulate is violated. You might argue that lines of longitude are not really straight. It is true that they are curved, but they are still geodesics of the geometry in which they exist. A geodesic is the shortest distance between two points. On a plane, this is a straight line. On a sphere, this is a great circle, and lines of longitude are classic examples of great circles (airplanes fly along them, for instance). Therefore, we have here a congruent set of parallel geodesics, something Euclid claimed could never be.[2]

There is one other terminological issue with non-Euclidean geometry. A straight line to Euclid becomes a geodesic when we wish to be general. Similarly, the term *flat* no longer refers to two-dimensionality. When we discuss whether our four-dimensional universe is Euclidean, we often say that if it were Euclidean, it would be "flat." Here, flatness refers not to the number of dimensions (obviously, we live in a four-dimensional spacetime, not in a two-dimensional space) but to the property of being Euclidean, which is to say having parallel lines that never intersect. If we draw a set of lines parallel to each other in our solar system and if we send spaceships flying along them, would the spaceships ever converge at a distant point or would they truly never meet? This is the question Einstein asked himself about cosmology in 1916 after finishing his theory. He answered that he thought the spaceships would converge; therefore, we live in a spherical universe. In doing so he answered an age-old paradox about whether the universe is infinite in extent (or "open") or whether it must have a boundary or edge. Einstein showed that it could be both finite (or "closed") without an edge in the same way that it is impossible to sail off the edge of the world, which is not infinite in extent. We could be living on a four-dimensional hypersphere. As long as the radius of this hypersphere is sufficiently large, it will seem flat locally, but parallel geodesics will still converge in the far distance. How could such a universe be? In general relativity, this is easy. It simply needs to contain sufficient mass. The gravitational field of all those massive objects would warp or curve the spacetime away from flatness, giving it the curvature of a hypersphere.[3] Even if the whole universe remained flat (*asymptotically flat*, meaning that it is still infinite in extent), it would still have areas of nonzero curvature locally, around massive objects like the Sun.

So what is the geometry of spacetime like close to the Sun? It is not flat, and this has certain implications for physics in this region. For one thing, circles are not quite what they were before. Since the geometry of a circle changes, it follows that as Mercury makes its orbit of the Sun, it does not cover a full twice-pi radians in one orbit. It actually moves through less than that. In trying to orbit, it consistently overshoots by a tiny amount—exactly that found by astronomers to be anomalous. It was this calculation, performed in November 1915 as soon as he had discovered his field equations, that finally convinced Einstein that his theory was correct. He said himself that he had "heart palpitations," such was his excitement (Pais 1982, 253).

What about straight lines, or geodesics, as we now should call them? An easy way to investigate this is to consider light rays, which

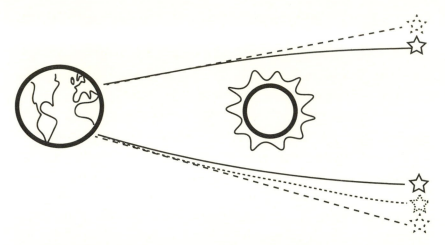

FIGURE 5. How light deflection happens as starlight passes by the Sun on its way to Earth. Above the Sun we see the so-called Newtonian prediction made by Einstein in 1911. The actual path the starlight follows is the solid line, and the solid star is the real position of the star. The path is curved because the light falls toward the Sun when it is close to the Sun. An observer on Earth presumes that the light has traveled along a straight line, which is the dashed line. The observer therefore thinks the star is at the position of the dashed star at the end of the dashed line. Below the Sun is the situation predicted by Einstein in 1915 after developing general relativity. The straight line and the apparent position of the star are still illustrated by dashed lines. The actual path of the starlight and the actual position of the star are still solid lines. In between, illustrated by a dotted line, is the geodesic traveled by starlight through the curved spacetime of Einstein's theory. This path would be followed by light even if it did not fall in a gravitational field. In this scenario both the tendency of light to fall toward the Sun because of the Sun's gravitational attraction and the curvature of space caused by the Sun's gravity influencing spacetime itself affect the path of the light. The total deflection is twice that of the "Newtonian" case. (Reprinted with the permission of the author.)

follow special geodesics of spacetime called *null geodesics*.[4] Light rays from a distant star as they pass close to the Sun on the way to Earth are almost parallel, diverging only very slightly. As they pass by the Sun, they enter a region of curved spacetime in which parallel geodesics are converging. Thus, the light rays stop diverging and begin slowly converging. If the region of positive curvature were large enough, the light rays would eventually converge as a set of congruent geodesics. The gravitational field would have focused the light rays! This is what is known as a gravitational lens, and Einstein was the first to come up with the idea. The Sun is a very poor lens. The amount by which the light rays converge before they leave the region of noticeable curvature is so tiny it is hard to detect. Nevertheless, this effect led Einstein to change his light deflection prediction, once he had the full version of his theory.

The original prediction had been entirely due to the equivalence principle. This states that light has weight and should fall in a gravitational field. Thus, it no longer follows a straight path because gravity deflects it from its path just a little bit—enough to change the apparent position of a star right up at the limb of the Sun by a small angle. But general relativity also says that gravity actually changes the shape of the geodesics themselves. The shortest path from here to there is no longer a straight line. This change in what *straightness* means also acts to focus light. In fact, it has the same effect as the falling effect does: it shifts the apparent position of a star at the limb of the Sun away from the Sun by an amount equal to that of the falling effect. Thus, Einstein had doubled the size of his light deflection prediction, which had the benefit of making it easier to measure the effect. And suddenly, the stakes were higher too. It would no longer be simply a test of whether light had weight but a test of Einstein's new theory of gravity versus theories, like Newton's, that were comfortable with the hypothesis that light had weight but could hold no truck with the notion of curved space and time. The stakes for future eclipse watchers had just been raised considerably.

4

Europe in Its Madness

By 1911 Einstein was actively trying to interest astronomers in testing his still-unfinished theory. But, in spite of the many influential papers he began publishing in 1905, it took time for Einstein to emerge from deep obscurity. However, in 1911 he took up a prestigious position at the German University in Prague. By then he had finally acquired an academic position in his adopted home of Switzerland, and one biographer has openly wondered why Einstein even accepted the position in Prague. He never really settled there and moved back to Switzerland within two years. But we can imagine that this man, who had always been ambitious in physics, was flattered by the offer from abroad. It tangibly confirmed that he enjoyed an international reputation in his field. The move to Prague did serve to set in motion a process that eventually gave birth to the 1919 expeditions. Einstein was thinking hard about gravity in 1911, and once in Prague, he met an astronomer named Leo Pollak and told him of his hopes to have his predictions tested. Pollak wrote to his fellow astronomers on Einstein's behalf (Hentschel 1997). One of those was a young man in Berlin named Erwin Finlay Freundlich, who was the only astronomer to leap at the chance to work with Einstein.

Freundlich was twenty-six years old in 1911. He had studied at the University of Göttingen in Germany, where one of his teachers was the great mathematician Felix Klein. Although he studied astronomy as well as math, he was taken aback when Klein found him a position as an assistant at the Royal Observatory in Berlin. Freundlich objected that he knew too little about actual observing to accept such a position, but Klein replied, "You did not come to a University to learn everything, but to learn *how* to learn everything. You will go to Berlin" (Batten 1985, 33; Hentschel 1997, 6). He did indeed learn how to observe but perhaps never how to get along with his superiors. Given his strong mathematical background, he was a natural fit for the role of Einstein's astronomer. He had training in subjects such as differential geometry, which most astronomers of the time lacked, in order to understand general relativity. The theory did not seem alien to him. He was also searching for something to take him away from the humdrum routine of most astronomical work. Pollak wrote to Freundlich that Einstein had proposed two tests of his theory, the light deflection and solar redshift tests, and urgently sought "verification through astronomy." Freundlich would devote much of his career to these tests. He at once wrote to Einstein, who responded enthusiastically: "I would be personally very pleased if you took up this interesting question" (Hentschel 1997, 12).

As a lowly assistant, Freundlich was not in a position to devote the observatory's resources, or even much of his own time, to a theory-testing program. He therefore began by studying existing eclipse photographs to see if he could measure the light deflection in stars on these plates. Einstein had published a paper earlier in the year giving a precise prediction of the size of the light deflection effect. Regrettably, it was not the full deflection he later calculated from the final theory. Even more unfortunately, he predicted a very small shift in the positions of the stars near the Sun. Freundlich found no plates in Berlin that were useful, so he began to make inquiries with astronomers at other observatories. These initial contacts sowed the seeds of many of the later expeditions. Before the end of 1911, he was writing to Campbell at the Lick Observatory. The letter is in his original English, which he spoke quite well, as his mother hailed from Cheltenham in the West of England (von Klueber 1965).

> I apply to you on account of a question of sur[e]ly high scientific interest, in which I depend from the kind support of astronomers. The modern theory of relativity of Mr. Einstein predicts an influence of any field of gravitation upon light passing near to the sun. The gravitation would have according to the investigation of Einstein the effect of deflecting the ray of the star, and Mr. Einstein asked me, if I would try to proof his results by observations. . . .
>
> Now eclipse-plates of the sun showing at the same time a few stars are the only material to be used till now, and I would be very much obliged to you for your kind support to get such plates. A few days ago I had the opportunity to talk to Mr. Perrine at Cordoba [Argentina] on his way through Berlin and he gave me the advice to apply to you on account of this question. (Quoted in Hentschel 1997, 13)

Both Perrine (the next year, 1912) and Campbell (1918 and 1922) would try to conduct the light deflection experiment at subsequent eclipses. For now, Campbell sent Freundlich plates from several earlier eclipses. Most of these plates had been taken as part of the search for the intramercurial planet Vulcan. As such, they had the advantage that the field of view was wider than the disk of the Sun itself. Unfortunately, because Vulcan would be moving along with the Sun, the plates had been taken with the telescope moving so as to keep the Sun fixed on the plates. Since the Sun moves with respect to the stars, the stars were not kept in position, and their images turned out a little streaky on the plates. For astrometry purposes this was obviously not good. Additionally, Vulcan was expected to be on the plane of the ecliptic (the plane in which Earth orbits the Sun, which the other planets also remain close to), so the photographs were taken not with the Sun at the center of the plates but instead off to one side. It would have been helpful to compare the positions of stars on opposite sides of the Sun, which could be expected to have shifted most noticeably with respect to each other. In the end, as Freundlich reported to Einstein, "The inadequate sharpness of the stellar images made any successful measurement of the plates illusory . . . I have therefore delayed investigating the whole question until really usable material is available" (translated and quoted in Hentschel 1997, 14).

Freundlich did, however, publish articles about his efforts, and this helped educate astronomers, at least in German-speaking countries, about the tests Einstein had proposed for his theory. Einstein was particularly appreciative of this aspect of Freundlich's efforts, writing, "It will be thanks to your diligence if the important issue of the deflection of rays of light now awakens an interest among astronomers as well" (Hentschel 1997, 15). Freundlich moved on to his next idea, trying to image stars close to the Sun without an eclipse or, failing that, determining whether Jupiter produced a measurable light deflection on stars in conjunction with it. Neither idea worked, although they did prompt Einstein to write to the great American astronomer George Ellery Hale in 1913 describing his prediction, still at that time only half what the full theory of general relativity would predict and accompanied by a little diagram of a ray of starlight being deflected by its path as it passes by the Sun. Hale replied that travel to an eclipse was unavoidable and referred the problem to Campbell, whose interest in the matter had already been piqued. It became clear to Einstein and Freundlich that a successful measurement would require traveling to an eclipse. Luckily, one was in the offing.

The eclipse of August 21, 1914, offered Freundlich a rare opportunity. The track of totality would run southeastward through Scandinavia and Russia, so he would not have to travel far from eastern Germany to reach a good station for observation. This was important because Freundlich lacked the seniority to organize an expedition from the Berlin observatory, recently relocated to Babelsberg on the outskirts of the city. Even worse, in addition to being junior in rank, he did not get along well with the director of the observatory, Hermann Struve. Freundlich was prone to personal antagonisms, but it is likely that his work for Einstein contributed to Struve's animosity. Struve came from a distinguished family of astronomers and understood the business of the observatory to be the careful, long-term study of the sky. Methodical and painstaking attention to routine and established tasks would add incrementally to the foundations

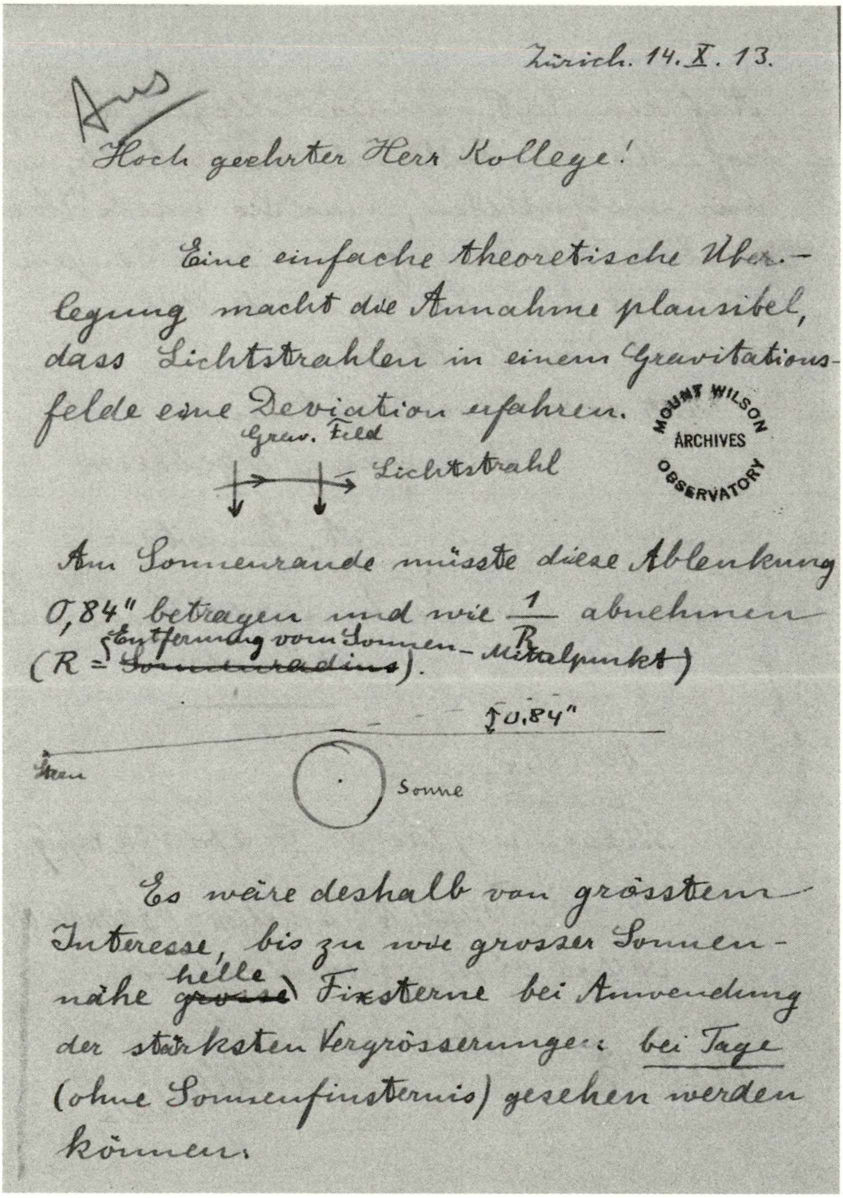

Zürich. 14. X. 13.

Aus

Hoch geehrter Herr Kollege!

Eine einfache theoretische Überlegung macht die Annahme plausibel, dass Lichtstrahlen in einem Gravitationsfelde eine Deviation erfahren.

Grav. Feld

Lichtstrahl

MOUNT WILSON ARCHIVES OBSERVATORY

Am Sonnenrande müsste diese Ablenkung 0,84" betragen und wie $\frac{1}{R}$ abnehmen (R = ~~Sonnenradius~~ Entfernung vom Sonnen-Mittelpunkt).

0,84"

Stern

Sonne

Es wäre deshalb von grösstem Interesse, bis zu wie grosser Sonnennähe ~~grosse~~ helle Fixsterne bei Anwendung der stärksten Vergrösserungen bei Tage (ohne Sonnenfinsternis) gesehen werden können.

FIGURE 6. A 1912 letter from Einstein to American astronomer George Ellery Hale asking if it was possible to image stars close to the Sun in daylight (see the underlined words *bei Tage*) without an eclipse. Notice the diagram in which Einstein illustrated how the bending of light by the Sun takes place. Hale replied that it was highly unlikely that a daylight scheme would work, concluding that "the eclipse method, on the contrary, appears to be very promising." He also noted that he had consulted William Campbell before composing his reply (*CPAE*, vol. 5, doc. 483).
(Courtesy of the Hebrew University of Jerusalem and the Huntington Library, Pasadena, California.)

laid down by earlier generations of astronomers. Theory testing had no place in his world, especially if it involved looking for tiny effects predicted by obscure theorists with newfangled mathematical ideas. Both Freundlich and Einstein were impatient with this approach to science. It had its merits. It was exactly this kind of endeavor that permitted astronomers to measure the tiny anomaly in the motion of Mercury, a process that took decades. Explaining this anomaly would be the first great success of Einstein's theory in 1915. It was made possible because astronomers had measured this extremely small effect so very precisely (it amounts to forty-three seconds of arc change in the perihelion of Mercury's orbit each century). At that time Einstein wrote to a colleague: "The result for the motion of the perihelion of Mercury fills one with great satisfaction. How we are helped here by the pedantic precision of astronomy, which I often secretly poked fun at!" (Einstein to Arnold Sommerfeld, December 9, 1915, *CPAE*, vol. 8, doc. 161). Einstein may have had Struve in mind when he wrote these words.

Germany had no equivalent to England's JPEC, engaged in central planning for eclipses, so without the support of his own observatory, Freundlich was stuck. But by 1914 Einstein had moved to Berlin to take up a very prestigious appointment that afforded him access to private and public sources of funding. He threw himself into the task of raising the money and succeeded. From the outset he assured Freundlich that he would find the money even if Einstein had to contribute it himself. His first stop was the Prussian Academy, which had just raised the money to fund Einstein's own position. He explained to Freundlich:

If the academy is not keen on becoming involved, then we will get that little bit of Mammon from private quarters. Right after the academy has informed us of the negative decision, I will with [Fritz] Haber's help touch Mr. [Leopold] Koppel, who, as you know, had parted with the money for my salary as academician. Should all else fail, I will pay for the matter myself out of my little bit of savings, at least the first 2,000 marks. So, after careful consideration, go ahead and order the [photographic] plates and don't let time slip by on account of money. (Hentschel 1997, 36)

The academy agreed to contribute, and some private funding (from the arms manufacturer Gustav Krupp) was also arranged. But Freundlich's position at the observatory was threatened; and Arnold Sommerfeld, one of the leading physicists in Germany, warned Einstein that his favorite astronomer might be dismissed from his post. Einstein replied that while he agreed that Freundlich was tactless, he thought the main problem was Struve's conservatism. Although he conceded that Freundlich was no scientific genius, he nevertheless valued his zeal and enthusiasm for the cause and argued that the young man should be permitted to continue his work (Hentschel 1997, 32).

Einstein's success in raising the funds permitted Freundlich to go to Russia to observe the eclipse from the Crimea, along with two companions, another Berlin astronomer and a mechanic from the Carl Zeiss optical firm. All their effort was for naught. Less than a month before the eclipse was to take place, with Freundlich and his collaborators already in Russia, World War I began. The three Germans were arrested and interned as enemy aliens, possibly even as suspected spies. Their instruments were confiscated, and they never got anywhere near to observing the eclipse. Freundlich eventually returned to Berlin in early September, after a prisoner exchange between the warring powers. His career as one of the unluckiest eclipse chasers ever was off to a flying start. He would be clouded out at least twice more before he ever got any data. After all the planning and effort, it was desperately disappointing. The war forced Freundlich to sit out several eclipses, and he ended up being scooped by the English, who had come so late to the game. But he made lemonade out of the lemons he was given. The enormous publicity surrounding the 1919 eclipse results enabled him to finally achieve some independence at the Potsdam Observatory (relocated again to an outer suburb of Berlin). Funding was raised for the Einstein Tower, a solar observatory designed to let Freundlich test the solar redshift prediction of Einstein's theory. He continued to work in this area until forced to leave Germany when the Nazis came to power, as he was of partly

Jewish ancestry. He spent the next war in Scotland, where he began using, as part of his surname, his middle name, Finlay, based on his mother's surname Finlayson, which is Scottish.

What would have transpired if Freundlich and his companions had been able to observe? This is one of scientific history's great might-have-beens. At this time, Einstein was making the wrong prediction because he had not yet finished his theory. Would Freundlich have remained unswayed by the temptation to agree with his mentor, the man who had made it possible for him to go to Russia and possibly saved his job in the bargain? There is every reason to expect he would have happily claimed that Einstein's prediction was too low, as he did eventually decide more than a decade later. But Sommerfeld seems to have had doubts about Freundlich's competence as an astronomer, and testing a theory against the wrong prediction would have been tricky for any experimenter. Whatever Freundlich's results, the publicity would have made Einstein's subsequent triumph in finding the right field equations for his theory seem like a retraction or a correction. He would have been forced to change his prediction after the experiment was performed. Ironically, as Freundlich set off, Einstein expressed his certainty about his then only partially correct prediction: "I have reconsidered the theory from all angles and can only say that I have complete confidence in the matter" (Hentschel 1997, 22). In the end, Einstein was spared all that, but at the cost of great disappointment at the time.

Scientists at War

After World War I, something extraordinary happened. German scientists were ostracized. The International Astronomical Union (IAU) was founded on July 28, 1919, in Brussels. Frank Dyson was one of the vice presidents of the new organization, and William Campbell was another. The IAU still exists as the premier organization uniting astronomers from different parts of the world. When the controversial decision was taken to exclude Pluto from the list of planets, it was the IAU that did it. Yet this body was founded with no German participation at all. The founding countries of the IAU were all members of the Triple Entente, or allied with it, during World War I. In the immediate aftermath of the war, the brave talk of the internationalism of science appears to have died. German scientists, with the exception of Einstein, did not receive invitations to speak at or attend conferences in former Allied countries. They were sharply criticized for their wartime support of German war crimes and for their participation in shocking uses of science, such as poison gas warfare. Many leading astronomers spoke as if this boycott of German science should persist indefinitely.

Why did this happen? Why were relations between scientists who had known each other well before the war now so strained? In the archive of the Greenwich observatory lie clues. Someone at the observatory had carefully preserved a London newspaper clipping discussing zeppelin raids on the Royal Observatory. The clipping took particular issue with the fact that the German press had justified the raid on the grounds that the observatory had provided the Admiralty with meteorological observations and was therefore part of the British war effort. Obviously, the astronomers at Greenwich felt personally targeted by the German war machine. The clipping, from the *Daily Telegraph* in September 1916, was authored by a special correspondent in Rotterdam, with access to the German daily papers, who reports a Captain Persius' effort to defend the usefulness of the zeppelin raids on England in the *Berliner Tageblatt*. The clipping gives a verbatim account of the German article:

> The chief use and object of the airship attacks on England consists in damaging military means and power of our most dangerous enemy. The idea of what are military forces is not a narrow one. Not only may bombs be thrown upon fortified places, warships, and workshops for making shells and ammunition of all kinds in order to destroy them, but they are also intended to destroy places of economic importance, which, if they remain untouched, would add more or less to England's power to continue the war. To the economic places which are looked upon as proper objects for bomb, such as railways, docks and wharves, may be added coal and oil depots, electricity and gas works, buildings which serve

for a meteorological purpose when they are in military hands, *such, for instance, as Greenwich Observatory*. All these are valuable targets, and the list could be continued ad infinitum. (Frank Dyson Papers, RGO Archive 8, Cambridge University Library)

Someone has underlined in pen the words italicized in the quotation.

Given the article's date, the bombing raid referred to was almost certainly the one carried out on the night of September 7, 1916. Interestingly, according to one version of events, the bombing of the Royal Observatory was carried out not by a zeppelin but by an airship built by the Schütte-Lanz company for the German army. These wooden-framed airships featured some advanced aeronautical technology but suffered compared to the aluminum-framed zeppelins in wet weather. *SL2*, captained by Hauptmann Richard von Wobeser, bombed the docklands area that night, and one of the bombs landed near, or maybe even caused minor damage to, the observatory (Castle 2015, 40).[1]

Though the debate about justifying the bombing gives the impression that the observatory was singled out for attention,[2] the airship in question was probably relieved to off-load its bombs anywhere in the vicinity of London's dockland. These aerial giants preferred to fly high to avoid aircraft and were not noted for their easy handling, especially in poor weather. On the same night as the observatory bombing, a zeppelin wandered around the countryside not far from London, doing its best to devastate agricultural production in the home counties (Castle 2015). Captain Persius seems to admit that the zeppelin raids were proving to be something less than an outstanding success from a military point of view, though *SL2* did succeed in orphaning one poor child in Deptford whose entire family was killed when a bomb landed on their house (Castle 2015). It must have been on the minds of the aviators that night that earlier the same week (September 3, 1916) the first German airship to be lost over England, the *SL11*, had gone down in Hertfordshire with all her crew after being struck with special incendiary bullets that ignited the hydrogen gas in her envelope.

Another document from the Greenwich archive that shows the war bitterness among scientists is a letter from Rudolf Moritz, a London barrister and amateur astronomer, to Phillip Cowell, a former Greenwich assistant, dated March 1, 1918. In it Moritz attempts to explain Einstein's relativity theory to Cowell. Most of the letter is taken up with an explanation of the special theory. But the author has reservations about the mathematical complexity of the general theory, based upon the differential geometry of the German mathematicians Bernhard Riemann and Elwin Christoffel. When he turns to the general theory, he has this to say:

> So much for the first theory of Relativity. I can follow it all analytically and physically and I believe it is true. The second theory of Einstein in 1914 [actually, 1915] is far more speculative and I fear only accord with observations can make me accept it. Besides the analysis is too beastly for words. I can well understand the compatriots of Riemann and Christoffel burning Louvain & sinking the Lusitania. (Frank Dyson Papers, RGO Archive 8, Cambridge University Library)

So here we learn that, in introducing differential geometry into physics, Einstein had, in Moritz' view, committed a war crime on a par with the most infamous so far committed by Germany during the war.

Inevitably, war provides many occasions to reflect on the atrocities of the enemy but is sparing when it comes to the wrongs committed by one's own side. For instance, refugees seeking aid in Great Britain were naturally from countries occupied by Germany. One such refugee was astronomer Robert Jonckheere, of the Lille Observatory in Belgium. In November 1914 he attended a meeting of the Royal Astronomical Society, where he recounted his experience of fleeing his country in the face of German artillery (Shears 2014). Accounts such as this will certainly have convinced many astronomers of the horribleness of this war and the German enemy in particular. Of course, we must remember that the feeling was mutual. In early 1915 the Dutch astronomer Kapteyn wrote to Eddington, informing him how to get in touch with the great German astronomer Karl Schwarzschild. But Kapteyn warned Eddington: "Though I know that you,

Dyson and perhaps some others are just as high in his esteem as ever, it might be well not to write to him direct. As with all Germans . . . his feelings against England—infinitely more than against France or Russia—are extremely bitter."[3] He offered to serve as an intermediary, reflecting the near total lack of communication between scientists in the belligerent countries. Some insight into the origins of Schwarzs-child's bitterness is provided by the following thoughts from a British astronomer, also in January 1915, about whether it would be possible to mount an expedition to observe the eclipse of 1916:

> Of course as regards such matters it was unsafe to prophesy, but his own strong conviction was that the war would be over some considerable time before the month of February 1916. He could not help thinking . . . that Germany was . . . getting very near starvation as regarded food supplies. He did not think it required a very sanguine man to imagine that there was a fair prospect of seeing the eclipse of February 1916, with the world more or less at peace. (Quoted in Shears 2014, 5)

So if only the Germans would hurry up and starve, the British would be able to get on with their eclipse observations without too much disruption.

By the middle of the war, even astronomers who were previously very involved in international collaborations had had enough. Herbert Turner (who would later serve on the 1919 eclipse subcommittee), writing in the *Observatory*, came out against the idea of including German astronomers in future endeavors of an international nature:

> At the declaration of War . . . the general attitude of scientific men was . . . that "Science was above all politics." . . . It would not be possible to adopt this attitude now. We have seen how engagements and relationships, which we all thought were "above all politics," and safe to be respected even in time of war itself, have, nevertheless, been broken and tossed aside in a moment if Germany took the fancy that it could thereby benefit itself. Many of us do not see how, after such an exhibition, we can face the

mockery of new understandings and undertakings with such a nation. (Quoted in Stanley 2007, 87)

That Turner was speaking for the majority is suggested by the lone voice of dissent, in the form of a letter from Eddington, to the same journal the following month:

> I think that astronomers in this country realize the disaster to progress which would result from dissolution of partnership, and there is no disposition to belittle the contributions of Germany. Some of the problems of our science can only be attacked by world-wide cooperation . . . the lines of latitude and longitude pay no regard to national boundaries. But, above all, there is the conviction that the pursuit of truth, whether in the minute structure of the atom or in the vast system of the stars, is a bond transcending human differences—to use it as a barrier fortifying national feuds is a degradation of the fair name of science. (Quoted in Stanley 2007, 88)

He urged his fellow scientists to concentrate on the personal rather than the national, inviting them to "think, not of a symbolic German, but of your former friend Prof. X, for instance—call him Hun, pirate, baby-killer, and try to work up a little fury. The attempt breaks down ludicrously" (quoted in Stanley 2007, 89).

But his efforts were in vain. The great physicist Joseph Larmor weighed in against him, and Turner replied in strident terms, arguing that his own opposition to reconciliation with Germany was based upon

> hard, horrible facts, such as we should not have believed possible before the war. He [Eddington] proposes to shut his eyes to these facts, and to test the situation by the play of our imaginations in connection with some individual. . . . Surely Prof. Eddington is here using his preconceptions formed before the war, and his own shrinking from horrors, to help him in ignoring actual hard facts? Is it not an actual fact that babies have been killed in ways almost inconceivably brutal, and not as a mere individual excess, but as a part of the deliberate and declared policy of the German army?[4] Is

it not a fact that the Lusitania was sunk with a national rejoicing that puts the cold-bloodedness of former pirates to shame?[5] Is it not a fact that German men of science have gone out of their way to declare their adhesion to these things, and that one of them who ventures some excuse still boasts of a "quiet conscience"? If we cast our memories back before the war, it is easy to recall that we should have vowed these things incredible; but that does not alter facts. (Quoted in Stanley 2007, 90–91)

Still, in war, atrocities are not uncommon. Why had German scientists become so associated with war crimes? Partly, as Turner mentions, because they stood up to be counted in support of them! In the early months of the war, many people, especially in the Allied countries, were outraged at the violation of Belgian neutrality, and the destruction of part of the town of Louvain during the conquest of that country had, as we have seen, become a staple of Entente war propaganda. A group of German intellectuals and academics, including scientists such as Max Planck, came together to sign the infamous "Manifesto of the Ninety-Three," in which they proudly stated, "The German Army and the German people are one." They responded to specific accusations of war crimes in a manner that they took for honesty but that shocked many in their audience and, certainly, their audience abroad. About Louvain, they had this to say:

It is not true that our troops treated Louvain brutally. Furious inhabitants having treacherously fallen upon them in their quarters, our troops with aching hearts were obliged to fire a part of the town as a punishment. The greatest part of Louvain has been preserved. The famous Town Hall stands quite intact; for at great self-sacrifice our soldiers saved it from destruction by the flames. Every German would of course greatly regret if in the course of this terrible war any works of art should already have been destroyed or be destroyed at some future time, but inasmuch as in our great love for art we cannot be surpassed by any other nation, in the same degree we must decidedly refuse to buy a German defeat at the cost of saving a work of art.

Here, the ninety-three intellectuals forthrightly defended the practice of collective punishment for civilian populations that harbored guerrilla fighters. They also pointed out that warring nations typically prefer to burn and destroy rather than risk defeat. But since the Belgians had not actually declared war upon Germany (the ninety-three insisted they had effectively done so by asking for Allied help in the event that they would be attacked), this robust defense of militarist realism rang particularly harshly on the ear. The Rape of Belgium, Louvain in particular, gained Germany a reputation for ruthlessness with which some Germans seemed unsettlingly comfortable. The defense of this harsh treatment of a country that had not declared war on Germany earned the signatories an unsavory reputation. Many regretted signing it afterward.

There was an attempt to counter the idea that German intellectuals placed military victory before all else, and Einstein was a part of it. One academic wrote and circulated a countermanifesto, called "Manifesto to the Europeans." It contained sentiments such as the following:

> It seems not just *a good thing*, but a dire *necessity, that educated men of all nations* direct their influence in such a way that the *terms of the peace not become the wellspring of future wars*—uncertain though the outcome of the war may now still seem. The fact that this war has plunged all European relations into an equally *unstable* and *plastic state* should rather be put to use to create out of Europe an organic whole. (*CPAE*, vol. 6, doc. 8)

Einstein was one of only three German academics (in addition to its author) who were willing to sign this manifesto, and it was not published. Meanwhile, his unease with the war only deepened as it dragged on. His new colleagues in Berlin were not, like him, horrified bystanders. They were patriots, anxious to do their bit for their country, autocratic government and all. Fitz Haber was, in the eyes of the British, the chief reason for the war dragging on. The optimistic assessment that "the war will be over by Christmas" (or at least by February 1916) was not just wishful thinking, at least from the British side. They had a plan to ensure that the war would be of short duration.

Their powerful navy, aided by the accidents of geography, was perfectly capable of blockading Germany's restricted access to Atlantic waters. This was a lifeline for Germany because she depended on imports of fertilizer, mostly guano from faraway countries like Chile, for her nitrates. These nitrates were necessary for two reasons. Obviously, the first was their use in agriculture, which became all the more essential as the blockade made importing food impossible. This put a terrible strain on the German agricultural sector's ability to feed the country's large industrial population. The second was the fact that nitrates were a key ingredient of high explosives. But, not long before the war, Haber had invented a method of extracting (or *fixing*) nitrogen from the air, where it is plentiful in gaseous form. This had then been industrialized in the Haber-Bosch process. At a stroke this discovery, arguably the achievement of twentieth-century science that has done the most to benefit mankind, undermined Britain's plan for a quick end to the war. Naturally, as the war dragged on and casualty lists lengthened, British patriots viewed Haber's ingenuity in a dark light, but it was nothing compared to their outrage over his follow-up contribution.

Einstein's other Berlin friend Walther Nernst, a chemist like Haber, was also anxious to help his country. He experimented with the use of poison gas as a weapon of war, but his methods reportedly inconvenienced the enemy so little that they were unaware they were the targets of a chemical attack. It was Haber who took up the research and decided to use chlorine gas. With his industrial experience, he was instrumental in making a battlefield weapon out of Nernst's idea. The result was one of the most horrific weapons of the war. Chlorine poisoning is a particularly painful way to die. As a weapon, poison gas turned out, like airships, to be less than totally successful. It can easily cause casualties on one's own side if the wind shifts at the wrong moment, and the successes that it did enjoy depended on the immobility of the front line in the Great War. But soldiers who saw a greenish cloud of chlorine coming toward them were terrified. The most vulnerable were the wounded, who would usually be lying down, the worst position to be in as the heavy gas stayed low and crept along the ground. The gas formed acid inside the lungs of those who breathed it, and victims drowned in the liquid created from tissue dissolving inside their own bodies. Allied opinion was outraged, even as their own militaries rushed to introduce new chemical weapons in response. The era of the scientist as an orchestrator of destruction on a massive scale had arrived. In the public mind, the nationality of these evil geniuses was German.

Oddly enough, governments did not anticipate the effectiveness of science in wartime. Most militaries, especially Germany's and Britain's, had plans to win the war quickly. They did not expect to have sufficient time for research to produce new weapons that would reach the front before the conflict ended. Accordingly, no effort was made to avoid conscripting scientists. Eddington's exemption was unusual. It was only during World War II that scientists were spared the front, even when they volunteered for it, and were instead assigned to research facilities. Important scientists—like Henry Moseley, who was shot in the head at Gallipoli while fighting for the British—died at the front during the Great War. He was the first to measure atomic number, the physical underpinning of the periodic table of elements, and might have won the Nobel Prize in 1916 for this discovery (no Nobel was awarded that year). A loss quite close to home for Einstein was that of Karl Schwarzschild, who responded immediately to Einstein's publication of general relativity, even while fighting on the Russian front with the German army, by finding the first solution to the theory in early 1916. He died later that year from a congenital skin condition that may have been exacerbated by conditions near the front.

So one can imagine that if scientists and their children were dying at the front, imperiled by weapons designed by former colleagues on the other side, feelings would be bitter in the war's aftermath. One has to remember that the scientists often went to the military with their ideas rather than the other way around. Nowadays, we have a notion that academics are liberal and antimilitary in their views. But Einstein himself often remarked on how isolated he felt in academia and that most professors seemed, to him, to be very nationalistic. Haber was anxious to demonstrate the intensity of his German patriotism. He did not see any contradiction between his Jewish heritage and his German nationality. But others did and, like Einstein, he had to leave Germany when the Nazis took power.

The experience of wartime for pacifists like Eddington and Einstein was one of loneliness and isolation. Many of their colleagues were at war or engaged in war-related research. The public had been enthusiastically in favor of war, and even when it turned out to be both long and bloody, they were committed to supporting the troops, which meant refusing to give in to war weariness. This was especially true when it came to questioning the need for the war. In the immediate aftermath of the war, to find that bitterness and resentment were being allowed to rule the day, even in relations between scientists, was terribly disappointing for men like Eddington and Einstein. The "Manifesto to the Europeans," which Einstein had been almost alone in signing, had warned of just this danger—that the seeds of future conflicts would be sown by the desire for vengeance after the current one. We know that Eddington and Einstein may have hoped to contribute to a change in outlook on this topic, but we must keep in mind that they must have had small hopes of any major successes, given their experience of the previous few years.

Einstein and Eddington Resisting the War

What were Einstein and Eddington able to accomplish, in terms of resisting the war, while it still raged? The answer is that they could not do much, but they did their bit. Einstein's response to the war is summed up in a letter to his friend Ehrenfest, written during the first month of the war: "Europe in its madness has now embarked on something incredibly preposterous. At such times one sees to what deplorable breed of brutes we belong. I am musing serenely along in my peaceful meditations and feel only a mixture of pity and disgust" (Holmes 2017, 3).

Einstein did more than just muse serenely. When his friend Georg Nicolai, who was also the author of the "Manifesto to the Europeans," founded the New Fatherland League, a group devoted to pacifism and the recognition of human rights, Einstein joined and was quite active within it. The authorities, however, suppressed the organization after it published its manifesto, and it was not able to resume its work until after the war. He also joined up with an international group called the Central Organization for a Durable Peace, serving on the national committee of its German branch. His letter of acceptance is fairly typical of him (Holmes 2017, chap. 1, 25): "I have absolutely no experience nor am I a competent person in political affairs. Nonetheless, I am happily ready to support this excellent cause. So do include me, if you consider it beneficial."

Certainly, his political activities attracted the attention of the authorities, who did place some antiwar activists in detention. In early 1918 Einstein's name appeared ninth on a Berlin police blacklist of thirty-one pacifists and socialists in the Berlin area (Holmes 2017, chap. 3, 2). As with Eddington, it is quite possible that Einstein would have ended up in jail but for his high academic standing.

Einstein did not say yes to every political adventure. He would sometimes fall back on the excuse that he was not a German citizen. In response to one proposal that he disliked the smell of, he said, "It is not appropriate for me as a Swiss to involve myself in local political affairs." But even in cases like this, he was forthright in his views. Later, in the same letter, he stated, "As one who was already a pacifist before the War, I have the right to express it now" (Holmes 2017, chap. 3, 10). This letter was written in 1918 when he obviously suspected that some Germans were being converted to the cause of peace now that they feared losing the war. He may have worried that pacifism could be perverted by those who unscrupulously used it as a weapon against their enemies but would later return to a warlike mentality when they once again held the upper hand.

While the war was still ongoing, and more effectively once it had ended, Einstein attempted to do something practical about uncovering the crimes that had been committed in its name. He worked with Belgians and Dutch who were trying to investigate German war crimes. For instance, his friend and colleague, the great Dutch physicist Hendrik Lorentz, was involved with a war crimes commission organized by private citizens. In July 1919 he wrote to Einstein regarding the progress of this and expressed the hope that Einstein would be able to meet with a man who had documentary evidence of the destruction of Louvain. Lorentz commented to Einstein on the extent of the postwar bitterness against Germany, exacerbated by the "deplorable" manifesto of the ninety-three. The result was that "it is clear that for the time being no Germans will be invited" to professional meetings. The only good news was that "there is no discussion of a formal exclusion; the door will be held open for them . . . in the future" (Holmes 2017, 20).

Einstein also tried to do good work within Germany. After the bloody suppression of the antiwar socialists and Communists during the Spartacus uprising, a petition was circulated to hold an inquiry into the murders of the Spartacist leaders Rosa Luxembourg and Karl Liebknecht. Einstein signed it. Of course, he was aware that doing these things antagonized extreme nationalists in Germany. Their view was that burning towns in other countries and murdering antiwar activists was not so much a natural part of warfare as it was the price of victory in war. Since warfare itself was part of the natural interaction between nations, it followed that nations with a conscience must inevitably succumb to nations without one. If men like Einstein represented the conscience of Germany, then a surgical metaphor suggested that the drastic action of physically amputating this conscience might be necessary to alleviate Germany's current suffering brought on by defeat. Sure enough, when former Freikorps troops operating as part of the infamous Organisation Consul assassinated the new Weimar Republic's foreign minister, Walther Rathenau, one of them was found with a list of other targets that included Einstein. By his own courageous actions and as a result of the fame the eclipse had brought him, Einstein had become a target. He spent some time traveling abroad in an effort to let the storm blow over but refused to leave Germany or be pushed out. The country was finally on the road to democracy, and Einstein did not wish to abandon it just when the struggle against antidemocratic forces showed some signs of success.

Einstein's opposition to the war involved more overt political acts than Eddington's, perhaps reflecting the fact that Einstein was more comfortable with being in the public eye. Nevertheless, he had it easier than his English colleague in one important respect: he was not subject to conscription. The two men were not very different in age. Eddington was only three years younger. But Einstein was helped in two ways. First, he had renounced his German citizenship as a young man, most likely precisely to avoid military service. Since conscription was universal in Germany before the war, once out of the system there seems to have never been any particular threat that he would be back drawn into it. This is in spite of the fact that when he won the Nobel Prize a few years after the war, the new German republic insisted he had become a Prussian citizen again when he had taken up his position as a Prussian state employee in 1914. It is worth noting that between the wars he became very publicly associated with a campaign to encourage pacifists and others to refuse conscription. So he certainly approved of what Eddington did during the war.

Unlike Germany, Britain had no military service before the war. There was considerable resistance to its imposition. Many people felt that the war was actually being fought against German militarism; therefore, it would be inappropriate to introduce conscription to fight it, thus violating the rights of the individual conscience. But in 1916 it was instituted since the supply of volunteers was drying up in the face of the horrible slaughter on the western front and in the Dardanelles. In Britain there was a tradition that Quakers, specifically, were entitled to plead exemption to military service through conscientious objector status. This status, though allowed for in the new conscription act, was more nominal than real, however. The state's main concern was to discourage anyone from claiming exemption, and the boards overseeing the program were openly hostile to anyone trying to do so. The issue was a central one of the era. For instance, the effort to introduce conscription in Ireland, then under British rule, played a key role in sparking the successful Irish independence movement that followed World War I.

Some of us are old enough to remember the derogatory term *Conchie*. The first court in which conscientious objectors were forced to stand was the court of public opinion. The government allied itself with public opinion against those who sought exemptions from military service. Eddington was very willing to declare himself a Conchie, but Cambridge University did not wish to be associated with one. It therefore obtained an exemption for him on the grounds of essential work. It was claimed that as the director of the observatory, he could not be spared. But the government decreed that a notice giving the names of exempt workers be posted on the exterior of any place of work that claimed exemptions. The Cambridge Observatory put such a notice up, with Eddington's name the only one on it (Stanley 2007, 136). He was not officially a Conchie, but he certainly stood alone and proud. Such a sign's ostensible purpose was to alert the public that a particular person had an official reason for not being at the front, but it also obviously served to ensure that people knew who was not at the front. Eddington's name was the only one on the posted notice because he was the only one of the observatory astronomers not in the military.

Once conscription was introduced in 1916, Eddington had to deal with difficulties on two fronts. The first was a private front on which he wrestled with his own conscience. When called up, men who wished to claim exemption, or c.o. status, had to go before their local Military Service Tribunal. Many Quakers did so and found it an ordeal. These tribunals felt it necessary to ensure that not going into the army was a more terrifying option than fighting at the front. They adopted tactics such as asking a man who claimed c.o. status if he would fight a German attacking his mother. If he said yes, he could not claim c.o. status. If he said no, he was clearly a coward and a traitor (Stanley 2007, 133–34). Probably the best that a Conchie could hope for, if his conscience allowed it, was to be assigned to a job at or near the front with an ambulance or in some other medical role. Many were sent to prison for refusing to be conscripted into a combat role—or into a noncombat role that too clearly aided the prosecution of the war. After a while the jails filled up, and camps were then opened up, such as an infamous one at Dyce in the north of Scotland (the site is now part of the city of Aberdeen). This camp was closed after one inmate contracted pneumonia and died. Of course, prison and camp guards also took it upon themselves to make sure Conchies did not have an easy ride. In a few cases, those claiming c.o. status were sent to the front anyway, where they could be shot if they refused to obey a direct order to take up arms. In at least some cases, people were allowed to think they were about to go before a firing squad before being told of a reprieve.

So Quakers (and others, such as some radical socialists, including the brother of future Labour prime minister Clement Atlee) were suffering for their cause. It seems that Eddington may have felt guilty about his exemption status. He did not feel guilt because he thought he should be at the front but rather that, in claiming an exemption, he was not showing solidarity with his fellow pacifists who were enduring harsh conditions for their beliefs. Fortunately, the government came to his rescue and opened up the second front in his wartime struggle. Dissatisfied with the number of people still avoiding military service, all existing exemptions were reviewed, and in 1918 Eddington had to go before his own local Military Service Tribunal in Cambridge. He now wished to claim c.o. status, but he found that it was not so easy. Because he had previously claimed exemption, the tribunal ruled that by definition he had no standing as a man of conscience. In general, it disallowed any claims that attempted to avoid service on more than one ground. Furthermore, his colleagues at Cambridge were still not keen on a prominent don seeking c.o. status (Stanley 2007, 135). Eddington was now being frustrated in his attempt to make a principled stand.

If anything, his Cambridge college, Trinity, became more adamant against c.o. status for their fellows as the war raged on because they felt shamed by one of them, Bertrand Russell. Russell, a future Nobel laureate, was one of the few public intellectuals of the twentieth century who rivaled Einstein in fame. He anonymously published a pamphlet highlighting the plight of the c.o.'s who had been sentenced to hard labor. When those distributing the pamphlet were prosecuted, Russell came forward as the author and was convicted, under the Defense of the Realm Act, of interfering with military recruiting. Upon his conviction he was dismissed from his fellowship, over the objections of many other fellows of his college, including Eddington.

His passport was revoked, and police looking for deserters in the audience disrupted a lecture he gave in Manchester (Stanley 2007, 144). He eventually went to jail for his outspoken pacifism. With the example of Russell and many of his fellow Quakers before him, Eddington was no longer disposed to hide behind his status as a scientist doing essential work. His employers and colleagues, however, were more determined than ever that he should.

The particular circumstance leading to Eddington's exemption being revoked was the great German Spring Offensive of 1918. After agreeing to the Treaty of Brest-Litovsk with the new Bolshevik government in Russia in early March, the Germans attacked with the aid of divisions released from the eastern front. They made large inroads into British lines in France, and it seemed possible that the British Expeditionary Force could be cut off from their French allies in what would have been an encirclement similar to that leading to the evacuation at Dunkirk in the next war. Casualties, as usual in Great War battles, were heavy, and manpower was now an urgent necessity. In response to this crisis, the British attempted to introduce conscription into Ireland, which in turn set off the process of the decline of the British Empire that characterized the mid-twentieth century. In these circumstances most exemptions were terminated by drawing their end dates forward (all were issued as ostensibly temporary). Eddington was notified that his would end on April 30. It is likely that he now wished to assert his status as a conscientious objector, and his college and university were determined to avoid having another public pacifist statement made by one of their dons. They wrote the following to the tribunal: "In making application for the exemption of the Director [of the Observatory], Prof. A.S. Eddington, it should be stated that in consequence of the death of the First Assistant in the explosion of the Vanguard,[6] and of the death of the Second Assistant in action in France, the Director is the sole remaining member of the Staff" (Stanley 2007, 145).

We now understand the sad reason why no members of his own staff accompanied Eddington to Principe. No one but himself was left alive. This appeal was successful. The local board agreed to extend his exemption by another three months. This would have been sufficient to see Eddington through to the end of the war because by then American troops arrived to reinforce the Allied ranks, and the German offensive collapsed. But the National Service representative on the tribunal, clearly irked by this local favoritism, appealed the decision (Stanley 2007, 145). Eddington would now have to appear before the tribunal to make his case. The college's worst fear had been realized. To the anger of colleagues such as Larmor and astronomer Hugh Newall,[7] staunch patriots with strong anti-German sentiment as a result of the war, Eddington intended to declare himself a conscientious objector.

Although Eddington wished to assert that he was a conscientious objector, he still hoped the tribunal would see that the alternative work to which he was best suited was astronomy. In fact, the tribunal was not disposed to see any such thing at all. Another Cambridge mathematical physicist, Ebenezer Cunningham, had already been before the tribunal as early as 1915. He had claimed c.o. status as a member, like Eddington, of a small nonconforming church. Since Cunningham was only a college lecturer and not the holder of a distinguished chair, he could not expect to receive an exemption. Cunningham had asked the tribunal to be allowed to work as a teacher but was sent to the countryside to do agricultural work, a typical verdict in such cases (Stanley 2007, 148). Cunningham was at that time the leading exponent of relativity theory (the special theory) in Cambridge. Eddington had largely ignored the theory until 1915. Though Cunningham tried to keep up his studies while working the land, he was obviously not in a position to hear news of Einstein's latest discoveries, since word of them filtered through to Eddington only in his role as the editor of important astronomical journals. After the war, when Cunningham returned to Cambridge, he attempted to teach the new theory of general relativity but found it so difficult to quickly grasp its concepts that he had to abandon the course. He never published in mathematical physics again (Warwick 2014, 479).

So Eddington had reason to be anxious about the effects of a sentence such as the one Cunningham had received. But Eddington's colleagues were not inclined to be all that sympathetic to his attempt to mix his status as a scientist with his status as a Quaker. He wrote to Oliver Lodge asking for a letter of support in these terms:

I should explain first that I am a conscientious objector. (No doubt you will deplore that, but I can only say that it is a matter of life-long conviction as a member of the Society of Friends from birth, and I have always taken a fairly active part in the affairs of the Society.) . . .

My position is that I should be willing to do work of that kind (not war work) if ordered [Eddington is referring to alternative work in industry or agriculture]; but I find it difficult to believe that that would really be for the benefit of the world even from the most narrow point of view. . . .

One feels reluctant to make much fuss about a particular case like this, when so many obviously far harder cases are being ruthlessly dismissed every day in order to supply the army. Still I think I ought to make the attempt to continue my work, provided that it is in the national interest.

I shall quite understand if you think it best in your position not to appear to be mixed up with a conscientious objector's case; and you may be sure that I should not take a refusal amiss. (Stanley 2007, 146)

Notice how apologetic Eddington's tone is. Modern commentators on the eclipse expeditions often talk as if the posteclipse outpouring of public enthusiasm for international reconciliation must have been Eddington's main purpose in going on the expedition all along. Personally, I find it difficult to believe that Eddington's experience of being a pacifist during the war led him to expect public approbation for his efforts. Note also that there is no evidence, according to Eddington's most recent biographer, Matt Stanley, that Lodge acceded to his request for a letter. By the end of the war, even mild association with a Conchie was socially undesirable. Eddington must have been surprised to find that his postwar efforts at internationalism met with some degree of public approval. That does not mean that he was not willing to try, but to suggest that he put international reconciliation ahead of his scientific goals for the expedition is to presume that he expected such a favorable reaction, which seems unlikely.

At the tribunal, Eddington's attempt to combine his request for c.o. status with his desire to continue scientific work received short shrift. Newall, appearing for the university, was quick to emphasize, when Eddington identified himself as a conscientious objector, that Eddington was a Quaker. He was determined that no one should mistake Eddington for another socialist radical like the disgraced Russell. The tribunal did not view it as its role to give Conchies their own free choice of war work, as Cunningham had learned. The whole point of alternative work was that it be as disagreeable as possible. Often, it was so boring and pointless that c.o.'s asked to be sent back to jail! Furthermore, they took the view that Eddington's exemption invalidated his plea for c.o. status. If he dropped the exemption, he could enter a plea to be an objector, but of course then the best he could hope for would be alternative work in industry or agriculture, which was what he was hoping to avoid.

After a private conclave, the tribunal concluded by voiding Eddington's exemption. Eddington would have to state his case before the tribunal if he wished for c.o. status. To the university's horror, his situation now was being reported by the local press. Worse, from Eddington's point of view, was that his application had been rejected. The tribunal took the view that his earlier exemption established that he was eligible for conscription, and therefore he could not now hide behind his conscience. Their only concession was to give Eddington a few weeks to try to procure some intervention from the government. This, presumably, was in recognition of the university's desire to do some maneuvering behind the scenes to get his exemption reinstated. This bore fruit. In a compromise apparently worked out by Larmor and Newall, the government supposedly agreed that Eddington could have his exemption, but he was expected to sign a letter agreeing that he would be working in a field of national importance (Stanley 2007, 148). The letter made no mention of his conscience. Fatally, Eddington signed it only after including words to the effect that he would, if denied the exemption, continue to try to claim c.o. status. With this rider, the government rejected the letter. We have this from the testimony of Chandra, Eddington's student (Chandrasekhar 1976).

At this point Dyson became involved. The planning for the eclipse expeditions was well underway. It was difficult to find astronomers who could go, since so many were either in the army or doing war work, and now his main collaborator was liable to end up in jail. Remarkably, someone managed to persuade the government to give Eddington a new hearing before the tribunal. Dyson himself now made a submission on Eddington's behalf in the form of a letter. He seems to have understood what note to strike:

> I should like to bring to the notice of the Tribunal the great value of Prof. Eddington's researches in astronomy, which are, in my opinion, to be ranked as highly as the work of his predecessors at Cambridge—Darwin, Ball, and Adams. They maintain the high position and traditions of British science at a time when it is very desirable that they be upheld, particularly in view of a widely spread but erroneous notion that the most important scientific researches are carried out in Germany. . . . I hope very strongly that the decision of the Tribunal will permit that important work to be continued. (Stanley 2007, 149)

So, Dyson assured the tribunal, letting Eddington continue his astronomy research would result in humiliations galore for the German enemy, so accustomed of late to taking the leading role in science. Furthermore, he had some unpleasant duties in mind for Eddington, which would take him out of the cozy Cambridge nest in which he had shirked up until now:

> There is another point to which I would like to draw attention. The Joint Permanent Eclipse Committee, of which I am Chairman, has received a grant of £1000 for the observation of a total eclipse of the sun in May of next year, on account of exceptional importance. Under present conditions the eclipse will be observed by very few people. Prof. Eddington is peculiarly qualified to make these observations, and I hope the Tribunal will give him permission to undertake this task. (Stanley 2007, 149)

It is perhaps just as well that Dyson left out the detail that Eddington's work at the eclipse would validate the work of

Germany's foremost scientist at the expense of England's greatest natural philosopher!

There is no way to know for certain, but it is quite plausible that Dyson was a key mover in the extraordinary (and very rare) turnaround of Eddington's reappearance before the board. The board now reinstated the exemption, for twelve months, until after the eclipse (Stanley 2007, 149). They even asked Eddington at the hearing to assure them the eclipse was of particular importance. They went further, going out of their way to mention that they did now accept that he was a genuine conscientious objector. This may be one reason why Eddington was willing to accept this compromise and not the earlier one worked out by his college. In support of the idea that Dyson played a key role in the final resolution, it must be remembered that he was quite a high official in British government circles at that time. The position of Astronomer Royal is part of the king's household, along with the poet laureate and the royal piper. It has, traditionally, a close association with the Admiralty and can be thought of as almost part of the military, as the Germans asserted in defense of their bombing raid.

The fact that he was willing to associate with a conscientious objector when others would not certainly tells us something about Dyson. Indeed, if Eddington and Einstein were publicly resisting the war effort, then Dyson, while not a pacifist, was engaged in keeping international cooperation going during the war in a very practical way. As the leading astronomer in Britain, he was responsible for his country's part in the international astronomical telegraphic system, which alerted astronomers to urgent discoveries, such as comet sightings. The rapid dissemination of such results would allow multiple observatories to focus their instruments on celestial events that might be quite transient. The system had been set up in 1882 by the German astronomer Wilhelm Julius Foerster and was headquartered in Kiel, an important naval base in northern Germany (Vinter Hansen 1955, 16). Foerster was still alive in World War I and was, along with Einstein, one of the four academics willing to sign the "Manifesto to the Europeans." This international telegraph system initially used what is called the Science Observer code to transmit positions on the sky, and from this it is sometimes referred to as the Science Observer system. After the war, when it was reconstituted without German participation under the auspices of the new IAU, it was called the International Astronomical News Service.

The Science Observer system went down at the very beginning of the war, when telegraphic cables linking the belligerent nations were physically cut. As a result, Dyson made an arrangement with Edward Pickering of the Harvard Observatory in the United States to retransmit American cables in Europe (the American system had always been somewhat separate from the European one). Meanwhile, the head of the Central Bureau in Kiel, Hermann Kobold, made an arrangement in late 1914 with Elis Strömgren of the Copenhagen Observatory in neutral Denmark to take over the duties of the Central Bureau (Vinter Hansen 1955). Dyson made an effort to transmit cables received by him to the Danes but encountered difficulties. The War Office was opposed to the sending of such cables in code. In 1916 Dyson received the following telegraph from the chief censor's office: "With reference to a telegram in code, addressed by you to Strömgren, Copenhagen, will you kindly let me know the exact meaning of the telegram, and whether you have authority to use private code to Copenhagen. The telegram was transmitted" (Stanley 2003, 60). Some astronomers, either out of a nervous fear of official disapproval, or patriotic impulses, objected to sending cables even to neutral countries such as Denmark, and it was not until some years after the war ended that the Copenhagen Observatory was accepted as the new Central Bureau. Even then, the German system was kept separate.[8] By 1918 international cooperation between astronomers was at its lowest ebb, and some voices argued against any resumption, even after the hoped-for outbreak of peace. Peaceniks like Eddington and Einstein, and their sympathizers like Dyson, may have been hoping to do their bit for reconciliation but must have had very low expectations of achieving any immediate successes. Still, they were no doubt pleased that the eclipse expedition, even if entirely mounted by British astronomers, had at least a flavor of internationalism about it, given the author of the theory they hoped to test.

Troubled Times—Einstein during the Great War

Einstein is one of the best-loved figures of the modern age. He himself was often astonished at the sincere affection with which people viewed him. "Why is it that no one understands me and everyone likes me?" he once asked rhetorically (Callaprice 2011, 17). He is usually portrayed in a very positive light, something he felt uncomfortable with, as when he said "since the light deflection result became public, such a cult has been made out of me that I feel like a pagan idol" (Callaprice 2011, 7). He could do nothing to alter the adoration that many people viewed him with. If he was silly, he was endearingly silly. If he had faults, they were adorable faults. His wisdom became legendary. Today, there are few negative qualities in the popular perception of him. Yet there were—and are—many people who hated Einstein. Nowadays, such people do not have much of a platform, but in the Germany of the twenties and thirties, Einstein dislike was a major industry—so much so that Einstein himself referred to his opponents as the antirelativity company GmbH, using the German acronym for a limited liability company. Of course, this infamy did not emerge until after he became a major public figure in 1919, but the seeds for it were sown during the war years.

Einstein had a difficult year in 1914. Every facet of his life was troubled. His marriage finally and irrevocably fell apart. The world fell apart with the onset of war and, in doing so, thwarted his plans to test his new theory. His main research project, the general theory of relativity, which he had worked on for years, ended in failure. Professionally, he had risen to the highest position he would achieve in science, the leading figure in the scientific life of the world's first city of science, Berlin. Personally, however, he was not thriving. For the moment, he still had his health, but before the war was over that too would be gone. Einstein may not have been on the front lines, but he did not have a good war.

Einstein married his first wife, Mileva Maric, over the objections of his parents. Although it was a love match between two aspiring physicists, the marriage ran into difficulties relatively quickly. Undoubtedly, the fact that Maric gave up her own hopes for a career in physics in order to be Einstein's wife contributed to the conflict between them. When Einstein moved his family to Berlin in 1914, it quickly became apparent to Maric that part of his motivation was to be close to his cousin and mistress Elsa. Relations between them had broken down so badly that in spite of the pain he felt at the separation from his two sons, Einstein seems to have pushed Mileva to return to Zurich. Because of the war, Einstein found it difficult to visit neutral Switzerland, and he became, to a considerable extent, estranged from his boys during this time.

If Einstein faced personal turmoil in his life in 1914, there was no respite to be found by throwing himself into his work. This was the hardest year in his work on the general theory, which had begun in 1907 and had kicked into overdrive in 1911. His effort to test the theory resulted only in the arrest of his collaborator Freundlich. Even worse, he had to accept that not only had he failed to reach his goal of finding the theory, he would never achieve it. When Einstein returned from Prague to Zurich in 1912 to take up a much more prestigious position than the one he had left there, he realized that his new theory must incorporate geometric ideas about curved spacetime. He sought the help of an old friend from his student days, Marcel Grossmann. The two cobbled together the Entwurf, or draft, theory of gravity that Freundlich set out to test in 1914. As its name implies, this theory was not the generally relativistic (or generally covariant, as Einstein called it) theory, which had been Einstein's goal. It did not permit a physicist to transform between any arbitrary coordinate frame. It was still somewhat restricted. Discouraged, Einstein came up with an argument, called the *Hole argument*, that such a theory was impossible. He not only had not achieved his dreams in 1914 but had declared them to be dead.

It must have seemed that things could get no worse, and indeed, late 1915 witnessed the sudden reversal of his view that the quest for a generally relativistic theory was futile. In a quick-fire burst of three papers in November, he presented the final form of his theory of gravity and showed how it had already passed an important test of explaining Mercury's anomalous motion. In 1916 he built on this success, inventing such important subjects of modern physics as cosmology and the study of gravitational waves (Kennefick 2007). But in early 1917, he fell very ill. This sickness affected his digestion and was exacerbated by the tight rationing of German life in the later stages of the war. If the English felt bitterness at the Germans over their methods of making war, imagine how the Germans felt about a nation whose aim lay in starving its enemies into surrender. Einstein suffered because he could eat only certain foods, but fortunately for him, relations in southern Germany and friends and family in neutral Switzerland kept him supplied with fresh foods unavailable to most Berliners. Without this help he might not have survived the war.

In November 1918 Einstein was soldiering on amid the difficulties of wartime Germany. His health had improved somewhat, though it was still precarious. He had managed to travel to neutral Holland in late 1916 but could not return there the following year because of his poor health. He had found comfort from the loss of his family in the arms of his cousin Elsa. She had insisted he move close to her when his illness struck, and she had nursed him back to health, preparing the foods his strict diet required. Of course, now that they were essentially living together, she wanted him to make an honest woman of her. But doing so would require divorcing Mileva, and when he had broached the subject with her, she had promptly suffered a mental breakdown and was hospitalized. He was now caught between the desire to protect his boys, who needed their mother, and the desire to please the woman who had possibly saved his life. As always, he sought refuge in work. His position in Berlin was with the Prussian Academy. He had been provided with his own research institute, the Kaiser Wilhelm Institute for Physics (now part of the Max Planck Institutes). Although he was also a professor at the University of Berlin, he had no requirement to teach. Nevertheless, he did teach regularly. In the winter semester of 1918, he taught a course on special relativity, the lecture notes for which still survive.

Einstein wrote his lecture notes in exercise books, with the date of each lecture as a heading, followed by his notes for that day. Things proceeded normally during this particular semester. His notes mention nothing to indicate the rigors of war that beset Germany as its army struggled to contain the advance of the Entente forces, now reinforced by American troops. Then on November 9, Einstein entered the phrase "Cancelled because of the Revolution" (*CPAE*, vol. 7, doc. 12). How many professors have written such a note for one of their classes? This day witnessed the abdication of the kaiser and the declaration of a German Republic. It was a happy day for Einstein, but it was also a day of revolution in the streets. Did he cancel class so he could hide in the attic? No. Instead, he went straight to the heart of the revolution and was even granted an audience with the first chancellor of the new republic, Friedrich Ebert.

This day, some revolutionary students had arrested certain administrative officials of the University of Berlin. Einstein was not alone in believing the university to be a center of reactionary politics. Now fellow professors called upon Einstein to intercede with the revolutionaries to obtain the release of the officials. As he wrote to his friend Michele Besso, "I am enjoying the reputation of an irreproachable Sozi [socialist or red]; as a consequence, yesterday's heroes are coming fawningly to me in the opinion that I can arrest their fall into the void. Funny world" (Einstein to Michele Besso, December 4, 1918, *CPAE*, vol. 8, doc. 663). Together with his friends Max Born and Max Wertheimer, Einstein set off for the center of Berlin, where he managed to find the revolutionary students in council at the Reichstag, where the German Parliament met. Those who had asked Einstein to intercede were correct in their view that Einstein had some good left-wing credentials, if not perhaps an *obersozi* or high-placed red. As Born later recalled, "I will not go into the difficulties we had in penetrating the dense crowds which surrounded the Reichstag building and the cordon of revolutionary soldiers, heavily armed and red-beribboned. Eventually someone recognized Einstein, and all doors were opened" (Born 2004, 147).

So Einstein and his friends were admitted to the meeting of the revolutionary students, who were debating new regulations for the university. They were a little taken aback when Einstein lectured them on the need to respect the traditional freedoms of the university faculty. But, whatever they may have felt about his arguments, they insisted that only the new chancellor had the power to release the prisoners. So Einstein and friends then went to the Reich chancellor's palace. As a sign that already, before 1919, Einstein was a well-known figure in Germany, he was actually admitted to see Ebert on this, one of the most momentous days in German history.[9]

Ebert saw Einstein only briefly. He had only become chancellor earlier that day, and it was only that afternoon that his Social Democratic Party colleague Phillip Scheidemann proclaimed the new German Republic from the window of the Reichstag, over the objections of the more conservative Ebert. He had to take his leave of Einstein to receive reports of Allied proposals for an armistice. But he did sign the release order for the academic prisoners, and Einstein's mission was complete. For the next few weeks, Einstein gave his class amid the ongoing turmoil of revolutionary Berlin. He did not find the time to make notes, as before, but instead merely recorded the date of each lecture with a one-line statement of the subject discussed. Then, on January 4, 1919 (still part of the same semester, which was due to continue into February), came another change: quite long, detailed notes that go on for several pages and continue at the back of the book. Indeed, they were written in Zurich, according to a comment in Einstein's own handwriting. What this means is that on January 4, students arrived at Einstein's classroom and were told to start writing—this was going to be the last day of classes, and they would have to cover several weeks' worth of material in one day. Einstein was leaving town. Why? The answer is twofold. The reason for leaving was personal—he wanted to get divorced and had to be in Switzerland to do it. Urgency was required because counterrevolutionary troops were massing around Berlin to suppress the worker's and soldier's (and student's) councils that had dominated the revolution until then. Einstein needed to get out of Berlin that day or he might not get out at all. If he didn't, the divorce wouldn't happen, and if the counterrevolutionaries didn't kill him, then Elsa would.

Einstein made it out just as Spartacus week was getting underway. During this week the Spartacists, the antiwar left wing of the German socialist movement, were bloodily repressed by Freikorps troops under the orders of Ebert's new Social Democratic government. These Freikorps were right-wing nationalist paramilitaries who opposed democracy and wished to see the imperial government restored and the German army victorious. Some began wearing swastikas around this time, and many went on to play roles in the rise of the Nazi Party. Meanwhile, the Spartacists had just taken on a new name, inspired by the name of Lenin's successful Bolshevik party in Russia, the Communists. In Berlin, Spartacus week ended with the bloody suppression of the Communists and their allies and the execution of their leaders. Einstein was surely glad to have escaped all this, as suggested by his postcard home to Max Born: "With brilliant sunshine and sweet chocolate, it's grand to read about Berlin events from up here. Monday the schoolmastering starts. This nest Zuoz is unbelievably attractive architecturally. Warm greetings to you all, from your Einstein" (postcard, Einstein to Max Born and his wife, January 15, 1919, *CPAE*, vol. 9, doc. 2).

In Switzerland the divorce was obtained, and although the decree forbade Einstein, as the guilty party in the proceedings, from remarrying for two years, he was already married to Elsa before Eddington returned to England that summer. Suddenly, everything was coming up roses for Einstein. He was surrounded by a new and loving family (Elsa had two daughters from a previous marriage), hoped to repair relations with his first family, and had witnessed the birth of democracy in the country of his birth. His theory, not only complete, held the prospect of a new confirmation that would surely establish it as an important contribution to physics. In the middle of 1919, Einstein, like most of Germany, was waiting anxiously for word from England. While everyone else waited for news that would decide what peace would look like, Einstein waited for news from England's astronomers.

5

Preparations in Time of War

When did British eclipse astronomers first hear of the light deflection experiment? The moment may be recorded in the minutes of the annual meeting of the Joint Permanent Eclipse Committee (JPEC) for 1913. Dyson chaired the meeting, which was engaged in planning for the ill-fated 1914 eclipse. "A letter from E. F. Freundlich was read, suggesting that the Committee should arrange for the photography of stars near the Sun during the eclipse, in connection with his investigation of the possible deflection of light rays by the Sun's gravitational field. It was resolved to inform Dr. Freundlich that the Committee was of opinion that a special equipment would be required to carry out this work satisfactorily, and to express the Committee's regret that no suitable instruments were at its disposal."[1]

Both the fact that Einstein's name is not mentioned and that nothing was done about Freundlich's proposal suggests that the theoretical importance of his program was somewhat lost on the committee at this time. Of course, in 1913 Einstein did not have a completed relativistic theory of gravity to test. There is an irony in the story of general relativity that in the early years Einstein was anxious to test the equivalence principle in order to know whether he was on the right track, but astronomers often did not see the use in testing a mere hypothesis. Only later, when Einstein had a beautifully coherent theory with a precise prediction to test, did astronomers rise to the challenge. By then, of course, the wonder of seeing the theory come together so beautifully and coherently had completely convinced Einstein that it was true. The agreement with the anomalous perihelion advance of Mercury made a powerful impression on him. He no longer worried that his original insight was misguided. He never lost his interest in experimental confirmation, but he no longer had much doubt that the theory was right and gave off a certain nerveless confidence in the outcome.

World War I imposed another difficulty in that when Einstein did finally finish his theory, in late 1915, news was slow to reach England. German scientific journals were not available in wartime Britain, and the conduit through which news reached English scientists was neutral Holland. Einstein had close ties with Dutch scientists and regularly visited there, especially after his friend Paul Ehrenfest moved to Leiden. Even in the midst of war, when obtaining permission to visit a neutral country was difficult, he paid a visit to Leiden in late 1916 and discussed general relativity with the astronomer Willem de Sitter, director of the Leiden Observatory. Out of this discussion, we can fairly say that modern cosmology, with its expanding universe, eventually grew. De Sitter, though, was already familiar with the theory from Einstein's papers and mentioned these in a letter to Eddington in England very soon after their publication. Eddington commissioned de Sitter to write articles specifically on the astronomical implications of Einstein's theory for the journal *Monthly Notices of the Royal Astronomical Society*. The first of these, which discussed the light deflection prediction in detail, appeared in June 1916 (de Sitter 1916a). A second one, which benefited from discussions with Einstein during the latter's visit, appeared in December (de Sitter 1916b). Eddington himself eventually took on the task of presenting the theory to an English-speaking audience, which he was to do to great effect over the years. That first meant understanding the theory, which took time, not least because Eddington may not have received a copy of any of Einstein's papers until the second half of 1916 (Warwick 2014, 463). Initially, he only had de Sitter's letters and articles to go on, combined with some knowledge of Einstein's prewar papers.

It is worth noting that learning a new theory solely from written accounts of it is not easy. Students, obviously, have teachers; tutors; and, at Cambridge, coaches, who explain difficult concepts verbally and interactively. Similarly, scientists often travel to visit more knowledgeable colleagues when they wish to master a new branch of their subject, or invite those colleagues to come and give a talk. This option was closed to Eddington in 1916. England had no experts on general relativity, a theory that had just been invented in a country off-limits to the English. England did possess at least one important physicist with expertise in special relativity. But Ebenezer Cunningham, as we have seen, was focused, much against his own inclinations, on bringing in the harvest in late summer 1916. Eddington seems to have had no one to talk to, though he saw Cunningham at least occasionally, such as at the meeting of the British Association for the Advancement of Science that year (Warwick 2014). As he attempted to master the general theory, he had precious little in the way of written material to learn from, either. It is a testament to his brilliance that he was successful, but it also reflects the thoroughness of his training as a mathematical physicist.

The modern expert on Cambridge mathematical physics at this time, Andrew Warwick, has given a detailed account of Eddington's encounter with general relativity. Warwick points out that Eddington, like Freundlich—and unusually for a student in England—studied differential geometry as a young man. Like other Wranglers who placed highly in their third-year math exams at Cambridge, Eddington benefited from the instruction of a mathematics coach, in his case, Robert Alfred Herman. In the nineteenth century, private tutors, known as coaches, helped prepare the best students for the mathematical Tripos. These tutors were often originally fellows of the colleges who might, for instance, have resigned their fellowships to get married (as was required at the time). Herman, on the other hand, was a fellow while he coached students like Eddington because instruction within the colleges had improved by then to the point that a student might expect to do well in the highly competitive exams without the help of a private coach. Herman has been described as "the last of the great coaches" (Forsyth 1935, 168). Herman gave lectures to his students, including Eddington, in differential geometry, which was not then part of the established curriculum at most universities. Thus, Eddington was well placed to overcome the main obstacle to understanding Einstein's new theory. In this respect, Herman plays a key role in our story. Yet, Warwick tells us, he himself could not accept that the geometry of the universe was non-Euclidean. For men like Herman, differential geometry was of interest in studying geometry on curved surfaces (such as the Earth) that are embedded in a higher-dimensional space that is, almost by definition, Euclidean or flat (Warwick 2014, 451). This prejudice was common. The reception of Einstein's ideas was hindered by the difficulty of accepting the reality of what he was proposing. He repeatedly advanced science by taking a mathematical or physical concept initially proposed as a mere trick or model and insisting that it be taken seriously as the actual way things work.

Eddington first wrote about Einstein's ideas about gravity in a 1915 article, in which he discussed them entirely in terms of the long-standing question of the speed of propagation of gravity. Indeed, he only gets around to mentioning Einstein's work at the end of a long review of historical attempts to measure or predict the speed with which gravity propagates. In this paper he describes the light deflection test as important because it provides a new way to test the question of whether gravity takes time to propagate. In Einstein's 1911 theory, the speed of light slows in a gravitational field. Thus, as a light ray passes close to the Sun its velocity decreases,[2] resulting in a focusing effect similar to refraction. At Cambridge, Einstein's earlier theory of special relativity was not really counted as a separate theory at all. Cambridge physicists, especially those working on the electronic theory of matter, viewed most features of Einstein's theory to be well understood in their own theories. They were highly skeptical of Einstein's axiomatic approach to physics. Although postulates had played an accepted role in math since the days of Euclid, many physicists were uncomfortable with raising any physical principle to the status of an a priori assumption. Surely, they argued, in physics one accepts only what has been experimentally demonstrated to be true. They had nothing against the principle of relativity (or the principle of equivalence) but rejected Einstein's approach, in which he assumed that certain conclusions derived from thought experiments must be valid because the results of the thought experiment were based upon these principles. Other physicists insisted that only a real experiment, not a thought experiment, could decide these matters. In this sense one of special relativity's weaknesses was that it was difficult to test because most objects simply do not move fast enough to tell if they are shorter, or more massive, or have clocks running more slowly. But general relativity was proposing tests that placed it in a different category from Einstein's earlier work. Of course, those tests were astronomical, so they were not of major interest to those physicists who had followed his earlier work. By contrast, most astronomers were not particularly au fait with the special theory, which seemed to have few implications for their subject. English astronomers occasionally drew attention to implications of special relativity, such as in the phenomenon of the aberration of starlight, which Einstein regarded as a critical test of the special theory. But when they did, they would usually mention Einstein only in passing, or not at all, since they were influenced more by Lorentz and others than by Einstein's own writings (Warwick 2014, 452–53). De Sitter's influential 1911 paper on the astronomical implications of what we would now call special relativity regards Lorentz, Poincaré, and Minkowski as the authors of that theory and neglects to even mention Einstein's name. Even Eddington, in 1915, does not talk as if he thinks Einstein is the *author* of the theory of relativity:

> There is another result of the principle of relativity of remarkable interest. Einstein (in 1911) has shown that on this theory a wave of light will travel more slowly when it enters an intense gravitational field of force. This must lead to a refraction of waves of light passing near a massive body. For instance, stars seen close to the limb of the Sun would be apparently displaced 0".83 away from the Sun's limb. This result is a natural (though, I understand, not

actually an inevitable) deduction from the principle of relativity. It would be extremely difficult to detect this deflection even during a total eclipse; but attempts are being made to find it. A decisive result, whether positive or negative, would be of remarkable importance—the first definite advance in our knowledge of gravitation for over two hundred years. A positive result would mean that gravitation has been pulled down from its pedestal, and ceases to stand aloof from the other interrelated forces of nature.

So at this stage, Eddington regards Einstein's work as interesting, primarily because he is a bold thinker prepared to show how gravity can be treated as simply another aspect of physics that must be made to obey the relativity principle. Einstein is not yet the author of a theory, as such. Eddington does not even mention the Entwurf theory, which Einstein had published the previous year and which predicted the amount of light deflection quoted by Eddington. Perhaps he was unaware of it. We cannot even be sure that it was Eddington who first drew Dyson's attention to Einstein's prediction. It could have been the other way around, since we know that Freundlich wrote to Dyson, the chairman of JPEC, about this in late 1913. Eddington could have heard about it from Perrine while in Brazil in 1912, but it seems likely the two men never met during their respective expeditions. Perrine wrote about his thwarted effort in 1912 only in 1923, when he wished, in the wake of the enormous publicity surrounding the 1919 test, to assert his priority as the first person to attempt to test the theory. If he and Eddington had met in 1912, he probably would have mentioned the fact. Of course, even if they did not meet, Eddington could have heard news of the experiment that was being conducted only thirty miles away. All we know for sure is that by 1913 at the latest, Dyson was aware of Einstein's proposed tests. The same was probably true of Eddington, who was sufficiently interested in them to discuss the light deflection test in favorable terms in 1915. But it is notable that he does not, at that time, mention any specific intention by English astronomers to involve themselves in the effort to make that test.

Now, since Eddington seems to have been very enthusiastic about general relativity by 1919, it is natural to imagine that other theorists in Cambridge had been receptive to Einstein's ideas all along. That was not the case at all. Most theorists insisted upon the reality of the ether, and Einstein's approach displeased them. Traditionally, physicists took a mechanical view of nature, trying to explain phenomena in terms of systems like the ether, which they took to be real mechanical entities, even if invisible to our direct senses. Principles and laws had a role in their science, but not as postulates. They could only be accepted provisionally pending further empirical proof. Einstein seemed to invert the normal practice of physics. He made the principle into a law and demanded that the ether be done away with because it could not be seen, thus leaving the electromagnetic field with no material basis at all. This is why one sees so many complaints about Einstein rejecting common sense and proposing obscure concepts. Making principles real and resisting mechanical explanations of fields seemed to lack common sense for those who believed the goal of physics was to construct a mechanical explanation of reality. Of course, Einstein regarded it as very commonsensical to decline to give a mechanical explanation for an ether that he did not believe actually existed.

So Einstein was quite well known by the time he published general relativity, but not so well known that he had an easy time persuading astronomers to test his theory. Did Eddington react so positively to the new theory because challengers to Newton were so rare and wonderful? No, the nineteenth century had seen many alternative theories of gravity proposed, especially toward the end of the century. At that time an outburst of unification efforts occurred, where physicists attempted to explain both electromagnetism and gravity in the same theory. In fact, many physicists, including Einstein, suspected that gravity might simply be an aspect of the electromagnetic phenomenon. Since the field concept had taken hold in explanations of electromagnetism, this meant recasting gravitational theory in a similar way. Indeed, general relativity is an example of a field theory of gravity. Most astronomers, reacting to general relativity, seem to have taken it to be another of an interminable series of rival theories to Newtonian gravity that would, as all the others had been, turn out to be wrong. Newton's theory had a highly impressive record of success in passing experimental tests. Occasionally there had been moments when there seemed to be a problem, as with the anomaly in Uranus' orbit discovered in the nineteenth century. But these had often turned into triumphant successes leading to new discoveries, such as the planet Neptune, which completely vindicated Newton's laws.

Eddington's interest in Einstein's theory stemmed from a problem that was obvious to a number of theoretical physicists at this time but not widely appreciated even among other physicists. Newton's theory was no longer a consistent theory of physics. Physical theories are not even worth testing if they fall victim to one of a number of pathologies. A theory, in physics, is primarily a calculational tool, so pathologies are serious when they undercut the theory's utility for performing calculations. An obvious example of a pathological theory is one that is incomplete. Such a theory lacks enough equations to inform the calculations that it might be called upon to inform. Some naive theories, usually produced by amateur physicists, lack any equations at all. A series of unquantified claims or concepts does not constitute a physical theory in modern science. Theories can also be ugly, poorly motivated, or underdetermined, but as with the problem of incompleteness, Newton's theory of gravity did not suffer from these problems. Theories are also prone to the pathology of being inconsistent. A theory is inconsistent when there are multiple ways to do a calculation, and these different approaches produce answers that do not agree with each other.

But how can a theory that had never failed any test suddenly become inconsistent? How had this happened? A structure that appears very soundly built may, owing to a shift in its foundations, crumble almost overnight. That is what had happened to Newton's theory of gravity. The foundation that had shifted was the subject of kinematics, the study of motion. The culprit was Einstein, who in 1905 had transformed physicists' understanding of relativity, as it applies to kinematics.

The problem that confronted physicists at the turn of the twentieth century was that Maxwell's theory of electromagnetism obeyed different relativistic transformations (known as *Lorentz transforms*) than those used in the study of Newton's theory, often referred to as *Galilean transformations*. A good deal of confusion reigned until Einstein showed that the Lorentz transformations actually applied to all of physics. It had never been noticed before because for objects that move very slowly compared to the speed of light, the two systems of transformations are indistinguishable. But Einstein's discovery raised a troubling point of principle. While Newton's equation for gravity was invariant under the old Galilean transformations, it was not invariant under Lorentz transformations. While this might not matter much when studying slow-moving bodies like the Earth and Moon, it could matter when studying how light itself behaved in a gravitational field. In that context, one might do the calculation in different ways and obtain different results. How could you compare what two different observers could see without some mechanism for transforming between their different frames of reference?

A theory may also be problematic if it is inconsistent with other established theories. If the answer produced by a calculation in one theory creates problems of consistency with the predictions of another theory, then it is possible that physicists will treat the new theory as being falsified without even the trouble of an experimental test. If introducing the new theory would mean throwing out an already established theory, then the new theory may be ignored completely. Newtonian gravity was inconsistent, in this way, with Maxwell's theory of electromagnetism.

So when Eddington received news of Einstein's success in finding a generally covariant theory of gravity, he must have pricked up his ears at once. A generally covariant theory is one that can be written in any set of coordinates, taking it beyond the limitations of the Lorentzian invariance of special relativity. In one stroke, theoretical physics would go from the uncomfortable position of dealing with two inconsistent theories of the fundamental forces to a theory that integrated all of physics into a single viewpoint-independent framework. What a marvelous tool for research that would be. There would be everything to be gained in hoping that this theory could supplant the existing theory. So in considering the perspectives of the main actors in 1919, we must keep in mind how radically divergent their appraisals of Einstein's theory would have been. Most astronomers thought of Einstein as another presumptuous pretender to Newton's crown, with the added deterrence of having seemingly complicated and highly technical ideas. But for a few theorists like Eddington, Einstein's theory was potentially the answer to a serious dilemma confronting theoretical physics.

The principle of relativity proposed two specific challenges to gravitational theory as it was understood in the early twentieth century. The first concerns the question of how fast the influence of gravity propagates through spacetime. Newton proposed that this happened instantaneously, which has greatly troubled physicists then and since. His theory suggested action at a distance, since it was as if the Earth reached out and tugged upon the Moon directly by some mysterious means. In the late eighteenth century, the great French physicist Pierre Simon de Laplace answered the question of whether the theory would make different predictions if gravity took time to travel from one place to another. Laplace showed there would be an effect because the Earth and the Moon would change their positions slightly in the time the gravitational influence took to travel between them. This would mean the force would not be precisely perpendicular to the motion (assuming a circular motion), as Newton assumed. The result would be a braking component of the gravitational force that would gradually draw the Moon and the Earth closer together. Since no such effect is observed in reality. Laplace concluded that the speed of gravity was at least one hundred million times that of light (Kennefick 2007, 28).

Throughout the nineteenth century, Laplace's calculation was thought to have proven that gravity propagates instantaneously. Even when Maxwell's field theory of electromagnetism became established, most physicists did not call for Newtonian gravity to be overthrown. But it did seem odd to some that two such similar forces as electromagnetism and gravity should be governed by such different mathematical laws. Furthermore, as special relativity (and its forerunners) emerged in response to Maxwell's theory, it dawned on some theorists that Laplace's calculation had been invalidated. He had assumed that corrections to the motion of a planet or a satellite would be of order v/c, where c is the speed of light, and v is the velocity of the planet. Since planets move far more slowly than light, this is a small quantity but still possible for astronomers to measure over long timescales. But in special relativity, the largest corrections to the motion of bodies are of order $(v/c)^2$, which is a very small quantity indeed—too small for astronomers to measure.[3] Therefore, there were reasons to suspect that a new field theory of gravity, different from Newton, would be both desirable and possible. The theory would have to be different because Newton's theory was not invariant under the kind of transformations used in special relativity. Therefore, either Newton was wrong and a field theory must be constructed to satisfy the new kind of relativity, or Newton was right and yet somehow gravitational effects propagated faster than light. This was troubling because Einstein's theory demanded that nothing can travel faster than light. But since Einstein's ideas were not, at first, widely taken up, most people beyond the man himself were not too worried.

The second issue emerged not only from relativity but from other theories, such as the electronic theory of matter, that were related but distinct. This was the idea that moving bodies have more mass than bodies at rest. In Einstein's theory this is merely a result of the famous relation $E = mc^2$. A moving object has kinetic energy, and kinetic energy, like all other forms of energy, has mass associated with it. In other theories of the time, this is a direct result of the fact that these bodies are moving with respect to the ether. Motion through the ether causes mass in a way that, numerically, is similar to Einstein's law. The key difference between Einstein's theory and the electronic theory of matter is that in Einstein's theory the change in mass is always relative because motion is relative. As the solar system moves through space, all the planets and the Sun share in that motion and therefore do not notice it. Their mass is completely unaffected, as far as anyone in the solar system can measure, because we are all moving along through space together. But if the solar system moves through the ether, then the electronic theory of matter demands that the masses of all the bodies in it should change depending on the state of motion of the whole system. Thus, strictly speaking, one has to know how the solar system is moving through interstellar space to do any calculations on, for instance, the perihelion advance of Mercury. This is unfortunate, because working out the state of motion of the solar system is no easy thing.

A similar situation faced Aristotle in classical Greece. One would like to acknowledge that motion is essentially something we measure in a relative way, yet in practice it is not always easy to actually perform calculations between multiple bodies if they are all in motion. Therefore, Aristotle provided us with a theory in which the Earth is defined to be at rest. Similarly, it is tempting to do all calculations within the solar system as if it were at rest in the ether. But is it really? In fact, in the eighteenth century, the great astronomer William Herschel showed that the solar system was moving through the system of nearby stars in the direction of a point in the constellation Hercules called the solar apex. The speed of this motion is quite fast, at some twenty kilometers per second. Would it affect the masses of the planets and the Sun significantly enough to be measurable in some fashion? Maybe not, but is it even the full speed of the stars through the ether? As we shall see, astronomers were just becoming aware that it probably was not. How did one do gravity without knowing something so essential as the speed of one's own motion through Newton's absolute space? Einstein offered a way out of this. He found a way to take the great principle of relativity and embody it, after two thousand years, in a set of equations that actually worked for any observer regardless of the observer's state of motion. Crucially, when using these equations, the speed of the solar system is completely irrelevant if one only wishes to do calculations within the system.

Now for an interesting question: How do you explain to people that you have just found the answer to a question waiting two millennia for a solution when they have been walking around the whole time pretending the problem does not exist? The answer is that it will take some time to get through to them. This explains the crucial nature of the anomalous perihelion shift of Mercury. It got the ball rolling. Up until then, Einstein, despite Freundlich's best efforts, was having trouble bringing the people who could help him most on board. Obviously, Einstein needed empirical tests to convince people of the benefits of his theory because most people did not even know the old theory had problems. He needed astronomers, but finding astronomers with the correct mathematical background was not easy. Someone like Eddington was the ideal person to persuade the rest of the world. He was a theorist with the right mathematical training who worked on the topic of celestial mechanics, which was suffering from the incompatibility of gravitational law with relativity theory. In addition, he had done extensive work in astrometry, the skill required for actually carrying out the observational text. It is true that Eddington and Einstein shared pacifist ideals and internationalist sentiment, but their common scientific interests are what brought them together.

6

The Opportunity of the Century?

Planning for the 1919 eclipse began in earnest on the morning of November 10, 1917, at the Royal Astronomical Society (RAS) in Burlington House, Piccadilly, London. The occasion was the annual meeting of the Joint Permanent Eclipse Committee (JPEC), which was in charge of all planning for eclipse expeditions mounted by either the RAS or the Royal Society. These societies were so influential that the committee controlled all expeditions of this type launched from England. A new, more egalitarian Europe was about to be born, and the most violent throes of that event were taking place in faraway Russia at just this time. For the moment, however, JPEC still reflected the old order of things that would so soon begin to pass away. The minutes recording those in attendance shows only one plain mister on the committee. Apart from this Mr. Jones,[1] there were three knights (including Sir Frank Dyson in the chair), a doctor (Crommelin, of the Greenwich observatory), a professor (Alfred Fowler, the man taking the minutes in his role as secretary), a priest (Father Cortie, a Jesuit astronomer), an admiral, a colonel, and a squadron commander. Obviously, the navy was well represented, as, traditionally, it had provided logistical support for these expeditions and was closely associated with the Greenwich observatory. Two more knights were unable to attend, but Professor Eddington of Cambridge University was there by invitation of the chairman, Dyson.[2]

At this meeting it was decided that "in view of the continuance of the war," there would be no expedition mounted from England to observe the 1918 eclipse. Since it was known that American astronomers would observe it, given that the path fell right across the United States, the committee felt easier about coming to this decision. The minutes then got straight down to the main order of business: "Attention was drawn by the Astronomer Royal [i.e., Dyson] to the importance of the Eclipse of 1919 May 28–29, as affording a specially favourable opportunity for testing the displacements of stars near the Sun, which are predicted by the Theory of Relativity. It was pointed out that such favourable opportunities are of rare occurrence, and there would certainly be no equally suitable Eclipse for many years."

Notice the odd fact that the eclipse is described as occurring over two days, May 28 and 29. How can the eclipse have taken place on two different dates? The answer is that Greenwich time used to end each day at noon, not at midnight. This was for the benefit of astronomers, since a night's observing would all take place within the same calendar date. A day that begins at noon is known as the *astronomical day*, in contrast to the civil day. It was not convenient for everyone else in society, and as Dyson worked to make Greenwich time available to nonastronomers, it became necessary, a few years after the eclipse, to make the calendar date change at midnight. But in 1919 the total phase of the eclipse took place on May 28 in Sobral and on May 29 for Eddington on Principe.

The especially favorable opportunity mentioned by Dyson was the position of the Sun, in late May, in the star field of the Hyades cluster. As this is the closest stellar cluster to Earth, this would mean the presence of unusually bright stars close to the Sun at this time. Additionally, as pointed out by Dyson's assistant at Greenwich, Crommelin, in a letter to *Nature* in 1919, the duration of totality would be unusually long, over five minutes at the stations eventually selected and nearly seven minutes in the mid-Atlantic, where it would, unfortunately, be of no use to anyone. Dyson had already, in March 1917, taken to the pages of the *Monthly Notices of the RAS* to discuss the eclipse in the context of the Einstein test. He reported on experiments undertaken with another of his staff members, Charles Davidson, to determine how many stars could be imaged in 1919. Davidson and Dyson had been present in Sfax, in modern Tunisia, during the 1905 eclipse, when Davidson had taken photographs with an astrographic lens with a wide field of view. The goal at the time had obviously not been to test light deflection, but Davidson and Dyson now took out these plates and found that only two stars could definitely be made out. They went so far as to take a new image of the eclipse field to serve as a comparison and then compared the two plates (eclipse and comparison) "film to film" to try to measure any apparent shift position. They found it impossible to measure the deflection in this way with the old plates. But Dyson pointed out that in 1919 there would be thirteen stars within the field of view of the astrographic that would be brighter in magnitude than the third brightest star in the Sfax field, which seemed just barely discernable on the eclipse plates. He concluded that

> this should serve for an ample verification, or the contrary, of Einstein's theory. Unfortunately, the track of the eclipse is across the Atlantic and near the Equator. Mr. Hinks has kindly undertaken to obtain for the Society information of the stations which may be occupied. I have brought the matter forward so that arrangements for observing at as many stations as possible may be made at the earliest possible moment.

Mr. Hinks was Arthur Robert Hinks, an astronomer turned geographer with perfect credentials for the task at hand. He had obviously been hard at work between March and November because by the date of the JPEC meeting, Dyson came prepared with the names of three possible observing stations. The first two were those eventually chosen, northern Brazil and the island of Principe. The third possibility was near the western shores of Lake Tanganyika in Central Africa. Hinks seriously considered sending one expedition on a journey to the middle of Africa because he devoted a considerable part of his published report in the *Observatory* (Hinks 1917) to this station. As he pointed out, the track of the eclipse was very unfavorable.

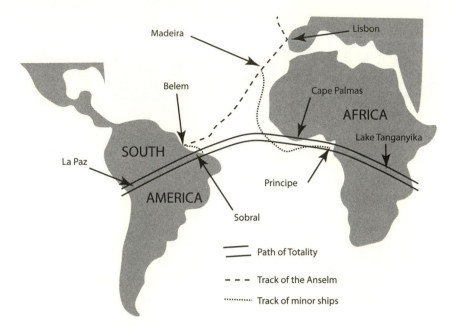

FIGURE 7. Map showing the track of totality and the voyages of the two expeditions. The sites considered as possible eclipse stations (either by the British or other expeditions) are indicated on the map, as well as ports visited by the expeditions' ship, the RMS *Anselm*. (Reprinted with the permission of the author.)

Most of it crossed either the Atlantic or the world's two great jungle basins, the Amazon and the Congo. The area around Sobral was the only other region, besides Lake Tanganyika, with a dry season in May. But Sobral had a rail link to a nearby port. Accessing Lake Tanganyika would have required the use of African porters—almost unheard of for an eclipse expedition, especially one already struggling with logistical problems. In any case Lake Tanganyika was ultimately rejected because the Sun would have been too low in the sky at that station, causing issues with the shifting of star positions due to the longer path of light through the Earth's atmosphere.[3] The atmospheric refraction of stellar positions varies with a star's height in the sky, a phenomenon known as *differential refraction*, and this would have been undesirable given the precise astrometry the data reduction would call for.[4] Hinks' article also clarifies why a much more accessible station on the coast of West Africa seems never to have been entertained. In his 1919 letter to *Nature*, Crommelin notes "a rather serious error in the maps of the eclipse printed in the ephemerides; they indicate the track of totality as lying to the south of the Liberian coast, but totality will, in fact, be observable on that coast, and the duration of totality and height of the sun are greater than at any other land station." We see here that even in the twentieth century, making maps to show the track of an eclipse was no routine matter. It is possible that in 1917 Dyson was misled by the ephemeris he was using into not even considering a station in Liberia. At some point he clearly realized the true situation. Unfortunately, Crommelin continues, "The weather prospects are not favourable [in Liberia], and it is not proposed to occupy a station there." Hinks also dismissed Liberia for climate reasons. Principe was chosen instead. However, as we shall see, weather conditions were not good there, and as reported by the American geologist Louis Bauer, observing conditions were ideal in Liberia on the day of the eclipse.

There is a considerable irony, or even pathos, in Hinks' role. In 1914 he had been the senior assistant at the Cambridge Observatory, where he had done important work measuring the astronomical unit, the distance from the Earth to the Sun and the most fundamental quantity in astronomy. He entertained hopes of succeeding to the directorship that year, but Eddington, the new Plumian Professor of Astronomy, was chosen instead. Hinks saw the malign influence of a preference for mathematicians over astronomers. He complained, "The whole trend of policy in Cambridge & England generally . . . is to take astronomical posts as sustenance for mathematicians" (Crelinsten 2006, 22). Rather than stay on under Eddington, he resigned his position, writing, "They must have been mad to imagine that a man who had had the ambition to do what I had been able to do would be content with an inferior position and no fun all his life" (Crelinsten 2006, 22). He moved into geography and greatly facilitated the planning of the expedition that would make his usurper famous. During the war he was invaluable to the British war effort as a geographer and also influenced the peace, with its many changes of national boundaries. He wrote to Campbell as the war was ending (Crelinsten 2006, 132): "Now that peace is in sight, I find my thoughts reverting to astronomy a little, and I hope eventually to finish off some things I had to leave incomplete in 1913." But he was daunted by the advent of relativity theory as a subject needful for astronomers to know. "The statistical stuff with its integral equations was bad enough. But relativity is much further beyond the limits of my comprehension, and I shall find when I start to make up my two years arrears of reading that I am hopelessly outclassed." Campbell sympathetically wrote back that "most astronomers could conscientiously make the same confession." He commented that astronomers were fortunate to have Eddington "rendering valuable service in keeping us posted on the applications and implications" (Crelinsten 2006, 132). For astronomers like Hinks, relativity represented the death knell of their hopes for a professional career in the subject, as the preference for mathematicians was shown to be a necessary course of action.

At the 1917 JPEC annual meeting, Dyson had placed before the committee the two essentials of eclipse planning, the date and possible locations. There remained three further key ingredients: funds, equipment, and personnel:

> On the motion of Sir F. Dyson, it was resolved that an application be made to the Government Grant Committee for an immediate grant of 100 pounds for the adaptation of instruments, and for a further sum of one thousand pounds, contingent on it being found possible for the expeditions to be sent out.
>
> A sub-committee was appointed to consider the details of the organization in connection with the proposed observations, the members being Sir F. Dyson, Prof. Eddington, Prof. Turner and Prof. Fowler.

Herbert Turner was the Savilian Professor of Astronomy at Oxford and also the director of the Oxford University Observatory.

The subcommittee met on May 10, 1918, by which time the award of the eleven hundred pounds had been made by the government. Lake Tanganyika had already been abandoned as a possible station, and it was decided that two observers would go to each of the other two stations. Since it would be difficult to take telescopes complete with their mountings, it was decided to use telescopes fixed in place, with coelostat mirrors as the only moving parts. It would be necessary to take exposures of many seconds in length in order to successfully image some of the stars in the field, and in this time the Earth would continue to rotate. Without some mechanism to move the telescope, the star images would be linear streaks rather than sharp points. Moving the whole telescope involved a large and complex mount, so the coelostat mirrors were a common instrument at eclipses. The name means "sky-stopping mirror," as its essential function is to stop the motion of the sky. The telescope remains fixed in place while the mirror at one end rotates in such a way as to counteract the rotation of the Earth, keeping each star image at the same point on the photographic plate. The minutes of the meeting note that "it was considered essential that the driving arrangements of these instruments should be greatly improved." Because a very delicate measurement of the positions of the stars would be required, it was essential that the driving mechanism be particularly reliable and accurate in producing a sharp image with no streaking.

Both expeditions would use astrographic lenses. These lenses were used in the *Astrographic Catalogue* project, which was an attempt to catalog the position of every star on the sky brighter than a magnitude of eleven. This effort was allied to another project, the *Carte du Ciel*, which was the first attempt to produce a photographic survey of the entire sky. In order to cover the entire sky, it was obviously helpful to use a telescope with a wide field of view; and the lenses for this purpose were specially made, mostly in France, which was the center of the project. The lenses used in Britain and Ireland, however, were made by the firm of Howard Grubb in Dublin. Two Grubb lenses were provided to the expeditions, one by Dyson from Greenwich and the other by Turner, director of the Oxford observatory. He would also provide the coelostat mirror to use with it, while the Greenwich team borrowed their mirror from the RAS.

The *Carte du Ciel* project was ill fated. It was never finished, and it has been argued that expending so much effort on it set French astronomy back, since those concentrating on survey work missed out on the period of enormous discovery in astrophysics during the first half of the twentieth century. In astronomy, as in physics, experts in the late nineteenth century foresaw that the great eras of scientific discovery were past and that the twentieth century would be the era of precise measurement. The *Astrographic Catalogue* focused on astrometry, the measurement of stellar positions. But instead, the first half of the twentieth century witnessed the discovery of nothing less than the expanding universe, the shape of our own galaxy, and the many other galaxies that exist (meanwhile, physicists discovered the inner universe of the subatomic world). However, in recent years the *Carte du Ciel* has proved very useful because its high-quality astrometry provides a baseline against which modern positions measured by the HIPPARCOS satellite can be compared to determine the proper motions of stars.

By the next meeting of the subcommittee on June 14, Dyson reported on his preliminary efforts to secure shipping facilities. More importantly, work was proceeding on preparing the equipment, and there was a long report on the driving mechanism of the Oxford coelostat. This report was provided by Edwin Cottingham, a clockmaker based in Northamptonshire. The importance of his work on the instrument is emphasized by the fact that at this meeting the teams were decided. Cottingham would join Eddington on one team. The other would consist of Davidson from Greenwich and Father Cortie. Cortie was accordingly invited to this meeting and reported that he had arranged for the Royal Irish Academy to place at the disposal of the committee a four-inch lens and accompanying coelostat that could be used by his team. This would, as we shall see, turn out to be decisively important for the success of the expeditions. He also reported on efforts to develop an electrically driven coelostat mechanism, but at the next meeting, it was noted that this had been abandoned because the engineer responsible was "occupied with urgent war work." This meeting, on November 8, also decided that Eddington would go to Principe and the other team to Sobral, in northern Brazil. By then, the 1918 eclipse had already taken place. The English planners may have wondered how the Americans had gotten on. They would have known that Campbell was one of the few astronomers who was interested in the Einstein test. They continued on, as often happens in science, with the knowledge that they could be scooped at any moment.

Difficulties—the 1918 Eclipse at Goldendale

Freundlich had accomplished one great thing before his arrest by the Russian authorities in 1914. He had interested William Wallace Campbell, the director of the Lick Observatory in California, in the Einstein test. Campbell was already involved in the search for Vulcan, so he was accustomed to selecting equipment that had a field of view extending beyond the body of the Sun. He came away from the Crimea in 1914 with the idea of performing such a test in the future firmly in mind (Crelinsten 2006, 113). There was a total solar eclipse visible in Colombia and Venezuela in 1916, but Campbell was unable to mount an expedition because of financial difficulties at his observatory. It should be recalled that the British system, embodied in JPEC, was exceptional at this time. In other countries, expeditions were dependent on the vagaries of each observatory's finances. In Britain, government largesse meant that expeditions were usually possible. But in 1916 the war put astronomy far from the mind of the British government.

The only professional expedition that might have performed the Einstein measurement in 1916 was one mounted from Argentina by Charles Dillon Perrine. Perrine, an American who had worked at Lick for a number of years, was then director of the Argentine National Observatory in Cordoba. Campbell urged Perrine to conduct the Einstein test in 1916, but Perrine, though he did go, lacked the resources to perform the experiment (Perrine 1923; Crelinsten 2006, 113). This would have been, in fact, the earliest possibility to test general relativity, since Einstein had only finished the theory late the preceding year, discovering his new, larger prediction of light deflection in the process.

Although Perrine may have missed his chance in 1916, he did play an important role in the story of the Einstein test. Back in 1911 Perrine had visited Freundlich at the Berlin Observatory (Crelinsten 2006, 57) not long after Freundlich had begun his association with Einstein. Learning about Freundlich's interest in using old eclipse plates to test Einstein's theory, Perrine recommended that he write to Campbell at Lick about the old Vulcan plates that he himself had helped take. Campbell pointed out that the Vulcan plates were not very suitable for the Einstein test because the Sun was usually positioned at the edge of the plates. A central position was preferable for the light deflection measurements. But he did lend Perrine the Vulcan apparatus to take to the 1912 eclipse in Brazil—the same eclipse that Eddington had gone to observe from Britain. Unfortunately, Perrine was completely rained out at his station in 1912. As he memorably put it, "We . . . suffered a total eclipse instead of observing one" (Perrine 1923, 283).

Once Perrine's chance had gone, the opportunity fell instead to Campbell himself. A total solar eclipse would be crossing the United States in the summer of 1918 (Crelinsten 2006, 114). The Lick team would only have to travel as far as Washington State to observe it. There was only one major obstacle, but it was a fatal one. Their equipment was still stuck in Russia. As neutrals the Americans had no difficulties leaving Russia in 1914, but their instruments were another matter. All shipping was commandeered for war work, and they were obliged to store it all at the Pulkovo observatory in Saint Petersburg. There it remained until after the Russian Revolution. The new Russian government under Alexander Kerensky was willing to ship the equipment eastward in August 1917. By December it had reached Vladivostok on the Pacific coast. By then the Bolsheviks had seized power, and all shipping was stopped. Only in April did the equipment set sail for Kobe in Japan. Campbell was now losing hope that the equipment would make it in time, though he chose his observation site of Goldendale, in Washington, because it was close to the Pacific ports, just in case. As Vesto Slipher of Lowell Observatory put it, "It is very discouraging [for Campbell], I am sure . . . after he has for so many years been observing carefully solar eclipses going to distant parts of the Earth and then through loss of his apparatus to be deprived of the opportunity to observe advantageously this one so near to home" (Crelinsten 2006, 114).

Campbell did find equipment he could use and chose his assistant Heber Curtis to take charge of the Einstein instruments (and those looking for Vulcan). The day of the eclipse was very cloudy, but miraculously, the clouds parted briefly just for the eclipse itself, and Curtis was able to take plates. For the first time, Einstein's theory could now be put to the test. The solar redshift measurements really only test the equivalence principle, not the full theory. The Mercury perihelion measurements were not specifically made to test general relativity. Einstein's calculation agreed with the measurements, not the other way around. For the first time, the theory of curved spacetime was on the line in an experiment in which Einstein did not know the correct answer in advance. However, Curtis still needed to take comparison plates of the eclipse field at night, with the Sun no longer obscuring the stars. This was necessary since the light deflection effect results in an apparent change of position of stars in the sky. Only by comparing the eclipse plate to the usual stellar positions can one measure the apparent deflection. Of course, the Sun moves only slowly against the background stars, taking a full year to return to its original position as Earth moves in its orbit. So it takes a few months for the stars of the eclipse field to rise early enough to be visible before sunrise (or after sunset) at the same position they were at in the sky when the eclipse took place. It is necessary to wait until they are in that position because of differential refraction, that phenomenon by which the atmosphere causes small shifts in position that vary with the altitude of the stars.

Unfortunately, Campbell chose not to take comparison plates before the eclipse because he was hoping his equipment would arrive from Russia. It had not, but Curtis now had to wait until some months after the eclipse to take them using the equipment that was actually employed. This needed to be done in the winter of 1918–1919, but Curtis was employed on war work during this time, and in the end the measurements were made in his absence (Crelinsten 2006, 126). It is interesting to note that the Lick team made no attempt to take the comparison plates from the Goldendale station, as ideally should have been done. Instead, they were taken with the same equipment remounted on Mount Hamilton, at the Lick Observatory itself. It has been argued that Eddington erred in 1919 by taking his comparison plates in England, rather than on Principe, in order to avoid having to wait months in the tropics. But Eddington was certainly not alone in judging that the inconvenience of leaving equipment in place for months at an insecure location outweighed the experimental benefits of using the same location and avoiding remounting the equipment.

At any rate, Campbell waited for Curtis' return in early May 1919 before actually beginning the data analysis. Thus, the next eclipse was about to take place before the Lick team had fairly begun their data analysis from the previous year. Although Campbell was already conscious of failings in the data, his plans to improve his equipment centered on the next North American eclipse of 1923 (Crelinsten 2006, 127). He did not try to observe the 1919 eclipse and instead concentrated on analyzing the data in hand. When he heard the news that the British would perform the Einstein test in May 1919, he suddenly realized that time was of the essence (Crelinsten 2006, 129). Accordingly, by the time Curtis did return, everything was in place, and the plate measurements proceeded with all speed in spite of difficulties with equipment. One particular problem that Campbell and Curtis encountered is worth mentioning since it affected the quality of the star images on their plates. According to Crelinsten (2006), the telescope mountings were subject to some movement during the eclipse, causing the star images to be doubled or to appear with tails, obviously compromising their suitability for careful measurement (126). In addition, Curtis was critical of the "second-rate driving clocks" (Crelinsten 2006, 134), which were all he had available since his best equipment had never arrived from Russia.

Nevertheless, they did have plates from two instruments used at Goldendale, and Curtis also used some plates taken during the eclipse of 1900. As we have seen, that eclipse had a very similar star field, as the Sun was in the Hyades at the time. The images from that eclipse, taken in the U.S. state of Georgia, were of better quality than those taken by Curtis himself at Goldendale. But, unfortunately, no one had taken comparison plates in 1900. So Curtis simply compared the stars' positions with celestial coordinates given by the *Carte du Ciel* project. All of his data convinced Curtis that there was no light deflection at all, and he came out strongly against the Einstein theory in his presentation at a meeting of the Astronomical Society of the Pacific in Pasadena. His final conclusion in that paper was that "there is no deflection of the light ray produced when the ray passes through a strong gravitational field, and . . . the Einstein effect is non-existent" (Crelinsten 2006, 137). In other words, Curtis rejected both of Einstein's predictions: the original one based solely on the equivalence principle claim that light, like everything else, falls in a gravitational field; and the later one, which includes the effect caused by the curvature of spacetime due to gravity. His framing of the theoretical issues was thus very different from that employed later by Eddington and Dyson, who chose to present Einstein's first prediction, what we might call the "half-deflection," as the "Newtonian" prediction. According to Curtis, not only was Einstein wrong but so was Newton himself, at least as understood by Eddington!

As the English continued their preparations, the war was helping them by frustrating the Americans. It delayed their work and steadily reduced the yearlong lead that the timing of the two eclipses had given them.

Further Difficulties—the Third Test and the Solar Redshift

Einstein's theory had gotten off to a great start with the Mercury perihelion advance calculation in 1915. This was a spectacular confirmation of the general theory in some ways, but it was not, strictly speaking, a prediction. It was more a sort of retrodiction. Note that Einstein had never discussed it publicly as a test until he discovered his theory passed it. While it is true that the eventual theory he developed allowed for no wiggle room in his prediction, this was not apparent to all scientists, so the theory could never have become established on the basis of this one success, no matter how great an impression it made on Einstein personally. Since the eclipse test could only be attempted every year or two, this left the solar redshift as the only ongoing research program engaged with the testing of general relativity. Unfortunately for Einstein, the solar astronomers engaged in this research were not very impressed with the way his theory matched up with the reality of what they were seeing.

The solar redshift prediction is based entirely upon the principle of equivalence. Unlike the light deflection test, it did not change at all when the final version of the theory was found. At the outset Einstein must have been quite hopeful that this test would lead to an early vindication of the principle because astronomers had already noticed the solar redshift phenomenon in the nineteenth century. One of America's first great physicists, Henry Rowland, working at Johns Hopkins University in Baltimore, Maryland, undertook a systematic comparison, beginning in 1887, of the absorption lines in the solar spectrum with the spectral lines observed in the laboratory. He noticed a general small tendency of the solar lines to be shifted a little bit to the red. Such a shift in wavelength could be explained by the motion of the source away from the observer, but obviously, this was not true for the Sun as a whole. While motions of the gas in the solar atmosphere could produce Doppler shifts of this kind, these motions must surely be about as often blueshifted as redshifted, since gas that sinks away from us (redshift) must surely be balanced by some gas rising toward us (blueshift). Rowland was inclined to think he had made minor errors in his measurement (Forbes 1963), but his research assistant Lewis Jewell did not accept this and repeated the measurements, establishing that there was a real, though small, redshift that varied from line to line so that a few lines were even blueshifted.

This was a known and active research problem in solar astronomy when Einstein came on the scene twenty years later. He must have thought the situation appeared promising. The only reason why Doppler shifting of rising and falling columns of gas had not been accepted to explain the shifts in the solar spectrum was that, although the amount of shift varied in an unpredictable way from one line to another (which would be consistent with motion alternately toward and away from the observer), it did not vary about zero. Instead, there was an overall tendency for a redshift. Since general relativity required all lines to be redshifted, it might appear that a mixture of the gravitational redshift and the Doppler redshift could explain the situation. All the lines were shifted to the red by an equal amount due to gravity, but there were variations about this point that left some lines even more redshifted, some less so, and a few actually blueshifted. This is, in fact, what eventually came to be accepted, but in the beginning the astronomers studying the problem were not impressed by the claims of Einstein's theory. Undoubtedly, the sheer complexity of the problem, with hundreds of lines that had to be carefully studied on an individual basis, hindered any decisive verdict in favor of Einstein. Although it played a very important role in the reception of general relativity, the redshift prediction did not possess the drama of the eclipse experiment.

One key player in this effort to test Einstein was an English astronomer based in India named John Evershed, director of the Kodaikanal Observatory in southern India.[5] He accepted the argument that the shifts in the Sun's spectral lines were due to the radial convection of gases in the solar atmosphere. To him, it naturally followed from this that the line shifts should only be visible at the center of the Sun and ought to disappear if one took spectra of the limb of the Sun. After all, if the rising and falling of masses of gas is responsible, then looking at the limb would eliminate the Doppler shift, since the columns of rising and falling gas would be viewed from the side. Instead of being motions toward or away, which produce Doppler shifts, they would be motions across the line of sight, which do not. Somewhat to Evershed's surprise, though there is a definite difference between the shifts of lines at the center of the Sun's disk and at the limb, the shift at the limb is not zero but is consistently to the red. This could easily have been taken as favoring Einstein, but Evershed chose not to. He even proposed that perhaps the Earth was using some mysterious force to repel the Sun's gases so that we always see a redshift! This idea was met with hilarity in some quarters, as when a colleague wrote to Einstein wondering if it was really the entire Earth—or just the British Empire, specifically—that was repelling gases at the limb of the Sun.[6] No one seems to have jumped up to offer Einstein's explanation as a solution.

Eventually, the American astronomer Charles Everard St. John, based at the Mount Wilson Observatory in Southern California, proposed the modern explanation of the solar redshift that it is composed of a superposition of gravitational redshift and Doppler shifting due to radial convection. This, however, happened only after 1919. Modern commentators, most notably the philosophers John Earman and Clark Glymour, have proposed that it was not until Einstein's theory had been vindicated by Eddington and Dyson that other astronomers felt comfortable using it in an explanatory way. Obviously, this shows that scientists can be strongly affected by social factors. Although the reliability and precision (and sheer volume) of solar line shift measurements had been increasing over the years, it does seem as if solar astronomers were not willing to go out on a limb in favor of the theory until someone else was willing to do so. But it must be kept in mind that the solar redshift never counted as what one might call a test of general relativity. Since the amount of the Doppler shift in each given line was totally unpredictable, there was no single case in which anyone could confidently state that the gravitational redshift had been independently measured. It was only by taking the totality of many line measurements into account, as St. John did, that anyone could say Einstein's theory contributed to a plausible understanding of the phenomenon, combined with other factors. It is a natural process in science for a theory to first be tested and then used to explain other phenomena. Since Einstein's theory figured in this explanatory role in the solar redshift problem, it is expected that it could be used in this different role only after passing other tests. In this sense the gravitational redshift did not count as an independent test of general relativity until much later.

The Antirelativity Company

Not everyone loves relativity theory or finds it beautiful. Einstein's theory has always had many detractors. Some people simply dislike his aversion to absolute space. Others dislike the mathematical complexity that goes along with coordinate transformations. Many people hated the cavalier way in which he banished the ether and, with it, the old dream of a mechanical explanation of all the forces of nature. Some philosophically minded people regarded him as an idealist who represented a threat to the realist view of nature. Some people just hated him because he was a Jew. Moreover, he was a Jew who proudly asserted himself. He became a Zionist as a result of his political awakening during the Great War. Yet he was even more defiantly a Democrat, a pacifist, and a Socialist, all political phi-losophies that were supposed to benefit from a great deal of Jewish support. In 1919 he was still safely anonymous outside the circle of those who followed the latest physical theories (though, as we have seen, he was a fairly well-known figure in Berlin). Only at the end of the year would he begin his journey toward becoming a target for all those who hated modernism, liberalism, and internationalism.

At this time racist prejudice was more open than it is today. A lead-ing Nobel Prize–winning German physicist such as Phillip Lenard could feel comfortable casting Einstein as an enemy of "German phys-ics." It is worth mentioning this, since an aspect of the movement for German physics was anti-English. Beginning during the First World War, scientists like Lenard began to object to using English terms in German physics papers. But there was an obvious anti-Semitic ele-ment to these views, as well. Einstein was one of their principal targets, especially for his work on relativity theory. Unfortunately for Lenard, he lacked the expertise to challenge Einstein's theoreti-cal work. Instead, he often employed the tactic of painting Einstein as a plagiarist—someone who stole the ideas of German scientists whose work had gone unrecognized. One example relates directly to the eclipse. In 1921 Lenard republished the work of a Prussian astronomer of a century before, Johann Georg von Soldner, in which Soldner calculated the light deflection effect based solely on the Newtonian physics of 1801. Lenard argued that Soldner obtained the same result as the half-deflection Einstein proposed in 1911 and that this very German physicist should be given priority for the enormous public success of the 1919 expedition.

Lenard's claims never amounted to much. Soldner's calculation was largely irrelevant to the physics of 1921, and he did not predict the result actually obtained by the 1919 team. Rediscovering his con-tribution did no more than highlight the fact that eighteenth-century physicists, like Newton himself, had been quite comfortable with the idea that light was affected by gravity. It was only in the nineteenth century that belief in the wave (as opposed to the particle) nature of light convinced physicists that light did not have weight and would not fall toward the Sun. By 1919 some theorists were so convinced that this view had been mistaken that Eddington admitted he would not, from a theoretical standpoint, quite know what to do if the eclipse discovered no light deflection. Yet an astronomer like Curtis could sincerely believe that the null result represented the status quo, which his experimental results were vindicating. So profound are the conceptual gulfs that can open up in the ebb and flow of scientific history, especially during periods of rapid change. Einstein's reputation among scientists suffered not at all from Lenard's barbs. Politically, with the rise of the Nazi Party, which Lenard joined, German physics was eventually in a position to expel Einstein and his theories from the German academy. It was a hollow victory in a double sense. Einstein left Germany and resigned his positions before they could fire him when the Nazis did come to power. After that, they could not succeed in doing more, while ignoring his theories and persecuting his fellow Jewish scientists, than ensure the complete takeover of English as the first language of science.

Politically based skepticism of relativity theory was not confined to the Right. The theory was controversial in the Soviet Union for allegedly being in conflict with Marxist thought. Closer to home was the case of Friedrich Adler, son of a very famous Austrian Socialist politician and himself both a politician and a physicist. Adler and Einstein knew each other. Indeed, one of the early academic positions that Einstein obtained might have gone to Adler, had Adler not decided to enter politics and nobly tell his would-be employers that Einstein was the better man for the job. Einstein and Adler shared antiwar views, which Adler acted upon with considerable vigor when, in 1916, he shot dead the Austrian minister-president (the head of government under the emperor). Sentenced to be executed, he was the subject of a widespread campaign for clemency, which Einstein joined. He was still alive when the revolution overthrowing the Austrian Empire freed him at the end of the war. He went on to a long political career, which includes the famous anecdote that someone once complained, as Adler gave a long speech, that "he shoots better than he talks" (Galison 2008).

The irony of Einstein's help in keeping his friend Adler alive is that in jail Adler used his time to launch an attack on relativity theory.

His father even attempted to use this quixotic attempt to overthrow the ideas of the leading physicist of the age as evidence that his son was insane, aiming at a legal defense of diminished responsibility. At least one other old acquaintance from Einstein's days in Switzerland gained minor prominence as a critic of relativity theory. Edouard Guillaume had worked in the Swiss federal patent office with Einstein. His correspondence with Einstein is replete with Einstein's patient efforts to explain how Guillaume's criticisms of the theory were based upon misunderstandings. Einstein seemed to genuinely wish to protect Guillaume from embarrassing himself publicly, ending one letter jocularly with the words "Repent, you hardened sinner!" Since Einstein certainly refused to respond in kind even when Guillaume tried to join the international movement of largely amateur physicists who opposed relativity in its early years, we may take it that he held his old colleague in some affection. Like Guillaume, many of these people simply did not understand the theory well enough to make a cogent critique of it, or even to comprehend efforts to set them straight.

At a higher level than the racists and the naive theorizers were men like Ludwik Silberstein and Dayton Miller, who, as we shall see, mounted a serious challenge to relativity by trying to show that it is possible to measure the speed of the Earth through space from inside a closed room. It should be kept in mind that the vast amount of activity against Einstein in the period before 1919 and his ascent to fame did not spring up out of nowhere. There was already a lot of unease among many physicists about relativity theory before 1919, and this was definitely true of many astronomers.

As we have seen in the letter from Moritz to Cowell, many people disliked general relativity because of its difficulty and mathematical complexity. Certainly, in the wake of the announcement of the 1919 results, there were many revealing comments from astronomers confessing that they did not understand the new theory. One astronomer who went so far as to join up with the antirelativity company was Curtis. He collaborated with Charles Lane Poor, a fellow astronomer who was one of the leading opponents of relativity in the United States. Poor was professor of astronomy at Columbia University and devoted a great deal of effort—and generated a lot of publicity—attacking Einstein's theory. In 1923 Curtis wrote to Poor: "Everybody seems to be 'falling for' the theory, without knowing very much about it. I am certainly still one of the irreconcileables [*sic*], and it does not seem at present that I can ever believe in it" (Crelinsten 2006, 243). Poor was a relatively rare example of a professional astronomer who was willing to attack relativity so vehemently and publicly as to reveal himself as a partisan rather than a suitably detached scientist. He even endeavored to find financial backing for antirelativity work and to create a network of skeptics. Curtis received a warning from Campbell at one stage not to join the "society" composed of Poor and another astronomer, Thomas Jefferson Jackson See. See, though a professional astronomer, was an inveterate controversialist who certainly did his reputation no credit with his attacks on Einstein's theory. Curtis, however, was a leading astronomer who is remembered today primarily for his role in the "great debate" with Harlow Shapley on the size of the Milky Way galaxy and whether there are galaxies other than our own. The fact that he was quite willing to associate with, and identify as one of, the irreconcilables suggests that there were deep wellsprings of hostility to relativity in the astronomical community. Certainly, as Dyson prepared to test the theory, he was aware of this animosity. He would certainly have been conscious of the importance of taking measurements fine enough to convincingly test the new theory against the old. Even so, as we shall see, he had to leave to Eddington's judgment the precise determination of exactly what the old theory was predicting!

7

Tools of the Trade

When Dyson went to Greenwich as chief assistant in 1894,[1] he found himself assigned to work on the *Astrographic Catalogue*, then still in its infancy. Like Eddington, his background up to that point was in mathematical physics, but he soon showed himself to be a very competent observer and remained primarily an observational astronomer for the rest of his career. The *Astrographic Chart*, also known as the *Carte du Ciel*, and the *Astrographic Catalogue* were two sides of a project launched in 1887 at the first important international astronomy conference, held in Paris. The project aimed to produce, for the first time in history, a photographic chart or map of the entire night sky with precise measurements of the positions of a large number of stars. The use of photography in astronomy was then still very new. The appearance of the comet of 1882, which was particularly striking when viewed from the Southern Hemisphere, had excited interest in it. Photography had become a popular hobby following the development of the dry plate method and the availability of inexpensive photographic film. Amateurs attempting to take photographs of the comet with ordinary cameras were disappointed to find that the movement of the Earth while their plates were exposed quite spoiled their images, since the comet shifted position in their field

FIGURE 8. This photograph of the comet of 1882, taken at the Cape observatory in South Africa by David Gill, showed that stars could be imaged with the new dry plate photograph technique and inspired the *Astrographic Catalogue* project and the birth of photographic astrometry. This image is from a print made in Gill's time.
(Courtesy of the South African Astronomical Observatory.)

of view during this time. A Scottish astronomer, David Gill, who was Her Majesty's Astronomer at the Cape of Good Hope in South Africa, invited one of these amateurs to attach his camera to the Royal Observatory at the Cape of Good Hope's equatorial reflecting telescope. Such telescopes have a mounting powered by clockwork that counteracts the motion of the Earth to keep stars in focus for long periods. Gill found that he obtained not only an excellent image of the comet but also of many stars in the background. This was an important discovery because many earlier experiments in stellar photography had, in the days of wet photographic plates, found it difficult to image any but the brightest stars. Early uses of photography in astronomy focused on the Moon and the Sun.

The Royal Observatory at the Cape of Good Hope was founded to precisely measure stars in the Southern Hemisphere and thus was primarily interested in astrometry. Gill's experience of imaging the comet, in which long exposure times were necessary to capture the dim tail, proved that hundreds of stars could be captured on one plate. Previously, astrometry was a laborious project involving the careful observation of individual stars using a transit circle as they passed overhead each night. Gill realized that dry plate photography combined with the clockwork-driven equatorial mount could revolutionize the subject. He began a project with the Dutch astronomer Jacobus Kapteyn to do astrometry by photography for the first time. Gill took the photographs in South Africa. Since he lacked manpower, Kapteyn, who lacked an observatory, made the measurements of positions from the plates supplied by Gill.

Major optical problems had to be overcome during the 1880s to make photographic surveys possible. First, early photographic plates were primarily sensitive to blue light, but telescope lenses were designed to focus yellow light, to which the human eye is most sensitive. Since the refractive index of a lens varies for different colors of light, this meant that ordinary lenses were unsuitable for use with photographic plates. Gill asked an instrument maker he had worked with before, Howard Grubb in Dublin, to supply him with a suitable lens for his photographic survey work. The telescope Grubb supplied had a lens that worked well as far as the color of light but revealed a further and more serious difficulty. Lenses were designed to focus light on the axis where the astronomer's eye was located. The eye is a small target, but a photographic plate is quite large. More importantly, it is geometrically flat. Light, as it travels from the lens to the eye, has a wave front that is spherical in nature. This means that the flat plate would only be properly focused at the center. There would be significant distortion at the edges and corners of the plate, a serious problem. The lens Grubb supplied to Gill suffered from this problem, so Gill ended up using a camera lens intended for use in landscape photography supplied by the London firm of Dallmeyer.

Fortunately, during the 1880s a pair of brothers in Paris, Paul and Prosper Henry, succeeded in developing a lens suitable for use in stellar photography that produced a relatively flat field on a photographic plate. Although they were amateurs, they received the support of Admiral Amédée Mouchez, director of the Paris Observatory,[2] and it was he and Gill who together came up with the idea of an international project to create the *Carte du Ciel*, the photographic map of the sky. Obviously, Gill could not cover the whole sky by himself, so the *Carte du Ciel* was to be an international collaboration involving over a dozen observatories in many countries. At the 1887 conference, each observatory involved in the project agreed to use similar equipment, and they decided to use a lens similar to that designed by the Henry brothers. This resulted in the manufacture of a large number of thirteen-and-a-half-inch diameter lenses, known as astrographic lenses. Two of these were used in 1919 because the large field of view, desirable for the all-sky survey, would permit more stars to be imaged during the eclipse.

Most telescopes focus on individual objects or small patches of sky to maximize magnification. The astrographic project had different requirements. A wider field of view was desired. Even with their wider field of view, the project's telescopes would require over ten thousand separate photographic plates to cover the whole sky, an enormous undertaking. This was indeed one of the first examples of big science and an early prototype of international cooperation in science. At the 1887 conference, differences in language, in culture, and in national interests had to be overcome to reach agreement on specifications for the standardized equipment (Turner 1912). In this respect it was the hosts, the French, who got their way, but it was still necessary for each observatory affiliated with the project to pay for its own instruments and reduce its own data. Observatories staffed by British astronomers played an important role in the project because Greenwich was responsible for the region around the North Celestial Pole (Oxford's zone lay farther south); while Melbourne, Sydney, Cape Town, and Perth were responsible for most of the southern sky. Probably, for nationalistic reasons, these observatories preferred not to buy from the Henry brothers, which opened the way for the Grubb firm based in Dublin to manufacture seven of the astrographic lenses, including both used in 1919.

More important than the actual photographs was the astrometry, which would produce the *Astrographic Catalogue* of stellar positions.

Indeed, the *Carte du Ciel* was never completed, but the catalog largely was and even exceeded its original goals in some cases. In recent times the original plates have become useful again. The *Astrographic Catalogue* itself had its star positions entered as electronic data in the 1990s. After that the United States Naval Observatory continued astrometric measurements from the original plates to create the *Astrographic Catalogue* 2000, which measures the proper motions of stars against modern astrometry measurements from the Hipparcos satellite.

Until the seventeenth century, astronomers paid little attention to the positions of stars. The planets were the main goal of their research, and the stars served only as landmarks against which to observe their motion. In order to solve the problem of longitude, it was proposed that sailors carefully observe the position of the Moon as a means of keeping time independent of local time. But in 1675 English astronomer John Flamsteed pointed out to the British government that even the positions of the bright stars were not known with sufficient accuracy to make using them as references in the sky possible. Accordingly, the Greenwich observatory was founded, and Flamsteed became the first Astronomer Royal, with a plan to remedy this defect in scientific knowledge. Further progress was made by James Bradley, the third Astronomer Royal (from 1742 to 1762), who was the first to measure stellar positions with sufficient accuracy to detect the motion of the Earth. His discovery of stellar aberration provided the first proof that the Earth really does move in the cosmos. Bradley was hoping to detect parallax, an apparent shift in the positions of stars between seasons caused by the change in our perspective as Earth orbits the Sun. He failed in this because parallax is only detectable for the nearest stars, and Bradley had no idea which stars were nearby. Stellar aberration is an effect caused by the relative motion of the Earth in its orbit across the path of the light rays reaching us from those stars. It depends not on the radius of the Earth's orbit, as parallax does, but on its orbital velocity as compared to the speed of light. It affects all stars equally, regardless of their distance. Stellar aberration was one of the key results Einstein relied upon in developing special relativity nearly two centuries later, since it shows

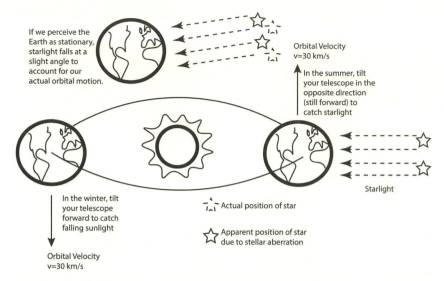

FIGURE 9. The motion of the Earth in the solar system affects falling starlight. As we move around the Sun at a speed of roughly thirty kilometers a second, we must angle our telescopes very slightly forward, just as we must tilt our umbrellas forward to catch raindrops as we hurry along in the rain. Since our motion reverses every six months, over the course of a year astronomers can detect small apparent shifts in the positions of stars due to stellar aberration. This small effect must also be accounted for in eclipse measurements, since comparison plates are typically taken months after eclipse plates.

(Reprinted with the permission of the author.)

how we can detect the motion of the Earth in space, provided we look beyond the confines of the Earth. Of course, this detection of the Earth's motion tells us nothing about the motion of the Sun in the larger system of stars.

Nevertheless, the work of William Herschel in the eighteenth century led to some clue of the motion of the solar system. He compared the positions of stars to those Flamsteed had recorded a century before. Stars behind us move closer together with the centuries, while stars ahead of us draw farther apart, simply because we are now closer to them (and farther from the ones behind us). Herschel showed that the solar system is moving, with respect to the nearby stars, at some twenty kilometers per second in the direction of a point in the constellation Hercules known as the solar apex. This is only a small fraction of the solar system's speed in its orbit around the galactic center. Herschel only measured the small amount by which our Sun exceeds the speed of the nearby stars as we all move together through the galaxy. He also roughly estimated stellar distances to come up with a description of our place at the center of a vast system of nearby stars that was not quite spherical but somewhat flattened, like a thick lens. This was the picture of the universe still held in the minds of astronomers when Dyson went to work at the Royal Observatory in 1894. Indeed, Kapteyn confirmed it in the early twentieth century when he estimated the width of this group of stars as less than twenty thousand parsecs (twenty kiloparsecs). A parsec, short for *parallax second*, is the distance at which a star, observed during adjacent seasons of the year (e.g., spring and summer), changes its apparent position in the sky by one arc second (1/3600 part of a degree). A parsec is somewhat more than three light-years, an enormous distance. It was, however, completely dwarfed by the universe that emerged in the era of the big telescopes after 1920.

Although Bradley provided very accurate stellar positions for a few thousand reference stars, they did not cover the whole sky equally because the original purpose had been to help take positions of the Moon, which follows the ecliptic in its motion across the sky. So when the work of the *Astrographic Catalogue* began, astronomers were confronted with the need to establish very accurate positions of reference stars throughout the sky. This was of particular importance to Greenwich, which had been assigned the North Celestial Pole as its sector of the sky, very far from the ecliptic. Each plate of the great sky chart would need to have some reference stars on it so that positions on that plate could be meaningfully compared to those on other plates. Furthermore, positions taken by Bradley could not be relied upon because some stars were known to have large proper motions. *Proper motion* refers to the actual motions of the stars themselves through the sky (Herschel had effectively measured the Sun's proper motion), rather than their apparent motion caused by the movement of the Earth (which includes both parallax and stellar aberration). It is typically very small, but the level of accuracy aimed for was great. As such, it was hoped to avoid any errors in the reference star positions, which would compromise the whole project. Dyson, accordingly, had to study proper motions, a topic that would dominate his career. Looking for any stars near the pole that had been measured accurately decades before, he found them in the work of an amateur astronomer named Stephen Groombridge, who had carefully measured many positions of circumpolar stars in the early nineteenth century, some eight decades earlier, using the transit method.

Groombridge had left very complete records, which enabled Dyson, working with another Greenwich assistant, to rereduce all of Groombridge's data to modern standards. They then made their own careful measurements of the positions of those stars as of about 1900 and calculated the proper motion of thousands. In so doing, Dyson succeeded in massively enhancing the knowledge of the motions of nearby stars and did so at a particularly important time in the history of astronomy. Kapteyn, using Bradley's data, had shown the existence of what were called *star streams*. It turned out that the stars were going places. On one side of us, the proper motions all pointed in one direction; on the other side of the solar system, they were going the opposite way. Thus, there were two star streams. When Dyson left Greenwich to go to Scotland in 1905, he was replaced by Eddington, who used Dyson's data to confirm Kapteyn's result. The reason for the star streams took a while to determine. It turns out that the twenty-kilometer-per- second motion of the solar system observed by Herschel is only a small part of our motion about the center of the galaxy. We orbit the galactic center at a speed of over two hundred kilometers per second, which cannot be measured by looking at the local stars because they are all orbiting along with us (relativity again!). However, those closer to the center are orbiting somewhat faster, so they are gaining ground on us ahead, while those farther away are falling behind. We are in the middle lane of traffic, watching cars in the slow lane and the fast lane as they gradually move away from us in opposite directions. This is the reason for Kapteyn's star streams. Dyson and Eddington's work persuaded everyone that Kapteyn's work had to be taken seriously. Thus began the hunt for the galaxy as we know it today.

With the positions of reference stars in hand, measuring the photographic plates themselves could move forward. Dyson's predecessor at Greenwich, Turner, had already developed a method of astrometry from photographic plates that involved measuring six plate constants that could then be applied to each star position on the plate (measured by a micrometer screw) to convert it to a position on the sky. It was left to Dyson to develop efficient methods of determining the plate constants from measurements of the reference stars on the plate. Carrying this work forward as an assistant from 1894 to 1905 also opened up a possibility for him when he returned as director of Greenwich. He could take new plates of the same fields twenty years later and look for the proper motions of even more stars. To do this, he took the new plates "through the glass" to produce images not reversed. These new images could then be compared directly back to back with the original plates and any tiny shifts in the position of individual stars measured with the micrometer screw. This is the same method of differential astrometry employed with the eclipse plates, and Dyson was working on this project at the same time because volume 4 of the Greenwich *Astrographic Catalogue*, published in 1921, contains this data (which is already referred to in a paper he published the previous year).

Dyson was accustomed to fine measurements of sub–arc second shifts in stellar positions in one other area as well. Ever since Bradley had failed to find parallax in the eighteenth century, astronomers had keenly felt their lack of any real knowledge about how far away the stars are. Since parallax would only be measurable for closer stars, they were in a bind because they had no idea which stars were closer. Measurements of the proper motions of many stars, such as those Dyson now had, solved this problem. A closer star will have much greater proper motion than a distant star, so Dyson and others could now measure the parallaxes of many nearby stars, finally putting numbers on the cosmic neighborhood depicted by Herschel. By 1919 Dyson had twenty years of experience in astrometry, measuring relative shifts of position, both in parallax and in proper motion, that were the same size as those Einstein had predicted to occur during the eclipse.

Even at this time, Dyson was at pains to emphasize to skeptics that he had complete confidence in his own ability to measure the shifts on the eclipse plates taken at Sobral. In recent decades much of the debate about the 1919 eclipse has focused on Eddington. Questions have been asked about whether he was biased, whether his data was good enough, and whether his claimed precision was trustworthy. If Dyson were alive to listen in, he would be rightly indignant. As Eddington himself took pains to point out in writing Dyson's obituary, it was Dyson who chose 1919 as the year of decision, who organized the expeditions, whose assistants took the essential data, and who handled the data analysis, with the aid of a man, Charles Davidson, with whom he had begun working nearly two decades before. Dyson had complete confidence in his own ability to achieve the precision required for the test and would have been, at best, amused by the modern assumption that he was the junior partner in the enterprise. In writing Dyson's obituary, Eddington said it best when he wrote, "Time after time, in scientific progress as in human fellowship, his was the inconspicuous yet indispensable part" (Eddington 1940, 171).

The Telescope Maker: Howard Grubb

The early years of the astrographic survey were very difficult. Everything was new: equipment, techniques, and the requirement for so much man power. The man-power problem was solved at many observatories by the application of woman power. The survey contributed to the rise of women in astronomy by greatly increasing the number of jobs, though admittedly at very low pay (Stevenson 2014). For astronomical instrument makers, also, the survey represented an opportunity. Obviously, the Henry brothers made many of the lenses needed by participating observatories, but one firm in Dublin ended up making astrographic telescopes for seven different observatories. Just as the Henrys benefited from their close relationship with Admiral Mouchez, so the Dublin firm of Howard Grubb (originally, the son in the firm of Thomas Grubb and Son) began by making an astrographic telescope for the other prime mover of the project, David Gill in Cape Town, South Africa. It was a difficult job. Gill was an exacting client, and the telescope was delayed repeatedly. But in the end, it was a success, and Grubb's firm went on to make more. In doing so, and through his prominence in making clockwork-driven coelostat mirrors, Grubb went on to indirectly play a key role in the 1919 eclipse.

The Grubbs were just one of a network of optical companies that rose up in the nineteenth century to provide telescopes and other equipment to the growing number of university and private observatories. In this respect Britain lagged behind countries like France and Germany, which had the leading instrument makers. In Ireland, the Earl of Rosse, a pioneer of large reflecting telescopes, built many of his instruments himself. But the sheer number of private observatories owned by wealthy Anglo-Irish landlords created an opening for Thomas Grubb to found a successful firm of astronomical instrument makers in Dublin in the mid-nineteenth century. The lenses used by the 1919 expeditions were manufactured by this company in the time of Thomas' son Howard. By the late nineteenth century, when Howard was running the firm, it was possible to prosper by adding newer university observatories to the clientele. University College Cork, my alma mater, fitted out its observatory in 1880 using equipment purchased from Grubb, including a lens similar to ones used on the eclipse expeditions.

The center of high-precision optics in Europe had moved to Germany by the early twentieth century. The preeminent firm was that of Carl Zeiss in Jena. Indeed, during World War I the British found they had no source for the optical-quality glass needed to manufacture lenses for high-precision naval periscopes. They were fortunate to have Grubb, who had great expertise in all areas of telescopic technology. For instance, he also built equatorial mounts for telescopes along with the clockwork to move the mounts. In 1900, he developed the reflector sight, an aid to targeting that permits a gunner to look through a small telescope into which the image of a target is projected by reflection. It is the forerunner of modern heads-up displays in military technology. Grubb was in high demand during the Great War, since he also developed and manufactured periscopes for the British navy. After the Easter Rising in 1916 raised the prospect that Ireland might gain independence, the British took no chances that they might lose this strategically important firm. The company was moved, lock, stock, and barrel, to Saint Albans near London before the end of the war. It did not prosper there once the war was over, and it later moved to the North of England.

Thomas Grubb was born in 1800 near the city of Waterford in southern Ireland. Like Eddington, he was a Quaker. He was an enthusiastic amateur astronomer, but his main business, initially, was manufacturing billiard tables in Dublin. Then, in the 1830s, he took advantage of the unusual level of demand for telescopic equipment for the private observatories of Ireland's wealthy Anglo-Irish landed gentry. He was aided in this by a friendship with Romney Robinson, one of Ireland's few professional astronomers, who was director of the observatory in Armagh in the north of Ireland. Grubb's expertise lay primarily in the complex mountings required to make the best use of large telescopes. A large lens or mirror was of little use without a stable mount that permitted motion, driven by clockwork, to follow the stars. He built the equatorial mounting for the refractor mounted at Markree Observatory in 1834. This telescope in a private observatory in County Sligo in northwestern Ireland was then the largest in the world. It was surpassed a few years later by the giant reflector built in the Irish midlands by the Earl of Rosse, for which Grubb also worked on the mounting. After these initial successes, Grubb branched out into providing the optics for telescopes as well, especially to equip new observatories in the Southern Hemisphere. He built a telescope for Melbourne, Australia, for which he took his youngest son, Howard, into the firm with him.

Howard Grubb enjoyed a very distinguished career as a maker of precision optics. Like his father he made a number of advances in telescopic technology. In many ways the astrographic project made his name, despite the difficulties in making the astrographic telescope for the Cape observatory. Although he continued to make telescopes into the early twentieth century, the naval arms race between Britain and Germany brought him more and more business, culminating in the move to Saint Albans. A visitor to the new site, having been shown around by Howard Grubb, discussed the reasons for the move:

[He] piloted me round his works and showed me two large telescopes which had been finished sometime previously for the Russian Government, but the war and revolution had prevented shipment. I noticed also that the entire production then in hands was for the British Admiralty and this fact was responsible for the loss to Dublin of the works. The British Admiralty insisted, for three reasons, on the transfer to England.

The reasons were: —(1) Transport delays and dangers—the 30 ft. steel tubes for the periscopes were made by Vickers' in the North of England; they were not really tubes, but steel rods that had been bored out in an immense lathe. (2) Proximity to headquarters for consultation and inspection and delivery of the finished periscopes (3) The growing political unrest in Ireland—an armed guard was placed at the works. (Glass 1997, 213)

So while the British and others waited to receive their telescopes back from Russia, the Russians were unable to take delivery of Grubb's telescopes intended for them!

As was mentioned, the Grubbs' original business was in telescopic mountings and their associated clockwork. One of their specialties was coelostat (or siderostat) mirrors and their clockwork, which provided a convenient and cheap way to construct a telescope with a view of the whole sky. Equatorial mountings, which permit the entire telescope to point in any direction on the sky and have a drive to keep the telescope pointed at a specific area in the sky as the Earth turns, are large and expensive. Furthermore, they require the construction of a dome that can open and rotate in order to view any part of the sky. A cheaper alternative is a transit telescope, as was traditionally used in astrometry. Such a telescope is mounted to move only in one direction, covering one entire track across the sky. It relies on the rotation of the Earth to move stars overhead, and the retractable roof is a simple device that opens up a long, narrow slit in the ceiling. Simplest of all, a siderostat mirror can be operated through a small opening or window since the telescope does not move. The mirror at one end is turned to keep the image fixed in place. Since the entire telescope did not move, the use of a coelostat mirror permitted the great telescope at the Paris exhibition of 1900 to be built. It had a refracting lens of nearly fifty inches in diameter with a focal length not far short of two hundred feet. This telescope simply lay motionless in place, and the giant coelostat mirror did the work. Unfortunately, no buyer could be found for the telescope after the exhibition ended, and it was broken up.

Dyson himself came away from the 1919 eclipse experience distrustful of coelostats. He decided to transport equatorial mountings to subsequent eclipses and urged others to do the same. Nevertheless, in 1919 he had little choice. Like the Americans, he never did recover his equipment from Russia in time, so even the range of coelostats he had available may have been restricted. Recall that he ended up taking one on loan from the Royal Irish Academy. He had planned to introduce electric driving (as opposed to using clockwork driven by the falling weights of a pendulum) on the coelostats in 1919, but this ambition was foiled by the wartime shortage of personnel. The desire for a wide field of view meant that he and Eddington each used an astrographic lens, but his use of two different instruments at Sobral, which turned out to be a fortunate decision, suggests that he harbored doubts about the ability of the equipment at hand to achieve the desired level of accuracy.

Thomas Grubb retired from the firm around 1870 (Glass 1997, 3). It was thus Howard Grubb who manufactured all the astrographic lenses. It was he who worked with David Gill in the 1880s in the early effort, which predated the *Carte du Ciel*, to produce a lens suitable for astrometry. When the *Carte* was launched, he certainly hoped that his experience and established relationship with Gill would help him in securing commissions to build the new telescopes that would be required for the project. He made a frank appeal to nationalism, writing to Gill in the immediate aftermath of the Paris congress of 1887, which established the *Carte du Ciel* project:

> I shall be glad to hear from you as soon as possible about the details of the photographic telescopes. Are there not one or two of them to be made for Australia as well as for Greenwich and the Cape? Of course, the greater number I have to make, the less

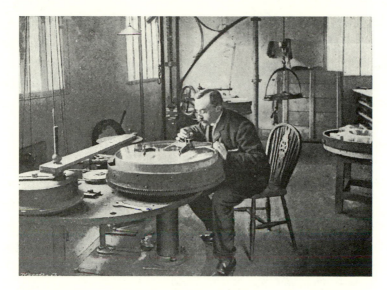

FIGURE 10. Howard Grubb in his workshop, using a spherometer to measure the curvature of a telescope lens during the final polishing process. He continued the telescope business begun in Dublin by his father, Thomas Grubb. His firm manufactured all three of the lenses used by the eclipse expeditions in 1919, as well as the coelostat mirror that produced the best photographic images.
(Reproduced from an article in the *Strand* magazine of 1896, vol. 12, 375.)

price I can make them for. . . . Now I dont think there can be any reasonable doubt that these instruments can be made in our own Country at least as well, probably decidedly better, than in France & my estimate is lower than that of the Frenchman. If therefore the Government send their orders to France they will have to be prepared with a good reason for doing so as I shall certainly take steps to have the question raised in the House [of Commons]. (Glass 1997, 134)

Grubb had apparently gotten wind of a plan to have all the instruments made by the Henry brothers, perhaps for reasons of standardization, and was prepared to take the matter all the way to the British Parliament. His early confidence that he could do better than the French eventually evaporated. Producing a lens with a really wide, flat field proved very difficult, and there were many delays in his delivery of the telescopes. Gill, in particular, though an old friend of Grubb's, was an exacting client, and Grubb seems to have often been driven to distraction in working with him. In 1882 (at the time when the photographic scheme first occurred to Gill), Grubb was even writing to his client complaining of the revolutionary fervor of his employees! It so happens that a member of the Irish Republican Brotherhood (IRB), Michael Lambert, was a key employee at Grubb's works. Lambert is even said to have made the keys used to liberate the founder of the IRB, James Stephens, from a Dublin jail in 1864 (Denieffe 1906, 123). It has been speculated that he may have used Grubb's machining tools for the job. This old Fenian, in spite of being "out" in the uprising of 1867 and in exile for a period thereafter, worked for Grubb for many years. In 1882 a Fenian revival took place as the organization was involved in the Land War agitation against Anglo-Irish landlords. In that year a spectacular terrorist outrage occurred in Dublin when the head of the British administration of Ireland and his chief civil servant were assassinated by knife-wielding attackers in Dublin's Phoenix Park. In a private note to Gill, Grubb wrote:

> The simple fact is (but one I did not like to have noted in our copying book) this: the minds of all our men are completely upset with the political situation they can think of nothing else but holding their Fenian meetings & when they are at work here they leave their heads behind them & we have had, oh such messes made of some of our work & such loss of time it is heart-breaking. (Glass 1997, 118)

In a subsequent letter the same year, Grubb specifically blamed Lambert, saying, "I have had to take all Lambert's hands away from him & put him to work in a room for himself & you can imagine what a state we are in when I tell you I have had to have some of the hands under Police Surveillance & I myself now never come down here at night to try an object glass without a loaded revolver in my hand" (Glass 1997, 118).

So one can imagine that when the Fenians came to a new and unprecedented prominence after the conscription crisis of 1918, Grubb may have been willing to move his business to England, since his political views were obviously unionist in sentiment.

In the end, Grubb made astrographic lenses for Oxford, Greenwich, Tacubaya near Mexico City, Cape Town, Sydney, Perth, and Melbourne, and another one of very similar design for Cork that was never actually used in the project. The Oxford and Greenwich lenses, although sister lenses, were by no means identical, as Grubb worked almost by trial and error to try to get the flattest possible field. The Greenwich and Cape lenses were constructed differently from the others, and a modern expert on Grubb's work, Ian Glass, an Irish astronomer working at the Cape observatory, reports, "The Greenwich and Cape telescopes have curved focal planes to this day and are usually focused for stars at 40 to 50 mm distance from the plate centre in order to get the best overall images. . . . Examination of the plates produced by the Cape astrographic shows slight elongation of the images situated towards the corners" (Glass 19, 154).

In one way the most interesting instrument was neither astrographic lens but the four-inch lens on loan from Dublin, which actually provided the data Dyson relied upon in making his decision in favor of Einstein. This equipment had originally been used to observe the eclipse of May 28, 1900—coincidentally, the first eclipse observed by Frank Dyson and an eclipse with a very similar star field to the one seen in 1919. However, the instrument was used to study the Sun's corona in 1900. Howard Grubb proposed that an Irish expedition be mounted to Spain, jointly sponsored by the Royal Irish Academy and the Royal Dublin Society. Grubb and a wealthy Irish amateur astronomer provided much of the equipment, but the two societies did purchase from him a coelostat mirror each, and Dunsink Observatory obtained a four-inch lens from Grubb (Joly et al. 1901).

The fact that an Irish instrument was used in 1919 suggests that British equipment was not widely available because of wartime circumstances. The fact that Grubb had manufactured all three of the lenses brought also suggests, of course, a generally high opinion of his work. The Irish connection to the eclipse is a surprisingly strong one and reflects the unusually strong position in astronomy that the country held in the nineteenth century. One more Irish connection would be added, with a final addition to the eclipse expedition personnel, in the months before departure.

8

The Improvised Expedition

The 1919 eclipse was one of the spoils of the Great War. In fact, since the war exhausted the economies of the victors as well as the losers, it might be said that it was the greatest prize of victory. As Dyson and Eddington made their plans in 1917 and 1918, a similar pair in Germany, Freundlich and Einstein, were setting their sights on the same eclipse. For a while it seemed as if Freundlich and Einstein had the pole position. It is true that, like everyone else, their equipment was stuck in Russia. It had been impounded in 1914, so at first there was no possibility whatsoever of its being returned. Since neither the friendly British nor the Americans ever got their gear back in time, it was, for most of the war, not clear that the Germans were much worse off. But Freundlich was really much more dependent on the 1914 equipment than the English because he commanded no resources of his own. So by January 1917, Einstein was writing to him proposing an effort to persuade someone in Russia to help them out. By then the equipment had been transferred to the observatory at the Black Sea port of Odessa. Einstein hoped that an appeal from one scientist to another might work. He obviously felt he was the right man for the job "since no bellicose odium weighs on me." For this reason, it would be "better if I use my personal influence concerning that instrument" (Crelinsten 2006, 123). But if Einstein was known as an antiwar German, there is no sign that this cut any mustard with the Russians.

However, by 1918 a solution to the problem that was more in keeping with the mood of the moment presented itself. It suddenly seemed an advantage for Germany to be at war with Russia when, in the wake of the breakdown of the Brest-Litovsk negotiations with the new Soviet government, the German army advanced on Odessa and occupied it in March. Freundlich and Einstein could now take advantage of their nation's military might, insofar as they would only have to negotiate with Germans to try to arrange for someone to retrieve their telescopes. Unfortunately for them, this still proved futile. In hindsight, the scientific honor of observing the 1919 eclipse may seem to have been one solid achievement of victory, but at the time no one imagined that the war was being fought for the benefit of astronomy. In the chaos of the German occupation of Ukraine, there was never an opportunity for anyone to go to Odessa to retrieve Freundlich's equipment.

Between March and November, the tide turned completely in the fortunes of war. In March the Germans captured the city where Freundlich's equipment lay, and the great offensive that followed on the western front, the famous Kaiserschlacht, almost had the effect of sending Eddington into captivity. Just when the Germans seemed to have successfully launched a war-winning stroke, the end came with shocking swiftness. Fresh American troops turned the tide of battle in the West. This precipitated the German revolution for which the Bolsheviks had been breathlessly waiting in the East. By the time the eclipse occurred, Odessa was once again under Soviet control after the Red Army's defeat of an Allied (French and Greek) expeditionary force that had replaced the Germans. But the Russian civil war was only just warming up, and Odessa would change hands multiple times before it concluded. Freundlich's equipment would not be returned until 1923, but it is highly unlikely that given the conditions of postwar Germany, he would have been able to mount an expedition in 1919 even if the instruments had been in his possession.

The Assistants

Wartime conditions in England made things difficult for Dyson and Eddington also, especially when it came to selecting personnel for the trip. Each of the two expeditions had a minimal complement of two observers, and no other observations were attempted apart from the Einstein test. It seems likely that Dyson himself would have gone in normal circumstances. He was greatly interested in eclipses and even wrote a book about them some years later. He saw six during his lifetime, and it is perhaps unfortunate that he did not accompany Eddington to Principe because he enjoyed uniformly good luck in his eclipse career, never being clouded out. Comparing his luck to Freundlich's, who traveled four times before he actually witnessed an eclipse, suggests that he had an eclipse fairy looking out for him. Of course, Dyson's luck seems to have traveled with his Greenwich team to Sobral, where the weather cleared at the last minute before totality.

Eddington had the most obvious difficulty finding a companion for the Principe stations, since both of his Cambridge Observatory assistants, as we have seen, died during the war. Edwin Turner Cottingham, who accompanied him, was a Northamptonshire clockmaker. As a young man, he had been apprenticed to a clockmaking firm, which he eventually had taken over, enjoying a successful career in astronomical timekeeping. He did work for both the Oxford and the Cambridge observatories, and he was a natural person for Eddington to take along given the reservations expressed about the coelostat clockwork before the expeditions set out. He had himself overhauled the Oxford coelostat and operated it on Principe. It is a testament to his skill that clouds were the only difficulty encountered by Eddington in his experiment.

If we have been impressed by the number of nonconformists in our story to date, the Sobral expedition seems to have been, quite by coincidence, a Catholic affair. Originally, the plan was for the senior figure there to be Father Aloysius Cortie, an experienced eclipse observer but not a member of the Greenwich staff. Father Cortie was a Jesuit astronomer who taught at Stonyhurst College in

Lancashire. He was a popular figure with his fellow astronomers, students, and the public, who all enjoyed his engaging character and outgoing personality. He was a frequently requested speaker at working men's clubs in Lancashire. He would annually deliver a sermon at the meeting of the British Association for the Advancement of Science and was considered a fine singer. His expertise lay in solar astronomy, which explains the interest in eclipses that led to him serving on the JPEC. Unfortunately, he died of complications from influenza in 1925, maintaining his cheery good nature to the last. When the doctor expressed his dissatisfaction that his patient was not recovering, Cortie replied, "Well, it is not my fault. I have taken all your medicines" (Turner 1926, 175).

It was he who arranged for the critical four-inch lens and coelostat to be loaned by the Dunsink Observatory and the Royal Irish Academy (RIA). Unfortunately, he was unable to obtain a leave of absence from his college, and Irish astronomer Andrew Crommelin took his place. Cortie was used to his vocation placing obstacles in his path at eclipse time. He had participated in the JPEC-organized 1914 eclipse, originally planning to go to Russia. However, he was refused permission, as a Jesuit, to enter the Russian Empire. Accordingly, he took up a station in Sweden, even though JPEC had originally rejected that country on account of poor weather reports for the time of year. As it happened he was eventually given permission to enter Russia but did not hear the news in time. This was fortunate because the weather turned out to be better in Sweden than it was in much of Russia on the day of the eclipse. His report of his experiences in 1914 (Cortie 1915) confirms the significance of his role on the 1919 subcommittee. In 1914 he used both of the coelostat mirrors that were taken to Sobral. One was a sixteen-inch coelostat owned by the Royal Astronomical Society (RAS) and used at Sobral with the astrographic lens. The other was the RIA's Grubb coelostat, used with the four-inch lens. His report of 1914 is highly critical of the performance of the driving mechanism of the sixteen-inch coelostat. At that time Cortie had already concluded that an electric driver would be superior, since the weight-driven clockwork typically required a second person (in addition to the person in charge of photography) to operate. In the end, the RAS coelostat was sent to Cottingham, who had already taken charge of the Oxford instrument (which, similarly, was a sixteen-inch coelostat used with a Grubb astrographic lens) to overhaul. The sixteen-inch coelostat at Sobral must be assessed as suspect. It was not looked over to the extent that the Principe coelostat was and yet had been the subject of intense criticism for its operation during a previous eclipse. Nor did Davidson benefit, as Eddington did with Cottingham, from having a second person to help him with the machine. It is worth quoting one of Cortie's specific complaints regarding the RAS sixteen-inch coelostat in 1914 (which would be used with the Sobral astrographic in 1919): "These errors we thought we had adjusted, as on the day preceding the eclipse, and on the morning of the eclipse day itself, the solar image remained perfectly steady for a period of fully half an hour. But our adjustments of the errors in driving and in the screw were completely upset by the drop in temperature during the eclipse" (Cortie 1915, 107).

In comparison, Cortie seems to have been very satisfied with the performance of the four-inch and its accompanying Grubb-made coelostat in 1914. This instrument, while owned by the RIA, may have been in Cortie's possession continuously from shortly after 1900. He used it at eclipses in 1905, 1911, and 1914. It seems clear that it was he who provided it to the expedition of 1919, and he had originally intended to use it himself. Given the concerns over the coelostat used with the astrographic, it is plain that the four-inch was given over to Crommelin to use as a backup in case the other instrument failed. It might be wondered why the suspect sixteen-inch coelostat was even used in 1919. Notice that it and the four-inch telescope had something in common: They were both with Cortie in neutral Sweden in 1914. Not being in Russia when the war broke out meant that this equipment made it back to England, though not without considerable adventure in dodging minefields at sea on the voyage home. It was available when a great deal of similar eclipse equipment was not. In this way Cortie's membership in the Society of Jesus played an essential role in the eclipse story. Had he not been banned from Russia because of the controversial character of his religious order, he would very likely not have returned with the equipment needed for the expedition of 1919.

The problems created for the English by the loss of equipment in Russia are revealed in Dyson's annual report for 1915 (published in 1918). In this he recounts the equipment that Davidson and Harold Spencer Jones ("Mr. Jones") took with them to Minsk in modern Belarus. They had a nine-inch coronagraph and two six-inch telescopes (in addition to a spectrograph), presumably, with coelostats or other ancillary equipment. None of this equipment made it back from Russia. In Dyson's words: "Owing to the outbreak of war, it was impossible for the observers to bring back the instruments with them, and these were therefore left at the Poulkova Observatory in charge of [the director] . . . [who took] charge of the instruments pending the termination of the war" (Dyson 1915, 14). Obviously, it was only Cortie's fortunate diversion to Sweden that prevented the same fate befalling the equipment he used in 1914.

Whereas Cortie was an English Catholic, born in London, his replacement, Crommelin, came from a prominent Irish Protestant family of Huguenot descent. It was his family who had brought the linen industry to Ulster in the eighteenth century (strictly speaking, his ancestor's family name was de la Cherois, the Crommelin being adopted from their relation by marriage to the famous linen family, whose male line had gone extinct). He was born in Cushenden in the glens of Antrim, a scenic part of northeastern Ireland. Though raised a Protestant, he converted to Catholicism as a young man. One of his sons became a priest. Years after the eclipse, two of his children died tragically in a climbing accident in the Lake District, not far from where Eddington was born (Davidson 1940, 236).

His specialty was comets and their orbits, and he was a master at predicting the time of their closest approach to the Sun. He joined the Greenwich staff in 1891, when the government approved an additional assistant for the observatory. His background was similar to that of Dyson and Eddington in that he had been a Wrangler at Trinity College, Cambridge, before joining the Greenwich staff as an assistant. His career was not as meteoric as theirs, however, and he had been a teacher for a time before this opportunity came

FIGURE 11. Charles Davidson and Andrew Crommelin in Sobral with Brazilian astronomers and American geophysicists. Davidson and Crommelin are in the center wearing white suits, with Davidson on our left and Crommelin standing to his left.
(Courtesy of Observatório Nacional, Rio de Janeiro, Brazil.)

along. According to an obituary written by his colleague Davidson, it was Crommelin who had operated the four-inch telescope that took the decisive data at Sobral. Admittedly, Davidson modestly did not mention his own work in preparing the instrument for use. Years later Eddington recalled, "The use of this instrument must have presented considerable difficulties—the unwieldy length of the telescope, the slower speed of the lens necessitating longer exposures and more accurate driving of the clock-work, the larger scale rendering the focus more sensitive to disturbances—but the observers achieved success, and the perfection of the negatives surpassed anything that could have been hoped for" (Eddington 1987, 118). Still, it is a curious fact that this English expedition was successful largely on account of an instrument handled by an Irish astronomer, on loan from Ireland, and manufactured entirely in Ireland. Crommelin was fifty-four at the time of the eclipse; he was likely included in the party because he was a member of the Greenwich staff not subject to conscription. He was thus available at short notice when Cortie had to pull out.

A key member of the party at Sobral was Charles Rundle Davidson. His background was quite different from the other members of the 1919 team in that he began his working career at Greenwich as a computer. He was not college educated and had attended what is known in England as a charity school. In fact, he had attended the original Bluecoat School, Christ's Hospital, near London. Eddington's best friend, C.J.A. Trimble, was the math teacher there for many years, although obviously not until after Davidson's time. He was therefore of a different social class than the other members of the expedition. He was hired in 1890 and became an established computer in 1896. By 1919 he had been promoted to the more prestigious post of assistant. He had been at Greenwich longer than anyone else involved with the expedition. Computers in those days referred to those hired to perform exacting calculations by hand, and they played an important role in the work of observatories like Greenwich. Many calculations had to be done using laborious numerical methods that today would be performed on an electronic computer, and astrometry using photographic plates provides an excellent example. In fact, two computers were used in the reduction of the Sobral data, as Davidson worked with another computer named Herbert Henry Furner on the plate measurements.

There was quite a strict division between computers and assistants at the observatory. Assistants were college educated and, frequently, Cambridge Wranglers, and the position was often a stepping-stone to the directorship of an observatory or a professorship. Most computers remained in a subordinate role throughout their careers, which were often quite short. Davidson was an exception but not a unique one, as Dyson made a practice of providing a path to promotion for computers. Davidson's obituary notes that "he established a reputation for ability to handle equipment which was so great as to elevate him to a recognized position of the Observatory's arbiter on all instrumental matters, and Dyson in particular relied on Davidson absolutely to supervise the later eclipse expeditions which Dyson made such a feature of the work of the Royal Observatory.

In all, Davidson went on eight eclipse expeditions" (Woolley 1971, 193). By contrast, Furner's obituary merely states that he was a "quick and accurate computer" (Melotte 1953, 306), though he too was eventually promoted to assistant. It is thus remarkable that a mere computer was charged with a leadership role on this important expedition. Partly, this reflects the peculiar conditions created by the war. Certainly, Davidson acted as Dyson's man on the spot, and the two developed great trust through working together on just this kind of task for many years. Indeed, Dyson's daughter reports that Dyson "learned much from experienced practical men such as Davidson . . . from whom he learned much of his skill" (Wilson 1951, 73, 63). Indeed, though a Wrangler himself in his younger days, he came to deprecate theory enough to be heard making remarks about theorists, such as "if he'd ever observed through a telescope he wouldn't say such things" (Wilson 1951, 73).

Just as Irishmen play an unexpected role in the story of the 1919 eclipse, so Davidson stands in contrast to the Trinity, Cambridge, background of the others involved. And yet it was he who was sent to handle the trickiest bit of equipment and he, undoubtedly, whose advice was central to the eventual result. It was the decision not to make use of the Sobral astrographic data that permitted Dyson and Eddington to find in favor of Einstein. A key part of this decision resulted from the claim that this instrument had not operated as it should. As Davidson was in charge of it, we may suspect that Dyson relied heavily on Davidson's judgment. As we discuss the decision to ignore the results from this instrument, we have to keep in mind Davidson's status as the practical, non-college-educated man who was nevertheless considered the master of instrumentation whose opinion could be relied upon. It is an important but not sufficiently remarked-upon fact of experimental science that only the operator of an experimental apparatus is in a position to understand how and whether it might go wrong. Even if operators cannot categorically say why it has malfunctioned, their tacit knowledge of its operation gives them a "feel" for it that enables them to say whether its data can be relied upon or not in a given instance. As such, Davidson played an essential role in our story.

In the end, the German military had one more part to play. This time it was the navy that took the lead. In the fall of 1918 the army realized that the war was lost. Rather than fight on across German soil and risk the loss of face that would have been their lot as the home population witnessed their defeat, the generals restored civilian government, which they had abolished at an earlier stage of the war. They did this so the politicians could do the dirty work of seeking peace. In response, the navy, which had cautiously avoided contact with the powerful Royal navy throughout the war,[1] decided to sail out in a suicidal attack on the British fleet. The sailors mutinied and the revolutionary movement spread throughout Germany, overthrowing the monarchy and canceling Einstein's class. This brought the war to an abrupt end in November 1918, just in time for eclipse planning to proceed. In the later report on the expedition (Dyson, Eddington, and Davidson 1920, 295), Dyson noted that in the summer of 1918, "Preliminary inquiries were . . . set on foot as to shipping facilities, from which it appeared very doubtful whether the expeditions could be carried through." Only in early November had conditions "changed materially." Had the war lasted much longer, it is unlikely that the expedition could have traveled, for lack of shipping.

9

Outward Bound

The Joint Permanent Eclipse Committee (JPEC) subcommittee on the eclipse met at the Royal Astronomical Society (RAS) at Burlington House in Central London on December 14, 1918. The war had ended, since the armistice had been in place for a month. No peace agreement had yet been signed. Indeed, Eddington would be writing home from Principe the following summer inquiring whether the peace had yet been concluded. But it was clear that the expeditions now had at least a chance of successfully reaching their stations. An hour later the full JPEC had its annual meeting at the Royal Society, which then also had its rooms at Burlington House (Sponsel 2002, 460). They reported on their plans, which were at an advanced stage. The war had, in its death throes, created one last obstacle for Dyson, with the withdrawal of Father Cortie. Too many of his teaching colleagues were already absent "in connection with the war." So although most of the personnel and the equipment had been decided upon, and with the difficult work of actually getting the teams to the observing sites before them, there was still a place to be filled.

Dyson moved quickly. By the next meeting of the eclipse subcommittee, again at Burlington House on January 10, 1919, he reported that Crommelin had agreed to join the Sobral expedition and that the

Admiralty (since the Royal Navy ultimately had responsibility for the observatory staff) had granted a leave of absence to both Crommelin and Davidson. He also reported that the equipment had mostly been collected at Greenwich. Plans were sufficiently far advanced that one impatient member of the subcommittee was thinking ahead to the data analysis stage. The minutes record that "Prof. Eddington drew attention to the desirability of each party taking out a micrometer, in order that preliminary measures of check plates and eclipse plates might be made at the eclipse stations."

There is no doubt that Eddington was hopeful that the expeditions could telegraph results home even before they had left their stations. Dyson never seems to have planned any such course of action. It seems clear that, even though he would not be present at Sobral, he always intended the data analysis of the images taken there to be undertaken under his supervision. In this respect, though the expeditions were planned by one committee operating under the auspices of the two learned societies, they were also separate expeditions launched by two different observatories. While the planning was a joint exercise, the gathering of the data and all of the subsequent analysis were to be done quite separately. Eddington did indeed make measurements from his plates on Principe, but he was forced to continue them over a considerable period back in Cambridge. The data from Sobral would be analyzed exclusively at the Royal Observatory in Greenwich.

For the moment the meeting returned to the immediate problem of getting to Principe and Sobral. Dyson and Eddington would "visit the Minister of Shipping to arrange for steamer passage." Dyson would take care of visas and passports. Eddington, like an excited school-boy, returned to the question of ensuring that the world know of their results immediately. He volunteered to develop a code by which the two teams could inform Dyson by telegraph "as to the weather conditions and general character of the results obtained during the eclipse." After this outburst of boyish enthusiasm, the meeting closed with a discussion of finances and the need for the observers to keep careful accounts of their expenses while traveling.

There was time for one last meeting of the subcommittee before departure, on February 14, 1919. This was really a meeting of the expedition members as they were packing to leave. Crommelin was now present, as well as Davidson and Cottingham, who had attended the previous meeting at Dyson's invitation. Apart from the four observers, only Dyson, as chairman, and the subcommittee's secretary, Fowler, were present. Final arrangements had been made. The instruments had been insured for the sum of three thousand pounds. Howard Grubb himself had been approached and had valued the astrographic lenses (or *object glasses*) at a thousand pounds apiece. Dyson had arranged for the instruments to be forwarded to Liverpool on the twentieth. They would then be put on board the steamship RMS *Anselm*, of the Booth Line, which would sail for Pará in Brazil on March 6. This may have been the first voyage resuming this service to Brazil after the war, according to one recent Brazilian academic paper on the subject, and the ship was greeted with considerable fanfare on arrival (Crispino and de Lima 2016). This emphasizes how the end of the war came not a moment too soon for the expedition to succeed. The same ship would bring Eddington and Cottingham as far as the island of Madeira. They had originally envisaged traveling by way of Lisbon, but "the political situation in Portugal" prevented this. Indeed, "no information was obtainable as regards steamers to Principe" from Madeira. Actually, the situation was ominous, as Eddington wrote to the director of the Lisbon Observatory the previous week, saying that "we find that all sailings of boats to Lisbon have been cancelled for the present—I suppose owing to the revolution. I trust that you and the observatory are unharmed" (Mota, Crawford, and Simões 2009, 256). Eddington and Cottingham would simply have to get there as best they could, dealing with the situation upon arrival in Funchal, the main town of Madeira. The minutes continue: "Under these circumstances he [Dyson] thought it desirable to approach the Admiralty with a view to obtaining the services of a warship to convey the Principe party from Madeira to their destination."

The use of Royal Navy ships had been quite common in previous expeditions. That nothing came of this proposal suggests that, in the immediate aftermath of the war, the navy had no ships to spare for taking astronomers to remote locations. The final meeting closed with a demand for more government money (250 pounds) and with one last contribution from the Jesuit order. Father Cortie had sent a letter informing the committee that a request had been sent via the Portuguese Jesuits asking members of the order in Brazil to welcome the expedition to Sobral. But it was clear that, from that point on, the expedition members would be largely operating on their own. As if the Great War had not been obstacle enough, a civil war had just broken out in Portugal following a monarchist insurrection against the republic declared less than a decade previously. This was the "political situation" referred to in the subcommittee's minutes. Although Madeira was a Portuguese possession, it was to be hoped that its great distance from the mainland would prevent the civil war from affecting the expeditions' plans. But there was little the Jesuits or the British government would be able to do if all Portuguese shipping was suspended. The expedition departed a couple of weeks later into the unknown.

The Outward Journey

The best sources for the expedition members' journey are Eddington's letters home to his mother and sister.[1] They lived with him at the observatory in Cambridge and formed a close family. The letters are typically signed, "Your affectionate son, Stanley." Eddington never went by the name Arthur. In any case it was customary at that time for British scientists to address each other by their surnames only. The first letter was not written until the *Anselm* was approaching Lisbon and recounted their experiences since leaving London. The first point to note was seasickness, to which Crommelin and Cottingham succumbed, even though the sea "has not been very rough." In his own published report, Crommelin (1919d) later mentioned that the cruise on the *Anselm* was "extremely enjoyable, save for a few hours in the Bay of Biscay."

Eddington reported their struggles with luggage at Euston station in London, where they were asked to pay extra for the transport of their fragile lenses. After that, "we got to Liverpool at 3.45 and then difficulties began." The hotels were full, and it took a long time to find

BOOTH LINE.   R.M.S. ANSELM

FIGURE 12. The RMS *Anselm*, pictured in a postcard intended for sale to passengers, was the ship that, on its first postwar voyage to South America, took Eddington and Cottingham as far as Madeira and carried Crommelin and Davidson as far as Brazil. The painting is by Norman Wilkinson in 1909. (Reprinted with the permission of the author.)

anyone from the baggage agency to take charge of their luggage for delivery to the *Anselm*. Eventually, "after three or four attempts," they found a hotel—and felt themselves lucky to do so, especially as it "was a howling wet night."

The next day they were down at the dock early, but the baggage did not arrive. They eventually had to go on board without it to present themselves to the immigration officer, who "was only there for a short time." After noon the bags turned up, "and we went down to lunch much relieved." The two teams had cabins next to each other on the *Anselm*, which Eddington found "roomier" than he expected. At 2:00 p.m. they set sail, moving slowly through docks down the Mersey. By 9:00 p.m. they could see the lights at Holyhead on Anglesey, an island in northwestern Wales, and there they dropped the pilot, a port official who steered ships in and out of the harbor. After that, Eddington reports, they had no further sight of land, and wartime regulations even prevented the crew from informing the passengers of the ship's course.

A curious note of life aboard the *Anselm* was that Davidson and Crommelin were privileged to sit at a table with the ship's captain so that, as Eddington put it, "our party was broken up." Obviously, representatives of the Royal Observatory ranked higher in maritime culture than a teacher and a clockmaker. Eddington was happy to find an amateur astronomer aboard whom he knew from correspondence, a Mr. Walkey who was on his way to Brazil "for the Bible society," traveling up and down the Amazon and its tributaries. "He expects to be out there most of his life." Chess seems to have been an onboard pursuit for Eddington, who concludes the letter, "I am quite glad to be having a long steamer trip again." The *Arlanza*, which he had traveled on previously, had not yet returned to civilian service. The *Anselm* had served as a troopship in the early days of the war, ferrying the British Expeditionary Force over to France. But she had later returned to civilian use and was thus available for the eclipse expedition.[2]

By the date of his next letter, March 15, Eddington had left the *Anselm* for good, having arrived on Madeira. Good fortune was with them in Lisbon during an all-day stopover on March 12. Monarchist forces had largely been defeated the previous month, and the city was no longer threatened by war. They were able to visit the observatory there, accompanied by its eighty-two-year-old director who looked, to Eddington's eyes, like a vice admiral. He reported, "Lisbon is full of soldiers. They have disbanded all the police, but the country seems pretty quiet." Thus, a part of the voyage that might have been perilous passed peacefully. The four men had a farewell lunch together in Funchal after their arrival on Madeira after checking the all-important baggage. Though they always kept the valuable lenses with them, they could not afford to lose the mirrors and other equipment consigned to the hold. Then Davidson and Crommelin went aboard the launch that returned them to the *Anselm*, and Eddington and Cottingham were left behind on Madeira to find some means of getting to Principe, still far away on the equator.

At first they had little else to do but enjoy the sights and the views from the island's mountains. Like the Canary Islands, which lie to the south and much closer to the African coast, Madeira is affected by hot, dry winds from the Sahara to the east, which Eddington mentioned in his letters. In ancient and medieval times, the islands were occasionally frequented by sailors. Plutarch's description of the legendary Islands of the Blessed in the Atlantic suggests he may have had reports of Madeira in mind. During the Portuguese voyages of discovery, they were claimed and occupied and became an important center of sugarcane cultivation. Their geographic isolation, over five hundred kilometers from the nearest coastline, was helpful to Eddington given the disturbances in Portugal. But he reported to his mother that the Great War had visited the town and that ships torpedoed by U-boats were still lying sunken in the harbor, their masts visible from the shore. Portugal had initially tried to remain neutral in the conflict, but her commercial and colonial ties with Great Britain brought her into conflict with Germany, resulting in a declaration of war in 1916. At the end of that year, a U-boat targeted Funchal, sinking a British ship and two French ships in the harbor. A year later, two more U-boats bombarded the town itself, killing several people. Eddington saw traces of the still-visible damage during his visit.

The letter reporting all of these details was dated March 27, and one can judge that at this date Eddington still had no idea how he was going to get to Principe. The tone of the letter is relentlessly upbeat, however, and Eddington kept himself entertained on the island. He even took in a football match played between the crew of a visiting British warship and employees of the telegraph cable station, very likely also British. It must have been disheartening to see a British ship in the harbor that he could not use. But, obviously, if Dyson, with all his contacts, had failed to secure a naval vessel for their use, then Eddington could hardly be expected to flag one down as if he were hitchhiking. He was still over four thousand kilometers from Principe as the crow flies and closer to six thousand kilometers by sea, and only a ship making a regular port of call on that island could be expected to take him there. Whether such a ship would turn up seems to have been quite up in the air at the end of March, two months before the eclipse.

On April 6, Eddington wrote home again, this time with hopeful news. "I think that our time here is nearly up. We are to go on by the steamer *Portugal* which is due here on Wednesday, April 9th and should reach Principe on the 23rd." He must have been relieved, as there had been previous disappointments: "The *Quelimane* which we had thought at first would be our boat was due here on the 3rd but did not arrive till yesterday; it was going direct to St. Thomé [São Tomé, a neighboring island to Principe, now part of the same independent nation], only a hundred miles from Principe, but did not call at Principe." By early April, Cottingham had had enough of Eddington's strenuous hill walks, and so the astronomer was unaccompanied for his last excursion. The island's steep hills were daunting; wheeled vehicles were considered unsuitable for the terrain, and sleds or toboggans were used instead.

In search of entertainment other than hiking and seabathing, the Quaker Eddington even frequented the island's casino, "generally go[ing there] for tea," having previously reported that the tea was better there than in his English-run hotel. "There is always a band there [at the casino]. Roulette is prohibited on the island; but the authorities pretend not to know that it goes on. Now and again they make a raid, but they always telephone up to say they are coming. One afternoon, I was wanting to come away and found the main doors, which lead out through the dancing saloon, were fastened, and we had to come out by a back way; the reason was that the Chief of the Police had come up for the dancing and he was supposed not to know what was going on the other side of the door." Madeira was an early center of tourism and had been a popular destination for wealthy or invalid Europeans since the mid-nineteenth century. Eddington's hotel housed one such invalid in sixteen-year-old Geoffrey Turner from Mumbles in South Wales. Eddington befriended the lonely young man and reported a trip to the cinema with him: "I have scarcely ever been out after dinner, but last night I went with Geoffrey to a picture-palace. The chief film was the funeral of King Edward VII![3] It was rather curious seeing it after so many years. After about 3/4 hour of pictures there was a short play of which we naturally could understand nothing. Then some recitations (chiefly serious) and some songs (chiefly comic). One of the comic songs was very amusing though one could not understand the words. It was a very crowded house, and very interesting to watch the audience."[4]

In addition to tourists and invalids, the islands had been of great interest to men of science since the eighteenth century. Eddington encountered a trace of the island's most distinguished visiting physicist. "I had a talk this morning with the English Doctor an old gentleman who has gone in for science a good deal. He is brother-in-law to the late Lord Kelvin, and told me a lot of stories about him. Kelvin met his wife at Madeira—Miss Blandy—the Blandys are the agents of most of the shipping companies here, and they saw after storing our instruments here."[5]

Finally, the letter concludes: "I expect my next letter will be from Cape Verde Islands. I shall be glad to be progressing again; but I have enjoyed the whole of my stay here immensely. It has been a splendid holiday."

Sure enough, the next letter home was written aboard the *Portugal*. In that letter, dated April 13, Eddington described his adventures with the exit formalities from Madeira. He received considerable aid from a local gentleman who he initially took to be a rather disreputable character but who turned out, on better acquaintance, to be a journalist! This man, editor of the local paper, was able to guide Eddington through the formalities, which included various, possibly informal, payments to the British consul and the civil governor. The chief of police, "for a wonder," made no charge! The journalist was even able to introduce Eddington to the governor of Principe, who was on leave in Funchal. Like the chief of police, the journalist surprised Eddington by not being on the take and "was merely helping out of politeness." He did take the opportunity to quiz Eddington about the expedition and his scoop, thus obtained, "duly appeared the next day."

Fortunately, the *Portugal* arrived on time, and hours were spent, as usual, getting the instruments on board. Eddington commented that she was "quite a decent ship about the same size as the Anselm. The cabin, which we share is large and airy. The food is good, but it is difficult to get used to the foreign meal times." After describing the ship's few English passengers for his mother, Eddington returned to his old shipboard pastime of chess, describing a long game with a doctor lasting hours. However, "I think he was not very pleased

FIGURE 13. Eddington playing chess with his mother as his sister, Winifred, looks on. This photo was taken in Weston-super-Mare, before Eddington's appointment as director of the Cambridge Observatory. Once he was director, the family moved to join him at the Observatory on Madingley Road.
(Reproduced from Douglas 1957, plate 5, 19.)

at being beaten, at any rate he has not given me an opportunity of another game." He confides that on Madeira his teenage friend Geoffrey had been his chief chess partner. He closes the letter by expressing the hope of receiving a letter on Principe. He had mentioned earlier that he would have asked his mother to write to Madeira had he known how long he would be stuck there.

His mother must have been relieved to read his next letter (if it indeed arrived home before her son), since it was datelined Roça Sundy, Principe, April 29, and began, "My very dear Mother: Just a month to the eclipse; and today we have all our belongings at the site selected and have started the work of erection."

Eddington's difficulties were far from over, but at least he had made it to the track of totality with time to spare. Moreover, he was, as he reported, "in clover," with word having reached the islanders from Lisbon of the expedition. "Everyone has been very kind, and they are not only anxious to give us a good time, but give us every help we need for our work." However, "I am afraid the weather prospects are not at all good, from what we hear, and we shall be lucky if we get a clear day." As with the 1914 eclipse, the climate reports, which then suggested that Russia was a better bet than Sweden, seem to have been unreliable. Eddington's hosts told him, accurately, that cloudy weather was the norm on Principe in the spring.

So eager was Eddington's host to be helpful that he came out by launch to the *Portugal* before the Englishmen could even disembark. Eddington introduced this gentleman to his mother thusly: "Mr. Carneiro is our host. He is rather a young man, and owns the largest private plantation. He has only been out here two years, but his family have had the plantation a long while. In Lisbon he was a well-known bull-fighter. (The Portuguese bull fight is not like the Spanish, the horse and bulls are not killed.)"

Written like a true Englishman! Other plantation owners also offered sites to the expedition, but Eddington found Roça Sundy, the Carneiro estate, more favorable. Language was a considerable barrier for Eddington, since few people on the island spoke any English. The governor (apparently not the same man Eddington had met earlier on Madeira, since this gentleman came out to the ship in the launch with Sr. Caneiro) "always collars my Portuguese dictionary when he sees me and hunts up things to say" in his "rudimentary English." He noted the absence of any women on the island, by which he presumably meant European women. Everything must have been oriented toward the plantation economy. Principe is one of the world's great producers of cocoa. A *Roça* (pronounced *rossa*) is a plantation, and the island had played a historic role in the slave trade, slaves being the original basis of the Portuguese colonial economy. Slavery was legal until 1876, and forced labor remained common for years after its official abolition. The period of the early twentieth century witnessed sensational revelations of corrupt and brutal colonial exploitation, especially in the rubber trade, and São Tomé and Principe saw their cocoa trade suffer as a result of similar exposés in the Western press. It may be that, in response, conditions had improved by the time of Eddington's visit. Perhaps, lacking any ability to speak the language, he was limited in his capacity to discuss conditions and pay with the local workforce, but he certainly did benefit from the cheap labor, based as he was in one of the large *Roças*. In his later report, he commented that "we used freely [the] ample resources of labour and material at Sundy." The location was chosen because the locals believed that the northwestern tip of the island, where the *Roça* was located, had the best chance of clear skies at that time of year, since it was away from the largest of the island's hills.

A few days after completing this letter to his mother, Eddington wrote to his sister. To her he admitted what he could not to his mother—that he had played roulette at the casino on Madeira, ending up "about a pound down," as he estimated, over the course of his visit. He confided that he got on well with Cottingham but found that the fifty-year-old could not keep up with Eddington's more vigorous exercise and so was "glad to find a more active companion in Geoffrey Turner, a very jolly boy keen on butterflies, on swimming and on chess, so we had several common interests." Eddington seemed to enjoy the company of the wealthy plantation owners on Principe, compared to the inhabitants of Madeira, where he had apparently socialized mostly with English people, saying "the Portuguese here are a very superior type to those we have met before." He still lamented his inability to converse with most of the locals, however, even in French. Nevertheless, the language barrier does not seem to have been a major problem in eclipse preparations. He somewhat implies that Sr. Carneiro spoke French (it was only in the main town of Santo Antonio that "they do not know French"); additionally, there were two "negroes," Lewis and Wright, who spoke English. These gentlemen are only mentioned briefly in the one letter to Winifred. Perhaps Eddington worried that, like roulette, speaking to Negroes might have seemed a bit shocking to his mother. One wonders if Messrs. Lewis and Wright, who worked for the telegraph cable station, according to Eddington's section of the expedition report, might not have played an instrumental role in facilitating eclipse preparations.

Meanwhile, Davidson and Crommelin were making their way to Brazil on the *Anselm* with much less uncertainty. They not only had continuous passage to their country of destination but received advice from the director of the observatory in Rio de Janeiro, a Frenchman named Henry Morize, about their site at Sobral and how to get there well in advance of the expedition. He had also facilitated the expedition of 1912, during which Eddington visited Brazil. They reached Belém, a large port at the mouth of the Amazon River in the state of Pará in northern Brazil on March 23. The Brazilian government allowed their packing cases through without an inspection at customs. They telegrammed Morize of their arrival and, feeling it would be pointless to go to Sobral without his aid, continued up the Amazon on the *Anselm* as far as Manaós (now Manaus). Such is the size of this great river that the ocean liner traveled a thousand miles upriver routinely. Crommelin (1919d) commented upon "luxuriant forests" along the river's banks, filled with exotic tropical birds.

Manaus lies near the confluence of the Rio Negro, coming down from Colombia, with the main tributaries of the Amazon. Upon arrival the captain of the *Anselm* took the two astronomers up the Negro by motor launch. The river was named for its tea-colored water stained by the abundant dead leaves that fall in it, so, as Crommelin said, he and Davidson were able "to see the forest at close quarters." Manaus is the center of the Brazilian rubber trade, and during the early twentieth century it was almost the only source of this commodity so vital to modern industry. But by 1919 the area was suffering from competition with other tropical regions to which European colonial governments had introduced Brazilian rubber trees. Wealth from the rubber trade had, nevertheless, ensured that this remote location, accessible at the time only by river, with no road or rail links to other regions of Brazil, had its own tramway. One of these lines, to Flores, ran straight out into the jungle; and Crommelin stated that he and Davidson twice took this route and marveled at swarms of leaf-cutter ants and exotic plants such as the mimosa, which "shuts up upon being touched." He also marveled at the floating wharves at

Manaus, necessary since the river could change its level by as much as sixty feet between seasons. At these wharves he reported the *Anselm* filling up with rubber, Brazil nuts, and cotton before returning downriver. They reached Belém again on April 8. Here, too, they had time to explore the jungle, having been given free passes by the English-speaking owner of the local tram company.

They left again on April 24 via the steamer *Fortaleza*, arriving in Camocim in the Brazilian state of Ceará, farther southeast down the coast from Pará, on April 29, the day that Eddington wrote to his mother from Roça Sundy on Principe. Here they found that the way had been prepared for them, and they were personally conducted everywhere as guests of the Brazilian government. The next day they took the train a moderate distance inland to Sobral. They found the region in heavy drought, which gave the landscape a "depressing aspect," to Crommelin. At Sobral they found a warm and well-organized welcome had been prepared for them by the civil and church authorities, primarily thanks to the efforts of Morize but also due to the intervention of Father Cortie. As Crommelin reported, "Several deputations were at the station to welcome us; it must be confessed that they were expecting Father Cortie, whose letter expressing his inability to go had never reached Sobral. However, the welcome was freely transferred to us" (Crommelin 1919b, 369).

Unlike Eddington, they had an English-speaking interpreter, Dr. Leocadio Araujo, throughout their visit. They were provided the use of a house owned by a cotton magnate. This was important because the house received water from a well used by the cotton plant. The drought was so extreme that most residents of Sobral were reduced to getting water from holes dug in the dry riverbed. Next to the house was a racecourse, and they unpacked their equipment in the cover provided by its grandstand and used the track itself as the site for their observations. The soil of the region was sandy, but the racetrack was covered with coarse vegetation to provide better traction for the horses. Thus, by the end of April, both expeditions were in place and able to begin preparing their stations for the eclipse.

Preparations

Each eclipse party now had a month in which to prepare their equipment. Since their goal was astrometry, they had to pay particular attention to the foundations of their telescopes, which had to be steady. In fact, at Sobral the first task (according to Dyson's later report in 1920) after the marking of a meridian line was to construct brick piers to hold the coelostat mirrors. Since these mirrors would perform the turning action needed to keep star positions in place on the plate, it was of utmost importance that they have a secure base. A meridian line is a line that follows a line of longitude, thus it runs from true (celestial) north to south. The most famous such line is the prime meridian at Greenwich, which is still a tourist attraction today. In this case the line served as a reference point for the astronomers to correctly orient the telescopes to observe the Sun at its exact position during the eclipse. The coelostat mirrors could not observe the entire sky from any given position, and indeed the smaller coelostat, constructed by Grubb for use with the four-inch lens, had not been designed to work in tropical latitudes. It was built for Irish astronomers who were unlikely to travel to an eclipse outside Europe, so although Grubb designed it to be adjustable, it appears it was not adjustable for Sobral's latitude. Accordingly, at Greenwich an inclined wooden base had been built for this coelostat mirror, and it was mounted upon this at Sobral, on top of its brick pier. The piers were carefully built, which including digging short shafts beside them to permit the weights to fall, thereby powering the clockwork that allowed the mirrors to turn during the eclipse. Given Cortie's difficulties in 1914, the aim was to allow the weights to descend freely for half an hour, so no adjustments would need to be made during the crucial minutes of the eclipse.

At Sobral a hut was also constructed to house the telescopes. Crommelin reported (1919d) that, during a lunch break while the hut was being built, a whirlwind sprung up and threw down the beams, which had not been secured at that point, and smashed them. Luckily, they had brought beams to construct a darkroom for photographic development if no suitable interior rooms were

FIGURE 14. The telescopes in position at Sobral in northern Brazil. On the left is the round tube for the astrographic lens. Note the larger coelostat in front, which is said to have caused all the problems with this instrument. On the right, with its square tube, is the four-inch telescope with its smaller accompanying coelostat, made by Grubb. The four-inch lens, with its smaller field of view, has been oriented on its edge to maximize the number of stars imaged on its plate. Notice also that the coelostat mirrors are mounted on secure piers to provide stability, as the turning of the mirrors was critical to creating precise images of the stars. The temporary hut is to protect the instruments from the heat of the Sun, to which they would not normally be exposed in nighttime observations.
(Courtesy of Charlie Johnson.)

available, so these beams were used instead, and the hut was finished as planned. But this event, apparently not uncommon in that part of Brazil, caused some apprehension that the sudden change of temperature at totality might stir up strong winds, and windscreens were constructed to protect the observing site.

A photograph of the scene (see figure 14) shows the hut with the two instruments visible at the front. The circular tube belongs to the astrographic lens. It was purpose-built out of steel. Father Cortie had loaned the square tube and had used it in Sweden in 1914 for the four-inch lens. As Dyson commented, the four-inch lens was taken as an auxiliary device. It lacked the wide field of view of the astrographic, and the orientation of the tube had to be carefully chosen in order to permit the maximum number of bright stars—seven in all—to impinge on the plate. For this reason, the square, four-inch tube was mounted on the pier so that one angle of the square pointed down, requiring the construction of V-shaped wooden supports. The photograph shows this unusual orientation of the tube. It might have been possible to use larger photographic plates with this instrument, but this would have required constructing a new tube, which was not feasible in the time available. The plates were a critical moving part of this experiment. They had to be replaced at each exposure while being securely held immobile, or the images could be ruined. It was undoubtedly for this reason that no modifications were made to Father Cortie's apparatus, and in his report Dyson described the plate holders on the steel tube for the astrographic, which were constructed with screws to hold the plates firmly in place. Problems with screws failing to hold the photographic plates sufficiently in place are said to have been an issue during at least one subsequent eclipse (Texas Mauretanian Eclipse Team 1976).

In early 1919 Cottingham was still overhauling the Greenwich coelostat, and Dyson stated that the unfavorable February weather had handicapped this work. Apparently, not only the clouds at Principe caused problems in 1919. The familiar English wintry weather did its part as well. Some were uneasy as to whether the astrographic telescope, untried in an eclipse setting, was up to the task—hence, Dyson's decision to play it safe and bring along a second instrument that had successfully been used before. Of course, no one had tried to do astrometry with its plates. In that sense, they were in uncharted territory. But everything that could be done given the time and the man power available had been done, and Crommelin and Davidson were hopeful as they checked the optics once everything was in place. They felt they were obtaining good star images as they checked both instruments by night in May.

Meanwhile, similar preparations had been underway on Principe. Having decided to avoid the clouds by fixing the site in the lowlying northwest corner of the island, Eddington faced the problem of transport. Luckily, the plantation owners had a tram that went through the jungle from the little port. It was still necessary to carry the instruments for a mile through the jungle using, as Eddington put it, "native carriers."[6] As at Sobral, a pier was constructed for the coelostat support, in this case built of stone. Again a shaft was dug to allow for the clockwork-driving weight to fall unimpeded for a half-hour run of the mirror's driving mechanism. They made sure that a particular sector of the coelostat's mirrored surface was used to image the eclipse star field, as this section was considered the most perfect reflecting part of the surface. At this time, the techniques involved in making mirrors were not as reliable as today, which is why refractors were preferred for astrometry even though, in our time, reflectors are considered superior. Eddington and Cottingham were concerned about shaking the telescope when putting in plates and made sure, during the eclipse, to wait to begin an exposure until one second after replacing plates.

A major issue with conducting astrometry during an eclipse concerns the change of temperature that often occurs as the shadow of the Moon sweeps over the land during totality. This can affect the instrument's focus, changing the scale of the image on the photographic plate. Such a change in magnification alters the positions of the stars on the plates when it comes to comparing the eclipse image with one taken of the same stars at night. To avoid this, screens were erected on Principe to shield the equipment from the Sun, which was also the purpose of the hut at Sobral. However, Eddington noted that in the unusual tropical conditions of Principe, the temperature remained unchanged by day and night. This also proved, exceptionally, to be the case during the eclipse. In this sense the tropical conditions were helpful to the experiment.

While the astronomers in Sobral feared the effect of dust on their clockwork, Eddington faced extremely humid conditions on Principe and chose not to expose his lens to it until quite close to eclipse day. Only then did he commence his program of taking check plates. These plates were taken of a star field near Arcturus, nowhere near the eclipse field. The purpose of the check plates was, as their name implies, to ensure there was no significant change of scale (i.e., magnification) in the plates taken in England and on Principe. In this respect the two expeditions had planned a somewhat different program of observations. The Sobral team intended to remain in Brazil long enough to photograph the eclipse star field at night when it was roughly the same height in the sky as during the time of the eclipse. This meant waiting for a couple of months until the Sun rose sufficiently far behind the Hyades that the star field reached the required celestial altitude before dawn. These plates, called *comparison plates*, could then be directly compared to the eclipse plates to see if the stars had changed position. Since they would have been taken with the same equipment at the same location, one could hope that nothing material had changed in the optics. They waited until the stars were at the same altitude because of the differential refraction referred to earlier. Refraction in the Earth's atmosphere marginally shifts the position of the stars. The amount of this shift depends on the amount of atmosphere the light traverses on the way to the telescope, which in turn depends on the stars' height in the sky when they are observed. Eddington, however, had a problem. The eclipse was to occur much later in the day for him, meaning he would have to wait months longer on Principe to take proper comparison plates. It seems clear he never intended to do so, not wanting to be away from home for so long. As he said to his mother at the close of his final letter, "I suppose I shall be back about July 10. I shall look forward to the strawberries, which are better than anything they have in the tropics." In Eddington's defense, the steamship strike that took place just after the eclipse ended any thoughts of staying longer, but it seems likely he never really thought it all that necessary. Dyson, after all, could afford to be more cavalier about having his team stay two months in Brazil since he, as the busy director of the observatory, was not with them!

So what was Eddington's plan? He had his comparison plates taken at Oxford before he ever left England. This was probably done before the lens was even removed from the astrographic telescope at the observatory there. Of course, care was taken to focus the instrument at each location before taking plates, but this does not guarantee that the magnification of the image will be identical. Photographers are familiar with the concept of depth of field, which refers to the fact that a camera focused on a person at one location will also sharply image the people standing immediately behind her. There exists the similar concept of depth of focus, which reminds us that there are a range of positions, all close together, at which the lens can be placed to produce an apparently good focus on the photographic plate. This was a problem for Eddington because two apparently equally well-focused plates might actually have a different plate-to-lens distance and therefore a different magnification, producing a different scale on the plate. The purpose of his check plates was to guard against this. Check plates of the field near Arcturus were taken by night both in Oxford and on Principe so that the scale change between the two optical setups could be directly measured for any significant changes. Initially, Eddington seems to have regarded this merely as a routine measure. As we shall see, he eventually came to make crucial use of the scale change measured from these check plates.

10

Through Cloud, Hopefully

May 28, 1919, was, according to some people, the day science turned 2,503 years of age. May 28 is science's birthday because on this day in 585 BC the first eclipse that was successfully predicted beforehand took place. According to the ancient Greek historian Herodotus, Thales of Miletus predicted the occurrence of a total solar eclipse on this date. The eclipse took place during a battle between the Lydians (neighbors of Thales in Anatolia, modern Turkey) and the Medes, an Iranian people then engaged in building the empire that would shortly be inherited by their relations, the Persians. Herodotus told us that the two armies stopped fighting and arranged a truce in response to the eclipse. If we can believe Herodotus' account that the eclipse took place during the actual battle, then this battle is the earliest historical event that can be dated to the very day it occurred in our calendar. But it is the claim that Thales predicted the eclipse in advance that has caught the imagination. Thales has often been described as the first scientist, and his supposed prediction would stand as the first great triumph of science, making the date of the eclipse the birthday of science.

The truth is that most modern historians doubt Herodotus' claim. Predicting a solar eclipse is difficult enough. Predicting that totality

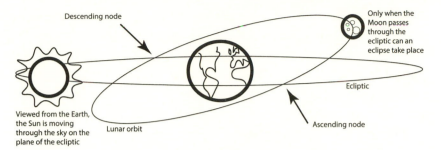

FIGURE 15. This figure illustrates the nodes of the Moon's orbit around the Earth. This orbit is inclined to the plane of the ecliptic, the plane of the Earth's orbit about the Sun (and thus the plane on which eclipses take place). Twice a month the Moon crosses the ecliptic, and the points where this happens are known as *nodes* of the lunar orbit. If the Moon is either new or full when it reaches a node, this means that the Sun, the Moon, and the Earth are all in line, creating the conditions for an eclipse. A solar eclipse occurs when the new Moon coincides with a node, because then the Moon is between the Earth and the Sun, potentially obscuring the Sun. The time it takes the Moon to return to a given node is a *draconic month*. The time to go from new Moon to new Moon is called the *synodic month*. After just over eighteen years, there will have been an even number of draconic months and an even number of synodic months, and a new eclipse in the same saros cycle will take place. (Reprinted with the permission of the author.)

would be visible at a given spot on the Earth became possible, as we know, only in the eighteenth and nineteenth centuries. It is not certain that Thales even understood that an eclipse is caused by the Moon blocking our view of the Sun. This can only happen, of course, when the Moon is new at just the moment it passes through a node in its orbit. Both these conditions are required to create an eclipse. The time between new Moons is called a *synodic* month, of roughly twenty-nine and a half days. The time the Moon takes to return to a given node is called the *draconic* month. It is just over twenty-seven days long. The draconic month gets its name from the ancient belief that dragons, who cause eclipses by devouring the Sun, live at the nodes in the lunar orbit (where the orbit crosses the ecliptic). After 223 synodic months (eighteen years, eleven days, and eight hours) have passed from one eclipse, the synodic and draconic cycles will again line up, and a new eclipse in the same series will occur. Thus, the eclipse of 1919 was part of solar saros 136 and was the successor to an eclipse that occurred in May 1901. The eclipse of 1919 was the longest of modern times when it occurred—one of the reasons Dyson was so anxious to seize the opportunity. The next two in the cycle were even longer, with the eclipse of 1955 the longest since the Middle Ages and longer than any that will take place in either the twenty-first or twenty-second centuries. Classical peoples' interest in eclipses is suggested by the claim that the Antikythera mechanism, the remains of an ancient Greek computer, may have contained machinery that permitted the calculation of saros cycles. The chief basis for this claim is that one major gear in the device has 223 teeth, the number of months in the saros cycle (Freeth 2009). But although the ancients knew of the saros cycle and could predict when an eclipse would occur, predicting that a solar eclipse would be observable at a given place was quite beyond their powers of calculation.

So if Thales did predict that the eclipse would be visible to people in his part of the world, he would have been very lucky indeed. But the story tells us that predicting eclipses in advance has been regarded as one of the highest aspirations of science since early times. Thus, on May 29, 1919, Eddington and the others were engaged in the ultimate scientific experiment. They were going far further than Thales could have dreamed. He was one of the founders of Greek geometry, and the theorem named after him depends on one of those facts that is true of Euclidean geometry but not curved geometries—that is, that the sum of the angles of a triangle adds up to 180 degrees. Now the members of the expedition, having successfully predicted the eclipse, would use it to draw a triangle of enormous proportions and test whether its angles really added up to 180 degrees, or something else, as Einstein predicted. In this fashion science had, two and a half millennia later, more than fulfilled the hopes held out for it at its birth. Even if their thoughts were not dwelling on Thales on the day before the eclipse, Eddington and the others were conscious of the importance of their task and aware that all their preparations would amount to nothing if the next day dawned cloudy.

Eddington later commented that it was "by strange good fortune [that] an eclipse did happen on May 29, 1919" (Eddington 1987, 113) because it is, he argued, not possible to pick a better day of the year to perform the experiment, given the number of bright stars close to the Sun at that time. The eclipse of 1900 took place on May 28, but of course no one knew to perform the test that year. A total eclipse in 1938 occurred on May 29, but in the Southern Ocean, close to Antarctica. The only land from which it was visible was the inhospitable South Georgia Island, and no one attempted to replicate the observations then. Eddington and the others found themselves far more comfortably situated in 1919. Eddington commented that "if this problem had been put forward at some other period of history, it might have been necessary to wait some thousands of years for a total eclipse of the sun to happen on the lucky date." He exaggerated somewhat, but the point is valid. The next eclipse to take place on May 29 will occur in 2310.

At both locations the dry season, fortunately, began before the date of the eclipse. In Sobral May is, according to the report, still part of the rainy season—the last month of it in fact—but the drought ensured there was very little rain in 1919. May 25 was an exception, when heavy rain fell to moisten the ground, to the delight of the astronomers, who had been worried about the effects of dust on their clockwork. On Principe the rainy season ended, as Eddington reported to his mother, on May 10; and the season of the gravana, a dry wind, set in. Unfortunately, Eddington discovered that this season was associated with heavy cloud cover in daytime. The nights were frequently clear, which facilitated taking check plates. The same was true, interestingly enough, at Sobral. Unfortunately, the mornings were typically cloudy, and the time of totality was before noon at that site.

After a dark, moonless night (since solar eclipses naturally take place only at the dark of the Moon), the light of day on May 29 brought bad news for Eddington. The sky was overcast, as it had been for days beforehand, and the cloud cover was complete. In fact, a thunderstorm broke out at midmorning and continued until nearly noon, "a remarkable occurrence at that time of year." In hindsight, Eddington commented to his mother, this may have been a blessing, "as it helped to clear the sky." Only half an hour before totality could the crescent Sun be glimpsed. This permitted them to align their telescope, though they had prepared a scheme to target the Sun even if they could not see it, using a landmark on the ground. Shortly before totality the Sun could be seen consistently through passing clouds. They began their program of taking plates "in faith," as Eddington confessed to his mother. He added, "I did not see the eclipse, being too busy changing plates, except for one glance to make sure it had begun, and another half-way through to see how much cloud there was." The clouds thinned as the eclipse progressed, and the last couple of plates provided the star images for Eddington's subsequent measurements.

At Sobral, Crommelin and Davidson were also dismayed to find the sky cloudier than usual. In fact, they estimated that it was nine-tenths covered with cloud at the time of first contact, when the Moon begins to partly obscure the Sun. From then until second contact, marking the onset of totality, was over an hour, so both expeditions were on tenterhooks, hoping the clouds would at least clear in the vital spot by the appointed time. The team at Sobral were the fortunate ones. As the report stated,

> There were various short intervals of sunshine during the partial phase which enabled us to place the sun's image at its assigned position on the ground glass, and to give a final adjustment to the rates of the driving clocks. As totality approached, the proportion of cloud diminished, and a large clear space reached the sun about one minute before second contact. Warnings were given 58s, 22s and 12s before second contact by observing the length of the disappearing crescent on the ground glass. When the crescent disappeared the word "go" was called and a metronome was started by Dr. Leocadio, who called out every tenth beat during totality, and the exposure times were recorded in terms of these beats . . . The region around the sun was free from cloud, except for an interval of about a minute near the middle of totality when it was veiled by thin cloud, which prevented the photography of stars, though the inner corona remained visible to the eye and the plates exposed at this time show it and the large prominence excellently defined. (Dyson, Eddington, and Davidson 1920, 299)

This prominence is a remarkable feature of the 1919 eclipse, best known from the photographs taken on Principe, which were mostly cloudy. The long exposures taken to image stars were unsuitable for such a bright object, which is overexposed on most of the Sobral

FIGURE 16. The great prominence visible during the 1919 eclipse is one of the iconic images of
the eclipse. This image was taken at Principe, since the cloud there prevented overexposure of
the bright prominence.
(Courtesy of Charlie Johnson.)

images. It is illustrative of the nature of professional astrophysics
that the observers did not spare a second to witness this remarkable phenomenon of nature live. Like everyone else, they enjoyed
it only in the images they took so skillfully. Prominences are enormous structures made of the Sun's plasma that extend out from the
surface in a bridge-like structure. Their formation is probably associated with eruptions in the Sun's magnetic field. Opportunities to
view them with the naked eye are rare, and the one seen during the
1919 eclipse was particularly large. Images of eclipses taken for the

Einstein experiment are typically a little dull. The Sun is small and the stars difficult to see. But the 1919 eclipse images that show off the prominence are among the most dramatic and reproduced of all eclipse photographs.

Eddington had been keen all along that the teams would telegraph news of the success or otherwise of the eclipse as soon as possible. This was done and Dyson was able to read out the messages at a meeting of the Royal Astronomical Society (RAS) in June. From Principe, Eddington cabled, "Through cloud, hopeful." Ironically, the Sobral team stuck with the code insisted upon by Eddington during the Joint Permanent Eclipse Committee (JPEC) meetings. At the RAS meeting, Dyson took the opportunity to elaborate because of the problems reported from Sobral:

> Both parties arrived safely, and we have had telegrams form both since the eclipse. We had adopted a code for the information to be sent, and the word "splendid" sent from Brazil was next to the word indicating a perfect eclipse. Prof. Eddington broke with the code in order to tell us that he was hopeful that something could be made of the photographs he had secured through clouds. A later telegram from Davidson stated that 12 out of the 13 stars had been secured on the plates taken with the Astrographic telescope. . . . The images are reported to be diffuse. This is probably due to the mirror or the coelostat in some way. The proper silvering of the mirror under the conditions prevailing before the expedition started proved very difficult. Also the coelostats, after many years of eclipse work, now need overhauling, a work which was impossible under the labor conditions during the war. The other telescope taken to Brazil had a 4-inch object-glass of 20 feet focal length, and was kindly loaned by Father Cortie. This had a smaller field, but on the plate taken by it 7 stars are shown. The observers are staying to take plates of the field after the Sun has moved away. That may keep them in Brazil a couple of months. (Jones 1919d, 261–62)

It is noteworthy here how Dyson and Davidson were already worried about the quality of the astrographic plates. Some modern commentators have claimed that the subsequent decision to discount the results from those plates was made because of their failure to agree with Einstein's theory. But already in May, before a single measurement had been taken, they were quick to blame the equipment in which they had never placed a completely wholehearted confidence. Years later Dyson published an extract from Davidson's diary, dated May 30 at 3:00 a.m.: "Four of the astrographic plates were developed and when dry examined. It was found that there had been a serious change in focus so that, while the stars were shown, the definition was spoilt. This change of focus can only be attributed to the unequal expansion of the mirror through the Sun's heat. The readings of the focusing scale were checked each day but were found to be unaltered at 11.0 mm. It seems doubtful whether much can be got from these plates" (Dyson and Woolley 1937).

One recalls here the wintry February of 1919. In the report Dyson tells us that the process of silvering the coelostat mirrors could not be carried out at Greenwich because the temperature could not be kept warm enough. The two larger mirrors, used with the astrographic lenses, had to be sent away for silvering. Only the coelostat used with the four-inch lens was silvered at Greenwich. Neither team seems to have been entirely happy with the silvering of the large coelostats. Eddington (in the report) refers to having to carefully position the Sun on the one sector of the mirror that was silvered best. Dyson was quick to blame the silvering of the large mirror at Sobral for the problems with the astrographic there. One might wonder why the silvering had to be done at the last minute during the month of departure. The answer is that the silver would tarnish within five or six months, so it could not have been completed any earlier for an eclipse in late May. At this time mirrors in reflecting telescopes had to be silvered twice a year. This is one reason why, even for the astrographic project, where they would have been advantageous, reflecting telescopes were not used. It was not until the development of the aluminizing process in the 1930s that reflectors began to take over in astronomy.

In the days after the eclipse, no one really knew if it had been a success or not. Eddington was merely hopeful. Davidson was worried about the quality of his images. Only Crommelin's device had worked pretty much as intended, but it could image little more than half a dozen stars because of its smaller field of view. Dyson, unable to influence events from so far away, must have been anxious. An extraordinary meeting of the RAS that took place in July could hardly have settled his nerves. Although the society normally did not meet in July, a large number of American astronomers were passing through London on their way to Brussels for the founding of the International Astronomical Union, to take place later that month. It is possible this Brussels meeting was one reason why Dyson did not travel to Brazil to observe the eclipse, as he did on six other occasions in his career. One of the Americans attending this special RAS meeting was none other than William Wallace Campbell, and the meeting opened with his discussion of the light deflection measurements carried out by Heber Curtis during the 1918 eclipse. He actually gave some numbers, none of them at all favorable to general relativity, and concluded with the statement that "it is my own opinion that Dr. Curtis' results preclude the larger Einstein effect, but not the smaller amount expected according to the original Einstein hypothesis" (Jones 1919b, 299). Thus, as Crommelin and Davidson prepared to take their comparison plates in Brazil, and with Eddington already departed from Principe, the news did not look good for Einstein. For Dyson, there was the fear that he would end up being scooped by the team at Lick. If the only comfort in Campbell's remarks had been his laments of missing equipment and large errors produced by poor-quality data, it seemed highly likely the same complaints would frustrate Dyson's program. It would not be until November that the results of the British expeditions would be released, and worries that there would be nothing significant to report persisted until the summer was well over.

11

Not Only Because of Theory

As promised, Eddington began to analyze data even before he left Principe. Both expeditions were developing plates immediately after the eclipse, only somewhat handicapped by the hot tropical weather. While Davidson and Crommelin were shocked to discover the poor quality of the images on the astrographic plates, Eddington was pleased to find that there were, after all, images of stars on the last couple of plates he had exposed during totality. Evidently, the clouds had thinned during the last few minutes of the eclipse and allowed the brightest stars to be imaged on the very last plates. Here, Dyson was vindicated for his urgency in taking advantage of this eclipse with its unusually rich star field (as well as the long duration of totality!). Eddington had brought a micrometer with him and had taken comparison plates before he left England, and so he set about taking measurements immediately.

A common modern take on Eddington, and Einstein as well, is that these two theorists hardly thought the experiment worth performing (Chandrasekhar 1976, 250), so sure were they of the result. But Einstein had spent years persuading astronomers to perform the test; and here was Eddington, on a tropical island, so burning with curiosity that he could not wait to be back in his observatory before beginning the painstaking work of data analysis. Perhaps their enthusiasm represented an overeagerness for good news, but no one can legitimately accuse them of indifference. A famous anecdote, told by Eddington himself, encapsulates his theory-led outlook. The incident in question took place when Eddington and Cottingham stayed with Dyson at Greenwich for a couple of nights before their departure alongside the Greenwich astronomers (Wilson 1951).

> As the problem then presented itself to us, there were three possibilities. There might be no deflection at all; that is to say, light might not be subject to gravitation. There might be a "half-deflection," signifying that light was subject to gravitation, as Newton had suggested, and obeyed the simple Newtonian law. Or there might be a "full deflection," confirming Einstein's instead of Newton's law. I remember Dyson explaining all of this to my companion Cottingham, who gathered the main idea that the bigger the result, the more exciting it would be. "What will it mean if we get double the deflection?" "Then," said Dyson, "Eddington will go mad, and you will have to come home alone." (Chandrasekhar 1976, 250)[1]

In typically English fashion, there was a riposte to Dyson's punch line. According to Eddington's biographer, after he had reduced the data from the first plate on Principe, he turned to Cottingham and deadpanned, "Cottingham, you won't have to go home alone" (Douglas 1957, 40). We have in Eddington's own handwriting the account he sent home to his mother from the deck of the steamship *Zaire*:[2]

> We developed the photographs 2 ea. night for six nights after the eclipse, and I spent the whole day measuring. The cloudy weather spoilt my plans and I had to treat the measures in a different way from what I intended; consequently, I have not been able to make any preliminary announcement of the result. But the one good plate that I measured gave a result agreeing with Einstein and I think I have got a little confirmation from a second plate. (Eddington to Sarah Ann Eddington, June 21, 1919, Eddington Papers, box 1, fol. Trinity/EDDN/A4/9, Trinity College Library, Cambridge)

It is certainly a point in favor of Eddington's critics that he adopted a definite stance at such a preliminary stage of his data reduction. As we shall see, he had to wait some months before any evidence came along to support this early statement. So it is interesting to ask whether Eddington's admitted interest in seeing Einstein's theory confirmed caused him to overlook contrary data. Many modern critics have argued that the 1919 team ignored data disagreeing with general relativity and that Eddington's bias was responsible. In essence, the charge is that Eddington failed to conduct a true test of the theory. I will argue that this version of the story is wrong on several counts, most importantly because it ignores Dyson's role. To begin with, Eddington's closing remarks in an article written before the expedition set out revealed that he was at least conscious of the possibility of an unexpected result:

> It is superfluous to dwell on the uncertainties which beset eclipse observers; the chance of unfavorable weather is the chief but by no means the only apprehension. Nor can we ignore the possibility that some unknown cause or complication will obscure the plain answer to the question propounded. But, if a plain answer is obtained, it is bound to be of great interest. I have sometimes wondered what must have been the feelings of Prof. Michelson when his wonderfully designed experiment failed to detect the expected signs of our velocity through the aether. It seemed that that elusive quantity was bound to be caught at last; but the result was null. Yet now we can see that a positive result would have been a very tame conclusion; and the negative result has started a new stream of knowledge revolutionizing the fundamental concepts of physics. A null result is not necessarily a failure. The present eclipse expeditions may for the first time demonstrate the weight of light; or they may confirm Einstein's weird theory of non-Euclidean space or they may lead to a result of yet more far-reaching consequences—no deflection. (Eddington 1919b, 122)

It is still worthwhile to note how theory-led Eddington's viewpoint was. Besides the null result, he admits only two possibilities, the two theoretical predictions Einstein made at different stages of the development of his theory. He has been criticized for his focus on this false trichotomy, as if only three numerical results were possible, instead of any amount of deflection. I think Eddington was conscious all along of the limitations of the experiment. He and his colleagues would have been utterly overambitious to have made very high claims of precision on the order of a repeatable experiment such as Michelson's. If they could decide between these three possibilities, that would be a good day's work.

The truth is that Eddington was conscious of the need for a new relativistic theory of gravity, but it did not have to be Einstein's theory. At the time of the eclipse, Eddington was already prepared to cast aside Einstein's theory if another one more in accord with ongoing experiments could be found. His interest in the unified field theory of the German mathematician Hermann Weyl reveals this fact. In 1919 it seemed likely that Einstein's theory would fail the solar redshift test. On December 16, 1918, only a few months before the planned expedition, Eddington wrote to Weyl:

> One reason for my interest [in your paper] is that it seems to me to reopen the whole question of the displacement of Fraunhofer lines, leaving the theoretical prediction unsettled. (Perhaps you will differ from me as to this). [Charles E.] St. John and [John] Evershed seem to be quite decided that experimental evidence is against the deflection, and this is rather a severe blow to those of us who are attracted by the relativity theory. I venture to think your theory may show a way out of the difficulty—but that is a guess. (Hermann Weyl Nachlass, Hs 91: 522, Swiss Federal Institute of Technology [Eidgenössische Technische Hochschule, ETH], Zurich; quoted in Kennefick 2012)

So although it is clear that Eddington saw a need for a new theory of gravity, he was by no means on a crusade to vindicate Einstein personally. He was undoubtedly worried about the solar redshift test. Writing to Walter Adams in 1918, for instance, he said, "St. John's latest paper has been giving me sleepless nights—chasing mare's nests to reconcile the relativity theory with the results, or vice versa. I cannot make any headway" (quoted in Douglas 1957, 42). The point here is that a relativistically invariant theory would be a good thing, but only if it worked! In that sense the eclipse test was perfect because it tested both the principle of equivalence (the half-deflection) and the idea that gravity altered geometry (the full deflection). Eddington's insistence that there were three possible results reflects his desire for the experiment to clarify the way forward for theorists such as himself wondering how best to proceed at an uncertain time of shifting paradigms. He wanted to know precisely whether and how Einstein was on the right track.

Eddington's emphasis on the experiment's implications for theory is understandable given that he was a theorist. The trichotomy frames the test in a way that emphasizes the lessons Eddington hoped he and other theorists would learn. The question is, does this somehow invalidate his participation in the experiment? Should theorists not be permitted involvement in experimental science for fear of bias? Keep in mind that Eddington was a perfectly capable observational astronomer, with years of experience in just this kind of work. There was no question of a Pauli effect, in which the theorist is such a klutz that he lays waste to the laboratory by his mere presence. As Dyson's successor as Astronomer Royal, H. Spencer Jones had this to say when later invited to comment on Eddington's career: "Observations of the determination of longitude are straightforward,[3] but require care and attention to detail to ensure accuracy. The results of the program show that Eddington was a careful and accurate observer. He is always thought of as pre-eminently a theoretical astronomer, and it is often overlooked that while at Greenwich he received training in observation and shared in various observing programmes, thereby acquiring a familiarity with observational astronomy that later stood him in good stead" (Douglas 1957, 17). The debate is not whether Eddington was capable of making the measurements, because he clearly was. The objection seems to be that he understood the implications of the measurements more clearly than his colleagues. Should this rule out his participation? It is a curious sort of "blind" experiment to demand that theorists not contribute because they understand the consequences of what they do too well!

Among the sharpest modern critics of Eddington's "trichotomy," the null, half, and full deflections, are the philosophers John Earman and Clark Glymour, who wrote about the eclipse expedition in 1981, in one of a landmark trio of papers on the three classical tests of general relativity. Earman and Glymour's papers raise a number of thoughtful points about the status of the light-bending test conducted during the 1919 eclipse. Their most important point concerned Eddington's careful framing of the theory test as a showdown between Einstein's theory and Einstein's first light-bending prediction, which Eddington labeled the Newtonian one. Earman and Glymour show how this played a critical and little-noticed role in the acceptance of the theory. The trichotomy forced people to read the results of the eclipse on Eddington's own preferred terms. This was a factor in the phenomenal success of Dyson and Eddington's campaign to win acceptance for their falsification of Newton's theory. Earman and Glymour further argue that this success prompted scientists who were hitherto skeptical of the theory, such as the solar astrophysicist Charles St. John, to reverse their previously negative verdict on the solar redshift test of the theory. The overall trend of their argument, as it has been read by many people, is that a carefully managed publicity campaign convinced physicists to accept general relativity on quite meager experimental grounds. However, nothing in their paper had the impact of their suggestion that Eddington's bias in favor of Einstein may have motivated him to throw out some of the data on dubious grounds.

Earman and Glymour's account of the eclipse has become widely known, at least in outline. Tracking the influence of their paper provides an interesting example of how a sufficiently compelling narrative can quickly evolve as it passes from the academy into popular culture (Partridge 2006). The carefully phrased version found in the original scholarly article is transformed into a bare-bones version as it enters popular discourse. It is stripped of all the caveats that originally hedged around the accusations against Eddington. Earman and Glymour's article was used extensively in an account of the eclipse expedition in Harry Collins and Trevor Pinch's *The Golem*, a book much more widely read than the original article. Collins and Pinch are very careful to avoid relying on the claim that Eddington fudged any data. Their argument is essentially that there is no such thing as a definitive experiment that resolves all doubts and proves one theory over another. This treatment in turn seems to have served as the principal source for a more popular book by John Waller (2002), whose aim is to debunk some of the well-worn anecdotes of scientific progress that modern historians of science have closely scrutinized in recent years.

When one finally gets to the reader reviews of Waller's book posted to amazon.com, one sees all of the scholarly analysis pared down to a headline in scare quotes. There,[4] it is stated that

> Eddington's observations of the eclipse over West Africa in 1919, which supposedly proved Einstein's theory, were worthless. Horror! . . . It's painful to discover that the guiding stories of one's lifetime are nonsense, but sometimes it just has to be done. This book has done it to me. John, you've broken my heart, but I've come through it stronger and wiser. Alas, I now see the old wives' tales everywhere, so I spend my days shuddering and shaking my head. As the motto of "The X-Files" put it: "Trust No-One."

> And we have pioneers like Robert Millikan and Arthur Eddington who made data fit a chosen theory, rather than the other way around. Yet, far from belittling such men, this book shows them in a new and more human light that transforms our understanding of scientific discovery.

> Remember learning in school how Eddington proved Einstein's theory of relatively by comparing the position of stars during and after an eclipse? Actually his images were so poor they proved precisely nothing except that Eddington was a dab hand at faking results. The book catalogues a series of famous scientists whose passion and belief in a theory blinded them to contrary evidence. In fascinating detail the book describes the circumstances surrounding the experiments both in the laboratory and in the wider social context. What links these scientists is that, as it turns out, the theories they were expounding happened to be right—just not for the reasons they gave. This compelling book should be

compulsory reading for all students of science and is delightful food for thought for anyone interested in science."

Thus, two main points emerge in the journey from scholarly article, through increasingly popular (though scholarly written) books, to the vox populi of the web. They are, first, that Eddington fudged, faked, or fit his own results to the theory he believed to be true and, second, that one must never take a story at its own evaluation, no matter how plausible it may seem (in this case, the story that Eddington proved that general relativity was true).

While I argue that the specific issue of the first point is quite wrong, I certainly cannot fault the attached moral. Indeed, it is true that the claim that Eddington's data were much dodgier than most people thought probably arose as a reaction to the wrongheaded belief that the 1919 expedition had somehow "proved" general relativity all by itself. One can confidently argue that the eclipse measurements were not very stringent tests of relativity because it was not an especially precise experiment. The compelling, but misleading, narrative that it somehow sufficed to prove Einstein right once and for all has now spawned, in reaction, a distorted myth of its own. In this counter-myth, the eclipse results were poor grounds on which to overthrow Newton's theory and replace it with Einstein's. I claim, on the contrary, that the eclipse results gave rather good grounds for believing that Einstein's theory was better than Newton's when dealing with the strong gravitational fields close to a massive body like the Sun. The lesson we should learn is that historians need to examine what scientists actually did very closely before deciding whether they did the right thing or not. Maybe we should never presume to tell them what is right or wrong, but at least let us look very closely at what they did. Let us now get down to that task.

Comparing the Plates

The method of comparing positions of stars on the plates was substantially similar for both expeditions. Plates taken during the eclipse were clipped together with comparison plates taken at night, so the star positions were as close to each other as possible. Ideally, one of the two plates would be a reversed image (taken with the use of a mirror as opposed to a lens), so the images could be compared face-to-face. A micrometer screw was then used to measure the separation between the positions of identical stars on the two plates. This would measure how far star A on one of the eclipse plates was positioned away from the same star A on a comparison plate.

In practice the Greenwich team faced the difficulty that both their eclipse and comparison plates were reversed (by the use of the coelostat mirrors in their instruments) and so could not be compared face-to-face. They made use of a third plate, which they called a scale plate, specially taken of the same field but not reversed (i.e., direct, using only lenses), which was placed against each of the eclipse and comparison plates in turn. The Cambridge team's comparison plates were taken using a telescope at Oxford (recall that the astrographic lens used on Principe was loaned to the expedition by the Oxford Observatory) and were thus direct because no mirror was involved. This permitted them to be placed face-to-face with the reversed eclipse plates taken via the coelostat mirror on Principe. At Greenwich the measurements of the Sobral data were made by two experienced computers, Charles Davidson and Herbert Henry Furner, under Dyson's direction (Dyson, Eddington, and Davidson 1920).

The Greenwich team had been able to take comparison plates of the eclipse star field while still in Brazil. This was possible at Sobral, where the eclipse took place in the morning with the Sun relatively low in the sky, permitting nighttime plates of the field to be taken two months later with the same instrument at the same location. Although the Greenwich team had originally entertained the possibility of leaving without waiting long enough to take comparison plates in situ, they decided it would be best not to do so after experiencing the aforementioned problems with astigmatism in the coelostat mirror used with their astrographic lens (Dyson, Eddington, and Davidson 1920, 298).

Even so, the task of the Greenwich team was not simply to measure the displacement of the star images between the comparison and eclipse plates and conclude that the resulting raw data were the

Einstein displacement. Regardless of the amount of light deflection found, there could be additional differences in star positions between the two plates due to three different kinds of misalignment between them. The first would concern whether the centers of each plate coincided when clamped together. The second dealt with the relative orientation of the two plates, either because of a rotation of the instrument, or simply because of the way the plates were clipped together. Finally, there might have been a change of scale (or magnification) on the plates, for instance, due to some change in the focus of the instrument between exposures.

If the star field was photographed at a different altitude in the sky, there would be differences in stellar positions on the plate due to differential refraction in the Earth's atmosphere. Even if everything else about the two plates was identical, the lapse of time between taking eclipse and comparison plates (two months, in the case of the Sobral expedition) meant that the Earth was moving in a different direction, relative to the direction of the star field, thus creating differences in stellar aberration between the two plates. *Stellar aberration* refers to a shift in the apparent position of a star due to the relative motion of the Earth compared to the line of sight toward the star. These last two kinds of change could be calculated theoretically. Both teams carried out these calculations during their data analysis, but other changes in scale and orientation between the plates, if they occurred, could not be predicted in advance. They had to be measured.

Fortunately, changes in the plate position and orientation behave differently from the purely radial displacement predicted by Einstein for the light-bending effect. The most important of the three changes, which must be determined in order to convert raw measurements from the plates into light-bending results, is the difference in *scale* between the two plates. This is because a change in scale between the plates can mimic the actual light deflection displacement because it shifts the stars radially away from the plate center. The scale can be distinguished from the light-bending deflection, as Eddington himself pointed out (Eddington 1919b, 120), because if measurements are taken with respect to the position of the Sun at the center of each plate, the light-bending deflection is greatest for stars nearest

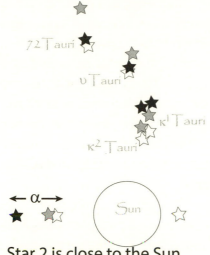

Star 11 (known as 66 Tauri) is far from the Sun and has a large scale shift (e) and a small light deflection shift.

Star 2 is close to the Sun and has a large shift due to light deflection (α) and a small scale shift.

☆ Actual position of star on a comparison plate

⭐ Star position shifted by scale change

★ Star position shifted by light deflection caused by the Sun's gravitational field

FIGURE 17. Stars close to the Sun in the 1919 eclipse, numbered as they were in the expedition's report, with names provided for all except for the dimmer, unnamed stars. The stars shown here are those visible on the plates taken by the four-inch lens at Sobral, with the exception of star 1, which was lost in the Sun's corona. Eddington's best plate from Principe showed only stars 3, 4, 5, 6, and 11. White stars give the actual positions of the stars, as shown on a comparison plate. Black provides the position on an eclipse plate caused by gravitational light deflection alone, greatly exaggerated. Close stars are deflected much more than stars far from the Sun. Gray stars show how a change in plate scale (or magnification) between the eclipse and comparison plates shifts the stars. Scale change shifts stars close to the center of the plate (where the Sun is located) much less than those far away. Only by carefully calculating the positions of at least six stars can the light deflection shift and the scale change be separated from each other, unless the scale change is independently determined.
(Reprinted with the permission of the author.)

the Sun, whereas the shift in position due to a scale change is greatest for stars farthest from the center of the plate, where the Sun is positioned. Of course, distinguishing scale from light-bending deflection in this manner requires measuring a variety of stars at different distances from the Sun. Measuring an insufficient number of stars would make it impossible to distinguish the scale change from the light deflection. For instance, in the case of the Principe plates, so few stars were visible that even the orientation could be difficult to distinguish from the deflection, thus causing some plates to be unusable (Dyson, Eddington, and Davidson 1920, 321).

In the jargon of the field, there are six plate constants to be solved for each plate pair (three kinds of possible misalignment times two dimensions on each plate). The method used is to set up equations that compare the measured displacements between stars on the eclipse and the comparison plates to equations based on the various plate constants and the sought-for displacement. Then, overdetermination of the plate constants is employed to derive values for all of the plate constants, including the scale factor and the light-bending factor. This works because there are only half a dozen plate constants, plus the light-bending factor itself, and seven or more stars exposed on nearly all plates at Sobral. Turner, Dyson's predecessor as chief assistant, had developed this method of data reduction at Greenwich in order to facilitate the Astrographic project. Dyson had taken over the work at an early stage and pioneered its implementation. There was no more experienced astronomer in the world than Dyson at doing this kind of work. It was a relatively new field for which he had helped establish the basic procedures.

As we have seen, there were problems with the images taken with the Sobral astrographic lens because, as was noted by the observers on-site, it had lost focus during the eclipse, possibly as a result of the change in temperature common during total eclipses. This meant the star images on these plates were not circular in form. This was problematic because determining the center of each image was essential to the measurement process. The shifts in position predicted by Einstein were sub–arc second in size. The sky has been divided by astronomers, since ancient times, into 360 degrees so that the Sun moves about 1 degree across the sky each day of the year. Happily, 360, which is so close to 365, has an unusually large number of factors, for easy computation. Each degree is further subdivided into sixty arc minutes. The full Moon is about half a degree in diameter, or thirty arc minutes across. Finally, each arc minute is divided into sixty arc seconds, the smallest division on the astronomer's sky. The Babylonians employed a sexagesimal counting system, based upon the number sixty. The saros cycle measuring the time between eclipses takes its name from the Babylonian word *saru*,[5] meaning thirty-six hundred. This number, equal to the square of sixty, is the number of arc seconds in a degree.

So the challenge of the eclipse expedition was to perform sub–arc second astrometry. Ideally, to do this one would like to have sub–arc second seeing. *Seeing* is the astronomical term for the quality of the atmosphere that permits sharp images of stars to form. A star is a point source. They are so far away that the light from the entire surface of the distant star seems to come from the same precise place in the sky. As the starlight passes through the atmosphere, it is refracted, and motions of our planet's air can cause tiny shifts in position, first one way and then the other. This is what makes the stars twinkle. It does not affect planets, which are not point sources but actually little disks in the sky. On the Moon the seeing is perfect, and stars do not twinkle because there is no atmosphere. Here on Earth, sub–arc second seeing is rare. Such still air is found only in certain places, and modern observatories are placed at sites where such seeing is relatively common—usually at altitude in dry climates. Among the worst places for seeing is at sea level in the tropics, where the air is warm and humid. The result is that images of the stars on the eclipse plates were three or four arc seconds across, as Dyson himself emphasized at a meeting of the Royal Astronomical Society (RAS) soon after the announcement of the results (Jones 1919a, 106). When placed next to the comparison plates, their disks would overlap, and the center-to-center distance of the two images had to be measured. But if the images were not circular and therefore the center was difficult to determine, this drastically decreased the precision of the experiment. Thus, the astigmatism in the Sobral astrographic images was a major problem for the Greenwich team.

In the end the Greenwich team decided to measure the star positions on the astrographic plates only in right ascension and not in declination (i.e., measuring in celestial longitude, east–west, only and not in celestial latitude, or north–south). The positions of the stars relative to the Sun meant that considerably more of the light-bending effect would be measurable in right ascension than in declination. Given that the data on the astrographic plates was considered "noisy," it seemed ill advised to try to measure in a coordinate in which the sought-for effect (the "signal") would be smaller than the noise. However, as we shall see, this may have resulted in an inaccurate determination of the scale, since so much information about the scale constant was lost by excluding the direction in which the scale would have been more visible, because it would have been larger relative to the light-bending displacement. An inaccurate determination of the scale would obviously result in an inaccurate determination of the Einstein displacement, since the two effects produce a similar radial displacement of star images.

As it turned out, the result obtained from the Sobral astrographic plates was discordant with the results from the other two instruments (the four-inch at Sobral and the Principe astrographic). Modern critics contend that the decision to discount the results from this instrument must have been largely made because the result was also discordant with the prediction of general relativity. But this seems highly unlikely for the team of astronomers at Greenwich, who did not believe that this theory was correct. The decision to measure in only one coordinate for these plates is clear evidence that the Greenwich team was unhappy with the quality of the data they contained. Indeed, problems were noted at the time of the eclipse itself, as we have seen.

The Greenwich team went further, however, as we learn from this report:

> The means of the 16 photographs [taken with the astrographic lens] treated in this manner [i.e., solving for all plate constants, including scale, and the displacement, from the same data] give $\alpha + 243$ e = $+0^{r}.0435$ [where α is the light-bending displacement, and e is the change of scale between the two plates] or with the value of the scale $+0^{r}.082$ from the previous table $\alpha = +0^{r}.024 = +0".93$ at the limb. It may be noticed that the change of scale arising from difference of refraction and aberration is $+0^{r}.020$. If this value of e be taken instead of $+0^{r}.082$ we obtain $\alpha = +0^{r}.039 = +1".52$ at the sun's limb. (Dyson, Eddington, and Davidson 1920, 312)

This last value is much closer to that recovered from the other two instruments (deflection of +1."61 on Principe, + 1."98 from the Sobral four-inch). It certainly suggests that the reason for the discrepancy is an inaccurate determination (and exaggeration) of the change of scale undergone by the astrographic instrument. Notice that the scale change as measured was four times the minimum expected scale change (0.082 vs. 0.2 turns of the micrometer screw). This alternative result appears nowhere else in the report, but the mere fact that it is mentioned suggests that the author (which means Dyson for this section of the report, written in his hand in the manuscript of the report [RGO Archive 8, fol. 150]) attached some significance to it. Furthermore, an earlier comment in the paper may be significant in this context: "These changes [in the focus of the astrographic lens during the eclipse] must be attributed to the effect of the sun's heat on the mirror, but it is difficult to say whether this caused a real change of scale in the resulting photographs or merely blurred the images" (Dyson, Eddington, and Davidson 1920, 309).

A straightforward interpretation would be that Dyson suspected the scale value was not accurately determined from the astrographic data, and he was therefore justified in ignoring any result derived from that data. But what about the possibility that the 1".52 deflection at the limb might be the true result from that data?

Two other publications mention this 1".52 deflection. Both are in the journal *Nature*, one by Dyson and one by Crommelin. Crommelin wrote a single-page account of the team's remarks to the famous joint meeting of the Royal Society and the RAS, in which he said, "This instrument [the astrographic] supports the Newtonian shift, the element of which is 0.87" at the limb. There is one mode of treatment by which the result comes out in better accord with those of the other instruments. Making the assumption that the bad focus did not alter the scale, and deducing this [scale], from the July plates, the value of the shift becomes 1.52"" (Crommelin 1919d, 281).

It is noteworthy that the apparently throwaway character of the remark in the published report (two sentences out of over forty pages) is contradicted by Crommelin's decision to devote two sentences to it in the mere page available to him in *Nature*. Both comments make it clear that this alternative approach was to calculate the value of the scale change theoretically and then, using that value, reduce the plate data to obtain the light deflection results. This is in contrast to the Greenwich team's normal approach to calculating the scale change and the light deflection from the same data. As it happens there is a parallel between the alternative approach and the method Eddington employed to reduce the Principe data.

According to the published report (Dyson, Eddington, and Davidson 1920, 317), Eddington was solely responsible for the reduction of the Principe data—what there was of it. It will be recalled that his comparison plates were taken in Oxford with a different instrumental setup, making a direct plate-to-plate comparison potentially problematic. Therefore, check plates of the star field around Arcturus were taken with both instruments in both places. Although these were originally intended merely as a safeguard against systematic errors arising out of changes in both instrument and location, his actual procedure was to take measurements on the check plates in order to calculate the difference in scale between the two instrument setups. He then *assumed* that the same change in scale applied to the eclipse and comparison plates taken in both places. In Eddington's words:

> As events turned out the check plates were important for another purpose, viz., to determine the difference of scale at Oxford and Principe. As shown in the report of the Sobral expedition, it is not necessary to know the scale of the eclipse photographs, since the reductions can be arranged so as to eliminate the unknown scale. If, however, a trustworthy scale is known and used in the reductions, the equations for the deflection have considerable greater weight, and the result depends on the measurement of a larger displacement. On surveying the meagre material which the clouds permitted us to obtain, it was evident that we must adopt the latter course; and accordingly the first step was to obtain from the check plates a determination of the scale of the Principe photographs. (Dyson, Eddington, and Davidson 1920, 317)

The material available from Principe was meager indeed. Owing to the cloud, which began to clear just as the eclipse was ending, only two plates with five stars on each were usable. This was insufficient to allow all six plate constants to be determined along with the light-bending displacement and barely sufficient, even if the data for those five stars were perfect (which was far from the case), to calculate four plate constants and the displacement. Thus, in Eddington's case the need for an independent determination of the change in scale was acute. But, of course, mere necessity does not provide any answer to the principal charge made by Earman and Glymour—that there was no justification for throwing out the results of the Greenwich astrographic while keeping the poor-quality data obtained on Principe by the Oxford astrographic.

First, let us quote Eddington's own attempt to justify the inclusion of his data:

> Our result [for the light-bending deflection at the limb of the Sun] may be written 1".61 ± 0".30. It will be seen that the error deduced in this way from the residuals is considerably larger than at first seemed likely from the accordance of the four results. Nevertheless the accuracy seems sufficient to give a fairly trustworthy confirmation of Einstein's theory, and to render the half-deflection at least very improbable. It remains to consider the question of systematic error. The results obtained with a similar instrument at Sobral are considered to be largely vitiated by systematic errors. What ground then have we—apart from the agreement with the far superior determination with the 4-inch lens at Sobral—for thinking that the present results are more trustworthy?
>
> At first sight everything is in favour of the Sobral astrographic plates. There are twelve stars shown against five, and the images though far from perfect are probably superior to the Principe images. The multiplicity of plates is less important, since it is mainly a question of systematic error. Against this must be set the fact that the five stars shown on plates W and X [from Principe] include all the most essential stars; stars 3 and 5 give the extreme range of deflection, and there is no great gain in including extra stars which play a passive part. Further, the gain of nearly

two extra magnitudes at Sobral must have meant over-exposure for the brighter stars, which happen to be the really important ones and this would tend to accentuate systematic errors [because it is more difficult to tell where the true center of the star lies when it is overexposed on the plate and thus has, in effect, become a large blob of emulsion], whilst rendering the defects of the images less easily recognized by the measurer. Perhaps, therefore the cloud was not so unkind to us after all.

Another important difference is made by the use of the extraneous determination of scale for the Principe reductions. Granting its validity, it reduces very considerably both accidental and systematic errors. The weight of the determination from the five stars with known scale is more than 50 percent greater than the weight from the twelve stars with unknown scale. Its effect as regards systematic error may be seen as follows. Knowing the scale, the greatest relative deflection to be measured amounts to 1".2 on Einstein's theory; but if the scale is unknown and must be eliminated, this is reduced to 0".67. As we wish to distinguish between the full deflection and the half-deflection, we must take half these quantities. Evidently with poor images it is much more hopeful to look for a difference of 0".6 than for 0".3. It is, of course, impossible to assign any precise limit to the possible systematic error in interpretation of the images by the measurer; but we feel fairly confident that the former figure is well outside possibility." (Dyson, Eddington, and Davidson 1920, 328–29)

This last paragraph is a close paraphrase of the closing part of a letter from Eddington to Dyson on October 3. Evidently, Eddington's first—and most critical—audience for his claims that the Principe result should be accepted as valid was his collaborator Dyson.

The key words in this part of the paper are "granting the validity" of the extraneous determination of scale. Should we, in hindsight, grant Eddington this validity? About this, Eddington himself said:

The writer must confess to a change of view with regard to the desirability of using an extraneous determination of scale. In considering the programme it had seemed too risky a proceeding, and it was thought that a self-contained determination would receive more confidence. But this opinion has been modified by the very special circumstances at Principe and it is now difficult to see that any valid objection can be brought against the use of the scale.

The temperature at Principe was remarkably uniform and the extreme range probably did not exceed 4° during our visit— including day and night, warm season and cold season. The temperature ranged generally from 77½° to 79½° in the rainy season, and about 1° colder in the cool gravana. All the check plates and eclipse plates were taken within a degree of the same temperature, and there was, of course, no perceptible fall of temperature preceding totality. To avoid any alteration of scale in the daytime the telescope tube and object-glass were shaded from direct solar radiation by a canvas screen; but even this was scarcely necessary, for the clouds before totality provided a still more efficient screen, and the feeble rays which penetrated could not have done any mischief. A heating of the mirror by the sun's rays could scarcely have produced a true alteration of scale though it might have done harm by altering the definition; the cloud protected us from any trouble of this kind. At the Oxford end of the comparison the scale is evidently the same for both sets of plates, since they were both taken at night and intermingled as regards date.

It thus appears that the check plate is legitimately applicable to the eclipse plates. But the method may not be so satisfactory at future eclipses, since the particular circumstances at Principe are not likely to be reproduced. (Dyson, Eddington, and Davidson 1920, 329–30)

The Greenwich team also underwent a change of heart in this respect, as shown by their preparations for the next eclipse, of 1922. Before the expedition departed, Davidson wrote a paper arguing that the independent determination of scale was not only a method superior to that employed in the reduction of the Sobral data but would be vital for the 1922 eclipse, which would lack the bright stars close to the Sun that were a unique feature of the 1919 eclipse (Davidson 1922). It was, indeed, this happy coincidence that had convinced

Dyson that the opportunity to test Einstein's theory in 1919 could not be passed up (Dyson 1917).

> In the Eclipse of 1919, the field of stars was unusually favourable for a determination of the Einstein gravitational displacement of light passing near the Sun—in fact, there is no other field on the Ecliptic with so many bright stars.
>
> In the Eclipse of 1922, if exposures are given of sufficient length to photograph faint stars near the Sun, there is grave danger of the images being drowned in the Corona. The brighter stars which are sure to be photographed are at such distances that the *differential* Einstein effect will be small, with consequent uncertainty in the result.
>
> If, however, one had an independent determination of the scale of the photograph, then two stars, each situated at 1° distance on opposite sides of the Sun, will show an increase in distance of $0''.88$, a quantity readily measured on good photographs. A scale may be determined by photographs taken the night before or after on a comparison field, as was done by Prof. Eddington in Principe. To this it may be objected that different conditions hold between the day and night observations. (Davidson 1922, 224–25)

Many efforts were made in subsequent eclipses—not always successfully—to get an independent measure of the scale without having to assume that it would be the same during the eclipse as during a later nighttime exposure. The most popular was taking check plates actually during the eclipse, as advocated by Davidson in his paper, even though this typically required pointing the instrument toward a different star field.

Thus, for most subsequent eclipses, an independent determination of scale was employed precisely because, as noted by Davidson, the smaller the observed effect when deprived of stars close to the Sun, the more danger in effectively halving the size of the thing to be measured by calculating the scale change from the same data used for the deflection.

It is important to keep in mind that the Sobral four-inch plates remain highly unusual, even after many subsequent eclipse expeditions to test Einstein's theory. They are almost unique in being taken with a working instrument, in clear weather, with several bright stars relatively close to the Sun. Even so, the experimenters were conscious of the difficulties of dealing with a possibly unknown change of scale. Much space is devoted to a discussion showing that the values adopted for the scale were consistent from one image to another across a plate (Dyson, Eddington, and Davidson 1920, 306–9).

In the case of the Principe plates, only the presence of unusually bright stars permitted any kind of measurement at all, given the cloudy conditions. The fact that, luckily enough, check plates were available made it possible for Eddington to derive a result that he, at least, was reasonably happy with. I accept, however, that his admitted biases might have made him especially anxious to extract a result from data that another experimenters would have been tempted to discard. It is, unfortunately, now impossible to analyze Eddington's work since neither his plates nor his data sheets seem to have survived.

In the case of the Sobral astrographic plates, the excellent conditions were compromised by the poor performance of the instrument. We should keep in mind something forgotten or ignored by many modern critics—that the poor performance of this instrument was known to Dyson before any data were ever reduced. It is not true that distrust of the Sobral astrographic plates only began when the measurements taken from them failed to agree with Einstein. Indeed, the experience with this instrument jaundiced the whole eclipse team against the use of coelostat mirrors in future eclipse experiments. And the reservations about the coelostats were articulated by Cortie even during the preparations. All along, there were worries about how they would perform. Eddington even put the troubles with the coelostats into verse, parodying the *Rubaiyat of Omar Khayyam*!

Ah Moon of my Delight far on the wane,
The Moon of Heaven has reached the Node again
But clouds are massing in the gloomy sky
O'er this same island, where we labored long—in vain?

And this I know; whether EINSTEIN is right
Or all his Theories are exploded quite,

One glimpse of stars amid the Darkness caught
Better than hours of toil by Candle-light

Ah Friend! Could thou and I with LLOYDS insure
For Gold this sorry Coelostat so poor,
Would we not shatter it to bits—and for
The next Eclipse a trustier Clock procure

———————————————————

The Clock no question makes of Fasts or Slows
But steadily and with a constant Rate it goes.
And Lo! the clouds are parting and the Sun
A crescent glimmering on the screen—It shows!—It shows!!

Five minutes, not a moment left to waste,
Five Minutes, for the picture to be traced—
The Stars are shining, and coronal light
Streams from the Orb of Darkness—Oh make haste!

For in and out, above, about, below
'Tis nothing but a magic *Shadow* show
Played in a Box, whose Candle is the Sun
Round which we phantom figures come and go

———————————————————

Oh leave the Wise our measures to collate
One thing at least is certain, LIGHT has WEIGHT
One thing is certain, and the rest debate—
Light-rays, when near the Sun, DO NOT GO STRAIGHT.

(DOUGLAS 1957, 43-44)

Note verse 3, where Eddington talks rather bitterly of the coelostats. Apparently, he was not happy with his, either. Indeed, he felt that only the cloud at Principe prevented his mirror from suffering the same fate as the very similar one Davidson used at Sobral, since Dyson and Davidson blamed the heat of the Sun for deforming the mirror's surface and causing the loss of focus. It has never been explained why the coelostat used by Crommelin did not behave this way. It may have been because it was smaller and less subject to a change of figure. Perhaps it was Grubb's superior manufacture. Maybe it was the fact that it was silvered at Greenwich, whereas the two larger mirrors were shipped elsewhere for silvering. Furthermore, Eddington, like other astronomers, looked forward to another eclipse and the opportunity to use better equipment. Finally, note how he also expressed his satisfaction at proving the most important point, that light does respond to gravity, even though he acknowledged continued debate about whether Einstein's theory was correct. On that score, he waited until after the 1922 eclipse to give his final verdict, this time quoting Lewis Carroll: "I think it was the Bellman in The Hunting of the Snark who laid down the rule 'when I say it three times, it is right.' The stars have now said it three times to three separate expeditions, and I am convinced that their answer is right" (Douglas 1957, 44).

The three separate expeditions are Principe, Sobral, and the Wallal, Australia, expedition of 1922. Eddington regarded his expedition as separate from the Greenwich one. At any rate both Eddington and Dyson came away determined not to use coelostat mirrors the next time. Dyson planned instead to go to the trouble of bringing along an equatorial mounting and erecting a real telescope at the eclipse location. But the trouble with eclipses is that the chance to benefit from your previous experience is not guaranteed. Neither Dyson nor Eddington went in 1922. Spencer Jones, Dyson's eventual successor as Astronomer Royal, led the expedition. As we shall see, it was clouded out and took no data.

Dyson and Davidson based all of their future planning on the realization that the Sobral astrographic data was really of very low quality. Although the Sobral astrographic plates showed more stars than Eddington's at Principe, in practice only the five brightest were used to determine the deflection (Dyson, Eddington, and Davidson 1920, 310–11). Recall also that these star positions were measured only in one coordinate, thus halving the amount of information available to correctly determine the scale (on top of the "halving" of reliability Eddington claimed as a consequence of not having an independent determination of the scale). In the circumstances, something might have been done had check plates been available to permit an independent determination of the scale. If one assumes the scale did not change (apart from what is theoretically predictable on the basis of differential refraction and aberration), then one obtains, as we have seen, a value for the deflection very similar to what Eddington derived from his data. More importantly, that calculation involved the assumption that the scale of the images did not change in the two months between the eclipse and the taking of comparison plates. The logic of accepting the "Newtonian" value obtained by the main data analysis was demonstrated by the fact that a large part of the deflection observed by the astrographic at Sobral was due to a significant change of scale *within the instrument itself*, presumably because the focus changed during the eclipse. Accepting the lower deflection result had as a corollary the issue of the instrument performing unexpectedly and perversely, which would tend to suggest the experiment was not terribly reliable. The Sobral astrographic data was internally inconsistent. One could perform the data analysis two different ways and obtain results that contradicted each other.

Ending the Experiment

Eddington arrived home in July, and there is every reason to believe he conducted the rest of the data reduction of his plates himself, though probably using a plate-measuring machine instead of a portable micrometer. On the other hand, with only a handful of stars on a few plates to work with, owing to the cloud over Principe, he must have been anxious to learn what was on the Sobral plates, since he knew the other expedition had had better luck with the weather. The Sobral expedition arrived back in England on August 25, 1919 (Crommelin 1919b, 281). Reduction of the data on their plates probably began almost immediately. Worksheets documenting the measurements made are preserved in the Royal Greenwich Observatory (RGO) archives (now housed in the University of Cambridge Library).[6] The first page, dealing with the plates taken by the astrographic lens, is headed "Total Solar Eclipse—1919 May 28–29–Sobral—Astro No. 1" and dated September 2, 1919 (RGO Archive 8, fol. 150). Some of the sheets are initialed C. D. and H. F., for Charles Davidson and

Herbert Furner, who made the measurements. The following year, Dyson commented, in a letter to the American geologist Louis Agricola Bauer on July 1, 1920:

> Dear Prof. Bauer,
>
> Your long list of "errata" rather alarmed me, though I could not believe that any serious error had been made in the reduction of the "Einstein" photographs, as both Davidson and I have dealt with some thousands of astronomical photographs in very similar fashion, in fact ~~almost~~ identical fashion except for the inclusion of the term α giving the displacement. (RGO Archive 8, fol. 147)

The displacement is that due to gravitational light bending. Note also that the word *almost* has been deleted by Dyson. This suggests that Dyson was intimately involved in the reduction process and, presumably, took the lead in it. Certainly, some of the key worksheets appear to be written in his hand. The data reduction proceeded throughout the month of September, while in Germany, Einstein himself waited with bated breath. He wrote to his close friend Paul Ehrenfest in Holland on September 12 to inquire whether the Dutch scientists, having closer contacts with their English colleagues, had received any news (Einstein 2004, doc. 103, 154).

On September 12, Eddington and Cottingham spoke before the British Association for the Advancement of Science at Bournemouth, discussing only briefly the significance of their endeavors for relativity theory. They focused instead on an interesting sidelight of their observations, the enormous solar prominence shown very clearly on the Principe plates and still today the most recognizable feature of the 1919 eclipse. All Eddington had to say concerning the actual results was described in the report on the conference as follows:

> Professor Eddington gave an account of the observations which had been made at Principe during the solar eclipse. The main object in view was to observe the displacement (if any) of stars, the light from which passed through the gravitational field of the sun. To establish the existence of such an effect and the determination of its magnitude gives, as is well known, a crucial test of the

theory of gravitation enunciated by Einstein. Professor Eddington explained that the observation had been partially vitiated
by the presence of clouds, but the plates already measured indicated the existence of a deflection intermediate between the two
theoretically possible values 0.87" and 1.75". He hoped that when
the measurements were complete the latter figure would prove to
be verified. Incidentally Professor Eddington pointed out that the
presence of clouds had resulted in a solar prominence being photographed and its history followed in some detail; some very striking
photographs were shown. (Eddington and Cottingham 1920, 156)

These photographs showing the prominence most clearly were precisely those taken through the thicker clouds at the start of the
eclipse, which were utterly useless for testing Einstein's theory.

Thus, Eddington cagily committed himself only to a value between
the two theoretical predictions Einstein had made. The result from
Principe, as it was later published, is much closer to the larger relativistically "correct" 1915 value, but in view of the limitations of his
data, Eddington was understandably unwilling to claim too much
in the way of accuracy in advance of the Sobral results. Nevertheless, he was confident at this stage that light bending, of some magnitude, was an established fact. He had managed to weigh light. As
we shall see, word of his remarks in Bournemouth quickly reached
Einstein in Berlin, who was elated by this vindication of his original
prediction that light was affected by gravity. It only remained to
test whether spacetime was curved. Eddington had been confident
of this when he had measured his first plate on Principe. On returning
to England, he developed several plates that were not developed on
Principe because of the high temperatures there.[7] One of these had
enough stars to be measurable, and this plate supported the results
from the first plate, which he had measured on Principe. But it was
clear that everything would depend on the Sobral data.

In early October, Eddington received the long-awaited news from
Dyson of the reduction of the final plates, those from the four-inch
lens. Eddington referred to them as the Cortie plates, after Father
Cortie. The four-inch plates were measured after the astrographic

plates because the size of the plates was unusual and required a mod-
ification to the Greenwich micrometer to hold them (Eddington
1987, 119). Now, at last, all the measurements were completed, and
Eddington replied to Dyson on October 3.

> Dear Dyson,
>
> I was very glad to have your letter & measures. I am glad the
> Cortie plates gave the full deflection not only because of theory,
> but because I had been worrying over the Principe plates and
> could not see any possible way of reconciling them with the half
> deflection.
>
> I thought perhaps I had been rash in adopting my scale from
> few measures. I have now completed my definite determination
> of A (5 different Principe v. 5 different Oxford plates), it is not
> greatly different from the provisional though it reduces my values
> of the deflection a little. (Eddington to Dyson, October 3, 1919,
> RGO Archive 8, fol. 150)

This suggests that the Sobral astrographic data had really worried
Eddington, as he felt his own data tended to support the higher result
predicted by general relativity. Indeed, he seems to have doubted that
his results could be made to agree with the lower figure, despite confessing to having redone his analysis more carefully. Thus, he greeted
the news that the Sobral four-inch results strongly favored the higher
value with relief.

There is much here that can be made to fit the model of Eddington as a biased experimenter, but one important detail throws cold
water on the overall data-fudging narrative of modern times. This letter strongly implies that Eddington was not involved in the reduction
of the Sobral data. The tenor of this opening paragraph indicates that
he is receiving his first news of the four-inch data reduction by letter,
and there is no suggestion that he had any earlier input, or even any
prior information, beyond hearing the results from the astrographic
instrument. Indeed, after receiving that earlier information, he occupied himself not with any analysis of data from Sobral but with a
reanalysis of his own data to see whether it could be reconciled
with the astrographic data. Coupled with Dyson's comment, quoted

1919 6ᵉ 3

Dear Dyson

    I was very glad to have your letter & measures. I am glad the Sobra plates give the full deflection not only because of theory, but because I had been worrying over the Principe plates and could not see any possible way of reconciling them with the half deflection.

    I thought perhaps I had been rash in adopting my scale from few measures. I have now completed my definitive determination of A (5 different Principe v. 5 different Oxford plates), it is not greatly different from the provisional though it reduces my values of the deflection a little.* The probable error seems sufficiently small. With this and some other minor corrections made in the definitive calculation, my four results are

$$1''.94 \, , \, 1''.44 \, , \, 1''.52 \, , \, 1''.67 \qquad \text{mean} \quad 1''.64.$$

    I do not think the ill-success of the Brazil astrographic affects the validity of these results, because the method of reduction makes the systematic errors of the images much less important in the Principe work. Granting that the method of determining scale is sound, the use of A enormously increases the weight of the determination. 5 stars with known scale give a weight 3·23 whereas 12 stars with unknown scale give weight 1·96. Thus the probable error should be smaller for the Principe results, though this

*The main part of the reduction is due to the fact that I now adopt independent values of "a" & "e".

FIGURE 18. The first page of an important letter from Eddington to Dyson, in which he acknowledged receipt of the news that the four-inch plates agreed with Einstein's prediction and with Eddington's own results from the Principe plates. This followed the earlier news of the disagreement between the Principe data and the result obtained from the Sobral astrographic plates. Note the key phrase "not only because of theory," in which Eddington acknowledged his theoretical prejudice but indicated that he saw no way to reconcile his measurements with those previously reported by Dyson, based on the astrographic plates.
(Cambridge University Library, RGO Archive 8, fol. 150, 138. Reproduced by permission of the Science and Technology Facilities Council and the Syndics of Cambridge University Library.)

previously, that he himself was responsible for the Sobral data, along with Davidson, we can conclude the Sobral data reduction was conducted independently of Eddington. The initials on the data sheets; the fact that the reduction was undoubtedly performed at Greenwich and not at Cambridge, where Eddington was; and the fact that Dyson was solely responsible for discussing the Sobral data in his written report all reinforce this impression.

It seems likely that Eddington was never present at Greenwich during the Sobral data reduction. In a letter to Dyson on October 21, 1919 (RGO Archive 8, fol. 150), he refers to having acquired a season railway ticket, rather suggesting he had not been traveling to see Dyson in the preceding weeks or months. We can be fairly confident that Eddington simply was not privy to the reduction of the Sobral data or to the crucial decision, recorded in the data sheets (in what appears to be Dyson's hand), to reject the astrographic data and accept the four-inch data as "the result of the Sobral expedition" (RGO Archive 8, fol. 150). The eclipse expedition appears to be a reasonably good case of independence being preserved between the two wings of the collaboration.

If Eddington did not make the decision to exclude the astrographic data, it is clear that it must have been Dyson who did. Eddington's motivations therefore turn out to be largely irrelevant to the central data-fudging claim. We must instead ask: What was Dyson's attitude toward war, peace, and relativity? Can we claim that he was biased or overly influenced by Eddington, a younger but more brilliant colleague?

In contrast to the views of Eddington and Curtis, both of whom had decided preferences in regard to Einstein's theory, Dyson's views seem to have been moderate. I suspect he was skeptical but not intransigent toward it. We have good evidence for this. Earman and Glymour have already noted that Dyson was skeptical of the theory and that he "thought it too good to be true" (Earman and Glymour 1980, 85). After the announcement of the eclipse results, Dyson sent copies of plates from the four-inch to a number of leading astronomers, several of whom politely replied about the clarity of the images, even though most were, at least privately, not well disposed toward general relativity. In reply to one correspondent, Frank Schlesinger of the Yale Observatory, Dyson wrote on March 18, 1920:

> We are planning to send an expedition to Christmas Island in 1922; & I hope it may be possible to send one to the Maldives; & that the Australians may do something. Is it likely that there will be an American Expedition? I hope so, in view of the importance of having the point thoroughly settled. The result was contrary to my expectations, but since we obtained it I have tried to understand the Relativity business, & it is certainly very comprehensive, though elusive and difficult. (RGO Archive 8, fol. 123)

We have confirmation of this from Eddington, who wrote to the mathematician Hermann Weyl on August 18, 1920:

> It was Dyson's enthusiasm that got the eclipse expeditions ready to start in spite of very great difficulties. He was at that time very skeptical about the theory though deeply interested in it; and he realized its very great importance. (Hermann Weyl Nachlass 525, ETH, Zurich; quoted in Kennefick 2012)

So I would argue that Dyson was free of the biases Eddington is alleged to have carried with him into the endeavor to test Einstein's theory. He was not a pacifist, like Eddington, though he was an internationalist. Significantly, his obituary in the *Observatory* reads, "After the Great War, when international co-operation in science had lapsed to a considerable extent, Dyson played a prominent part in the reconstitution of international scientific co-operation through the International Research Council (now the International Council of Scientific Unions) and in the formation of the International Astronomical Union. . . . Dyson took an important part in the initial deliberations that resulted in the formation of the Union, which owes much to his wise guidance" (Jones 1939, 186). Nevertheless, I doubt he was a crusader for the cause of pacifism, as Eddington was. It is worth recalling that the International Astronomical Union, at its birth, did not permit Germany and its allies to enjoy membership in the new body, even though Dyson played a leading role in it from the beginning.

Those modern critics who claim that Eddington's bias motivated the decision to reject the Sobral astrographic data have overlooked a major point against their view. It was not Eddington who made the decision to which they object. I can find no evidence that Dyson shared Eddington's biases. Personally, I highly doubt that bias in favor of Einstein played any role in Dyson's decision-making.

Newton versus Einstein?

Nevertheless, modern critics do argue that Eddington stands guilty as charged on the count of manipulating the theory-testing process to suit his own agenda. It is by no means apparent to me that this is unusual or sinister. I suspect that the theory-testing process more or less demands it. But let us assume that Eddington had an ulterior motive in arranging the theoretical outcomes in his "false" trichotomy and naming them as he did. The funny thing is that if he were trying to simply compose a narrative to heal the wounds of war, he should have approached the presentation of the results in another manner. Let us compare his presentation with Curtis' to see how he might have done things differently.

Both Curtis and Eddington spoke in terms of three theoretically possible results: the null deflection, the half-deflection predicted by Einstein in 1911, and the full deflection demanded by general relativity. They both agreed that Einstein would be triumphantly vindicated in the case of a measurement agreeing with the full deflection. This would prove not only that light has weight and falls in the gravitational field but also that spacetime is curved by the gravity of massive bodies like the Sun. Similarly, a null result would be bad for Einstein. But Curtis regarded the half-deflection, quite rightly, as also predicted by Einstein. It is clear from his presentation of the results in Pasadena that any definite light deflection measured at all would have at least partially vindicated Einstein. For Curtis the status quo was vindicated only by a null result. He did not frame the contest as one between Newton and Einstein but between Einstein and the expectations of most contemporary physicists that light is unaffected by gravity. As such, Curtis interpreted his result as irrelevant to the question of which theory of gravity is correct. His experiment did not test the gravitational law at all. Instead, it confirmed that light is unaffected by gravity. Obviously, to test a theory, one does not study the one thing that is impervious to it. If light was proven to be unaffected by gravity, then the eclipse experiment was irrelevant to the question of whose gravitational theory was correct. While the result was a blow to Einstein's belief in the equivalence principle, if he had come up with a version of his theory that could do without it, he could still have, as far as Curtis' result was concerned, gloried in a victory over Newton on the matter of Mercury's orbital anomaly.

Contrast this with Eddington's curious attitude toward the null result, which for him was the hardest to explain. For Curtis the null result was the most straightforward: defeat for Einstein, victory for the status quo. For Eddington it was a troublesome, though admittedly exciting, result. He behaved as if the status quo were untenable. It could not be returned to, and therefore, a result that would have been expected twenty years before now potentially undermined the whole of physics. It was of great importance. Eddington repeatedly framed the experiment as being first and foremost about weighing light. So we might say that Eddington and Curtis were in some agreement about the null result but viewed it from opposing standpoints.

It was in interpreting the half-deflection that Curtis and Eddington really differed. To Curtis the half-deflection was Einstein's prediction, yet paradoxically, Eddington insisted it was Newton's. I believe that Eddington had valid reasons for doing this. But whether he was justified or not, if Eddington's main goal was to heal the wounds of war by confirming Einstein's theories, then he certainly should not have tried to turn his experiment into a showdown over general relativity. He could have easily followed a path laid out by Curtis and undoubtedly acceptable to most astronomers. That path involved portraying Einstein as departing from nineteenth-century physics orthodoxy by predicting that light deflection existed at all. Then, any non-null result would confirm the genius of this German scientist.

Even if Eddington had wanted to portray the test as one between gravitational theories, he could have pitted Einstein against the Finnish physicist Gunnar Nordström, whose contemporary theory also predicted no deflection. Instead, he preferred to portray Einstein's equivalence principle prediction as being Newtonian in disguise. It seems clear to me that his reason for doing so was a sincere belief that this was the most scientifically interesting question at issue. It was certainly a silly thing to do if he merely wished to promote Anglo-German scientific friendship because it would reduce by at least half the chances of getting a clear verdict in Einstein's favor, since there is not a large difference between the half-deflection and full-deflection predictions.

It is worth noting that Eddington had already framed the test in this way before setting out for Principe. It was not a case of framing the narrative after the fact to give Einstein the maximum victory possible. Furthermore, Eddington knew he would not be performing the data analysis from the instruments on Sobral. If he was so anxious for Einstein to do well, he should certainly have been wary of the risk of Dyson ending up with a result that did not discriminate between the half-deflection and full-deflection results. His framing of the experiment was, however, completely consistent with his desire to test general relativity as a plausible successor to the suddenly problematic Newtonian theory. We can see why he was so anxious to drag Newton into the picture if we accept that he wished to dramatize a changeover from the old, inconsistent theory to the new relativistically invariant one. By framing things as he did, he left himself with two tests—one easier, one harder. The easier test would show that light had weight. He could have awarded all the glory to Einstein on that score but chose not to, as he knew well that Newton would have expected exactly the same result. Eddington was English enough to want to give Newton his due. The second and harder test would knock Newton off his perch and enthrone a new theory of gravity. Eddington can have had no certainty beforehand that this would be achievable. He clearly was willing to accept the verdict of the experiment only if Einstein had well and truly earned it.

The Shadow of Doubt

In recent decades many people have asserted that the 1919 eclipse experiment was not a very convincing test of general relativity. Eddington has been accused of partiality for Einstein, or of engaging in a crusade for peace at the expense of scientific truth, or of merely being lucky. The backlash against Eddington began as early as the 1970s. The last eclipse expedition team to conduct the light deflection test the old way, with photographic plates, was from the University of Texas in 1973 and did so in Mauretania in West Africa. They went to great lengths to do the experiment well and came away with a very thorough understanding of all the things that can go wrong on such an expedition. The experience left some members of the expedition with a definite skepticism regarding the 1919 team's claims. From this time on, there were quite a few relativists and astronomers who were skeptical of whether the 1919 measurements could really have been as precise as Eddington and Dyson claimed. Around the same time, a parallel but distinct claim arose among philosophers and historians of science. This line of skepticism is easier to trace. It is primarily due to the landmark paper written by Earman and Glymour in 1981. There is no doubt that their work popularized the notion that Eddington's bias had a detrimental effect on the experiment's status as a genuine theory test. With modern physicists often skeptical of the claimed accuracy of measurement and with the allegations of bias leveled by Earman and Glymour, it is not surprising that ordinary readers are left with an impression that Eddington behaved in a scientifically improper way.

A specific claim made by Earman and Glymour is that Eddington and Dyson prepared the ground for the reception of eclipse results by portraying the decision as a showdown between Newton and Einstein. They insist that Eddington had no legitimate basis for labeling the half-deflection result as Newtonian. They argue very convincingly that any of a number of assumptions might have been made as to how to do the "Newtonian" calculation in the context of twentieth-century physicists' understanding of the propagation of light through the electromagnetic field.

But precisely how Newtonian gravity, as a theory, is supposed to have lost out in the tricornered frame-up allegedly perpetrated by Eddington and Dyson is not quite apparent. After all, if the "true" Newtonian result was framed as being smaller than the half-deflection, this test would have been worse, from Newton's standpoint. For instance, the null deflection might have been chosen, in which case every single plate measured, except for one of the Sobral astrographic plates, would have falsified Newton. Alternatively, it might be argued that the true Newtonian result was greater than the half-deflection, which would make it closer to the value predicted by general relativity. If this is the argument, then no one should have gone on the expedition, since it would have been impossible to distinguish between two theories making such similar predictions. As for Cottingham's double-valued deflection possibility, no one, to my knowledge, has ever maintained that Newton's theory predicts such a thing. In other words, it can certainly be asserted that by choosing the Newtonian value he did, Eddington gave his old Trinity fellow the best possible chance of success.

Many reasons have been advanced to explain Eddington's bias. Principally, he is accused of a theoretical predisposition in favor of general relativity, perhaps because of the theory's inherent "beauty" or a personal agenda to advance the cause of international peace. I have emphasized how unconvincing I find the latter idea. It is true that he and Einstein were prominent pacifists, but this obscures the isolation that was their political lot at this time. The two men never met until after 1919, and neither had fellow physicists to speak with who shared their views (Eddington did have one until 1915, when Ebenezer Cunningham was hauled away from Cambridge as a draft resister). Some, like Dyson, were sympathetic, or at least prepared to be helpful, but it is hard to see how Eddington could have envisaged some master plan to achieve peace through scientific expedition. If he harbored hopes of furthering the cause of international reconciliation, then it must have been as a secondary goal of the whole project. If we accept that he therefore leaned toward general relativity as a theory, we must ask why. The answer lies in the very criticism that Earman and Glymour so cogently make. Newton's theory had become so inconsistent that none of Eddington's arguments and calculations make a clear-cut case that the half-deflection is really a prediction of the theory. But this is exactly why Eddington had hopes on behalf of general relativity! Even if he had been neutral to begin with, his experience as the theorist who actually had to do the calculations to frame the test left him with a decided preference. One theory was a pleasure to calculate with, making a clear and precise prediction that admits no ambiguity or wiggle room. The other theory had no clear way of tying itself into the other physical elements needed for the calculation, so debatable presumptions and assumptions had to be made to make it work at all.

Let us hear what Eddington himself had to say on this topic. In chapter 6 of his book *Space, Time and Gravitation* (Eddington 1987), he began by quoting Newton's famous remark about the great undiscovered ocean of truth that still lay before him after his discoveries and asked if "there [was] any reason to feel dissatisfied with Newton's law of gravitation?" He noted that "observationally it had been subjected to the most stringent tests, and had come to be regarded as the perfect model of an exact law of nature. The cases, where a possible failure could be alleged, were almost insignificant." Specifically, this was the anomalous perihelion advance of Mercury. He continued: "The most serious objection against the Newtonian law as an exact law was that it had become ambiguous. The law refers to the product of the masses of the two bodies; but the mass depends on the velocity—a fact unknown in Newton's day. Are we to take the variable mass, or the mass reduced to rest? Perhaps a learned judge, interpreting Newton's statement like a last will and testament, could give a decision; but that is scarcely the way to settle an important point in scientific theory. Further *distance*, also referred to in the law, is something relative to an observer. Are we to take the observer travelling with the Sun or with the other body concerned, or at rest in the aether or in some gravitational medium? Finally, is the force of gravitation propagated instantaneously, or with the velocity of light, or some other velocity?"

Eddington was in the position of someone charged with creating a set of boards mounted to permit the testing of a new saw being considered for purchase by a carpentry collective. The test will compare the new saw with the old tool the carpenters have used until now. The tester starts out using the old saw but has trouble and decides to use the new saw to construct the setup. It is a great help, cutting cleanly and reliably with little fuss. The other saw seems difficult to use, unwieldy when put to this newfangled kind of task. Obviously, by the time the job is done, the tester really hopes the new saw will pass the test and be purchased. His worst nightmare would be that because of nostalgia, the old saw will be chosen, and he will end up being stuck with it. It would be easy for him to imagine setting up the test to show off the new saw to its best advantage. Indeed, that would seem to be a sensible course to take. In doing so, he would be allowing the carpenters to take advantage of his superior experience with the two instruments. As the only one of them who had used the new saw and as someone who knew the limitations of the old saw, it would be his duty to do so. Is framing of this sort a case of manipulating the test or simply making it fit for purpose? Are you helping people make a wise decision or ensuring that they make the decision that suits you?

In essence, Eddington was one of the first people to realize that Einstein had finally solved the age-old problem of relativity. It is common to speak of relativity as a theory or a principle, but what Einstein created can also be understood as a tool. The philosopher Thomas Kuhn introduced to science studies the famous word paradigm. A paradigm can be a whole framework used to look at the world, something bigger than a theory, which also encompasses beliefs, knowledge, experience, and other things necessary to do science. He also sometimes used paradigm to refer to a research tool, a particular calculation that could be used as a model for many different sorts of problems requiring solutions. He eventually started using the word *exemplar* for this kind of tool. Exemplars play a large but often unappreciated role in scientific research. Not as all-encompassing as paradigms, they nevertheless make paradigms work. It is all very well to have a framework to inform your view of the world, but if you cannot actually do the calculations that framework demands, then it is really not that valuable.

The idea that one judges one's speed with reference to a particular object or landmark is an old one. Obviously, a ship at sea might be interested in its speed relative to another ship underway, or to the shore. We often have to deal with the reality that motion is relative. But transforming between frames of reference is not a trivial exercise, and it is not surprising that the ancients left us with a cosmology in which the Earth was absolutely at rest at the center of the universe. Since all our measurements of the planets and other celestial bodies are made from Earth, how easy it is to stick to a system in which there is never a need to worry about velocities with reference to a third body, such as the Sun. We know that at least one classical astronomer, Aristarchus of Samos, proposed that the Sun was the true center of the solar system. It is quite plausible that one reason his model never caught on was the difficulty of transforming all of our Earth-based measurements into a different, Sun-centered system. Certainly, one thing that set Copernicus apart was his willingness to do the hard graft of calculating. He actually produced a model of the solar system in which our measurements are referred to a Sun-centered system by converting every single observation from its original Earth-centered frame.

In other words, there has been too much focus on paradigms as big ideas and not enough on exemplars as the tools that make big ideas work. Let us take the example of Newton's universal law of gravitation. Newton enunciated the big idea that every massive particle in the universe attracts every other massive particle in the universe. This is a revolutionary and cleanly stated idea. It represents natural philosophy at its finest. On the other hand, it is useless for astronomy. Astronomers would like to treat planets as individual bodies attracting each other without having to separately calculate the force exerted by every rock on each planet on every rock on another planet. Fortunately, Newton also invented integral calculus. He used this tool to formally prove that two spherical bodies do behave as if they were two massive particles positioned at their respective centers. He did this by integrating all the forces exerted by all their individual parts. Thus, the tool of calculus enables the paradigm of universal gravity to work. He went further and argued that spherical planets should bulge at the equator because of their rotation and that the Sun's gravitational force on this nonspherical bulge explains the precession of the equinox. Thus, he explained a phenomenon that was previously mysterious, thanks to the interplay of his paradigm and his exemplars. This kind of empirical success was inherited by Einstein's theory, since Einstein was able to show that in the limit of low speed and weak gravity, the conditions of our solar system, his theory approximated Newton's. But he had also shown how gravity was to be understood in situations where modern relativistic effects came into play. Like Newton's, his theory is at once a big idea, a new paradigm, in which gravity determines the geometry of spacetime and, at the same time, becomes a calculational tool for showing how different observers can reconcile their respective measurements of spacetime quantities.

When people say that Eddington favored the new theory over the old, they imply that it was the paradigm's beauty that attracted him. I argue that it was also the tool's practicality that won him over. Of course, a tool has its own beauty, and Einstein applied the tools of differential geometry to physics in a particularly compelling way, but the attraction of the theory as a tool should be kept firmly in mind. This is important because you can read Earman and Glymour as making two major criticisms against Eddington. First, they observe that he labeled one of his trichotomy of possibilities as Newtonian without clarifying the ambiguity as to what Newton's theory in fact predicted. Then they accuse him of favoring Einstein over Newton. But sadly, they were not clear in explaining to their readers that one of Eddington's chief reasons for hoping that Einstein would be vindicated was the unfortunate ambiguity in Newtonian theory as it then stood! It is no good arguing that since Newton's theory was impossible to test, the attempt to test it was invalid. It is true that Eddington could have informed his audience that Newton's theory was not even included in the competition, having failed to meet the entry requirements, but this would have been absurd. You cannot expect people to calmly accept that the most famous theory in physics, which had been around for over two centuries, simply died quietly in the night with no one present to hear its last words.

So we can argue that Eddington was obliged to pick one of his three results to be the Newtonian one. Did he somehow pick one subsequently shown to be wrongheaded? No! In fact, it turned out to be a good choice. Later, it happened that in the old Newtonian era, that value had actually been calculated as the amount by which the Sun would deflect light (by von Soldner). In addition, modern relativists do have a suitable approximation of general relativity, which they label the *Newtonian limit* of the theory, and those modern theorists would typically agree with Eddington that the half-deflection represents that Newtonian limit's prediction. Therefore, Eddington's choice in labeling his trichotomy has been doubly vindicated since 1919.

The Joint Meeting

May 29, 1919, has become one of the most famous days in the history of science, but only because of what happened afterward. There were few witnesses to what took place that day. Few people could later tell of the observations that gave rise to such widespread excitement. Eddington's station on Principe was lonely indeed. Since Cottingham was a clockmaker by profession, Cambridge's Plumian Professor was, perhaps uniquely for a modern eclipse observation site, the only scientist present. Apart from the plantation owner and a few other local observers, no one saw what went on. Sobral was more like a typical eclipse site, though less busy than most. Photographs of the scene show quite a few instruments. The only other astronomers present were Brazilian, part of Morize's expedition from the Rio Observatory. There were a couple of American geophysicists from the Carnegie Institute, and that was it, apart from any interested locals.

The obscurity of Principe extended further than just the day itself. Eddington began his data reduction alone while still on the island. It is probable that he never enlisted anyone's help after his return to Cambridge. If he kept any notes or records, they were later lost. Even the plates he took have long since disappeared. They may have been disposed of when the observatory was cleared out after the Second World War. Eddington had passed away, and the university was anxious to remove Winifred, his sister, from the premises in order to rehouse needy academics in the postwar housing shortage. A fellow astronomer called Chubby Stratton was sent in to go through Eddington's papers and seems to have thrown out a great deal of material, judging by the relatively small amount deposited at the Trinity College Library. Apparently, no one thought the eclipse plates were worth keeping. No one now can examine what Stratton did and second-guess his methods and decisions. In this sense, it is not surprising that doubts have been expressed about Eddington's results. Apart from what he published, we have nothing else to go on. To be fair, he published in some detail, but it would be nice to have the plates. Apart from the fabulous published images of the prominence, which show no stars because of the clouds, we have no record of the eclipse at Principe other than the data and tables of calculations given in the report.

The Sobral expedition, on the other hand, began with witnesses and received more and more scrutiny as the years passed. The director of the Carnegie Institute, Louis Bauer, right from the beginning seems to have regarded it as his institution's duty to maintain a sort of oversight, apparently only because members of his staff were the only people from outside the Portuguese-speaking world to observe the Greenwich astronomers at work. Bauer himself witnessed totality from the Liberian coast, where he enjoyed the clear skies that eluded Eddington on Principe. He devoted a great deal of time to double-checking the data reduction of the four-inch telescope from Sobral and undertook a lengthy correspondence with Dyson about it. Dyson made sure to send prints of the four-inch images from Sobral to many astronomers around the world, so they could see the good-quality images obtained by that instrument for themselves. He arranged for the plates themselves and the micrometer screw used to measure them to be taken to an RAS meeting in December 1919 so that he and Davidson could demonstrate how the plates were measured to a group of British astronomers (Jones 1919e).

He and his staff published in many outlets and patiently responded to many objections. They entertained questions and listened as others attempted to find alternative explanations for their results—anything other than Einstein's theory. They attended meetings of learned societies and spoke at length to colleagues and the public. Of course, Eddington played a major role in all this, but his role was always oriented toward explaining Einstein's theory and the theoretical ramifications generally. Dyson and his staff discussed the experimental questions. Also, it was they who prepared for the next eclipse. Eddington seems to have felt no need to redo the experiment, but Dyson always intended to go back for higher-quality data, using better equipment and more careful and extended preparation.

The moment that really sparked the enormous wave of interest that enveloped them all—Einstein and the eclipse teams—was November 6, 1919, when the joint meeting of the RAS and the Royal Society took place at Burlington House (Sponsel 2002, 460). This meeting was held so that representatives of the Joint Permanent Eclipse Committee (JPEC), which had organized the expeditions, could make a report to the two sponsoring societies. It was part of the 1899 operating agreement between the two societies that governed JPEC that such a joint meeting would take place, and this had happened after a number of previous eclipses before the war (Sponsel 2002, 455). In 1919 many people were in attendance, and many of them reported afterward their impressions of what had transpired. The philosopher Alfred North Whitehead later said that "the whole atmosphere of intense interest was exactly that of the Greek drama" (from *Science and the Modern World*, quoted in Douglas 1957, 43). Minutes of the meeting, with what appear to be verbatim accounts of what each speaker said, were published in the *Observatory*, a publication of the RGO read by large numbers of amateur astronomers throughout Britain. Scientific work is typically done in the seclusion of a lab, and major results can be greeted by considerable publicity, but seldom is the contrast so extreme as between the isolation of Principe on May 29, thousands of miles from the nearest astronomer, and the crowded meeting room of the Royal Society on November 6 and the instant fame that followed it.

Dyson, Crommelin, and Eddington all spoke, describing the preparations for the trip and aspects of the voyage while giving thanks to some of those who had facilitated the expedition. They presented the results, and both Dyson and Crommelin discussed the reasons behind the most critical decision of the whole enterprise, to exclude the data from the Sobral astrographic instrument. They described its loss of focus during the eclipse and ascribed this to the Sun's light heating the mirror in the time leading up to totality, when the instrument was in readiness. They were followed by the president of the Royal Society, J. J. Thomson, who chaired the meeting. The famous physicist, discoverer of the electron, eloquently explained the significance of the result:

> I now call for discussion on this momentous communication. If the results obtained had been only that light was affected by gravitation, it would have been of the greatest importance. Newton did, in fact, suggest this very point in the first query in his "Optics," and his suggestion would presumably have led to the half-value. But this result is not an isolated one; it is part of a whole continent of scientific ideas affecting the most fundamental concepts of physics. It is difficult for the audience to weigh fully the meaning of the figures that have been put before us, but the Astronomer Royal and Prof. Eddington have studied the material carefully, and they regard the evidence as decisively in favour of the larger value for the displacement. This is the most important result obtained in connection with the theory of gravitation since Newton's day, and it is fitting that it should be announced at a meeting of the Society so closely connected with him.
>
> The difference between the laws of gravitation of Einstein and Newton come only in special cases. The real interest of Einstein's theory lies not so much in his results as in the method by which he gets them. If his theory is right, it makes us take an entirely new view of gravitation. If it is sustained that Einstein's reasoning holds good—and it has survived two very severe tests in connection with the perihelion of Mercury and the present eclipse—then it is the result of one of the highest achievements of human thought. The weak point in the theory is the great difficulty in expressing it. It would seem that no one can understand the new law of gravitation without a thorough knowledge of the theory of invariants and of the calculus of variations. (Jones 1919b, 394)

Ludwik Silberstein, a Polish physicist then teaching in Rome, played a prominent role at the meeting and became, over the years, the respectable face of antirelativity sentiment. He was unusual among the skeptics in that he understood the theory very well. Indeed, he was an acknowledged expert on it who wrote a well-used textbook on the subject. He cannot, because of this, be regarded as a typical anti-Einstein campaigner. Yet his familiarity with and even association with the theory did not prevent him from aligning himself with several efforts over the years to overthrow it. A hint of his motivation for doing so may be given at the joint meeting, when he pointed at a portrait of Newton and said, "We owe it to that great man to proceed very carefully in modifying or retouching his Law of Gravitation" (Jones 1919b, 397). He may have been a natural conservative who preferred Newton to Einstein and the traditional ether theory to the newfangled relativity. This may have been true even given his ability to make his way in the world of relativistic physics, unlike, for instance, Arthur Hinks. In spite of this, he would later boast of having killed Einstein's theory in language that leaves little doubt of his relish in playing the role of assassin. For now, Silberstein pointed out that of the three tests proposed by Einstein himself, general relativity had only passed two. In the years ahead, he devoted considerable effort to identifying tests that the theory would fail.

It was supposedly at the end of this meeting that Silberstein approached Eddington and congratulated him, saying, "Professor Eddington, you must be one of three persons in the world who understands general relativity." When Eddington appeared to demur, Silberstein chided him: "Don't be modest, Eddington." Though in many ways a private man, Eddington was also quick with a riposte. "On the contrary," he replied, "I am trying to think who the third one is," rebuking Silberstein, who obviously saw himself as the third man.[8]

The idea that only three men in the world understood relativity really took hold in the popular mind. It became a byword for the incomprehensibility of modern science. Obviously, it is true that the theory is mathematically complex and rather difficult. It is true that many professional scientists, especially in the early days, lacked the ability to properly comprehend it. But it was never as difficult as it was made out to be. There is a story told by Buckminster Fuller that his publisher did not want to let him include chapters on Einstein in his first book because he was not on the "list" of the nine people who understood relativity (Fuller 1975). Needless to say, there has never been a countable number of experts on relativity, and no such list of them has ever been made. The people who understood relativity at some functional level already numbered at least in the dozens in the theory's early days (including 1919), quickly rose into the hundreds, and must be a considerable number today, including a reasonable percentage of professional physicists. The American Physical Society currently has roughly fifty thousand members, so the total number of people who understand the theory at least numbers in the tens of thousands. Not exactly mainstream stuff but certainly a number bigger than three.

Dyson had the last word at the meeting. Silberstein closed with an argument for the superiority of the third test, the one Einstein's theory was failing. He pointed out that one did not have to wait "years or centuries" for that test to be repeated.[9] Dyson rose to reply, saying, "Dr. Silberstein is under a misapprehension with regard to my views on repeating this work at future eclipses. I think it most important that this result should be verified at the next two eclipses, only I should hope to avoid the use of mirrors. If necessary, a suitable equatorial mounting should be prepared. The fields are not very favourable, but results can be got with object-glasses similar to those used last May" (Jones 1919b, 397). Dyson was determined to try again, even without the bright stars of 1919, but next time he would ditch the coelostat mirrors that he blamed for his problems in May.

12

Lights All Askew in the Heavens

The report given by Dyson and Eddington at the joint meeting caused an international sensation. No one can have been prepared for the outpouring of public enthusiasm that greeted the news that Newton's theory of gravitation had been overthrown and replaced by a new theory developed by a little-known European scientist. Einstein was catapulted almost overnight to worldwide and lasting fame.

The day after the joint meeting, the *Times* of London ran an article about it under the headline "Revolution in Science. New Theory of the Universe." English scientists reported enormous interest from the public, which many were ill prepared to satisfy, knowing nothing about Einstein's theory. The *New York Times* followed a couple of days later with the headline "Lights All Askew in the Heavens." Einstein's most important biographer, Abraham Pais, has claimed that this article was the first time this American newspaper had mentioned Einstein's name. Actually, Crelinsten informs us that he had already been mentioned in a report by Campbell on the 1918 eclipse (Crelinsten 2006, 116). Not a year of his life was to pass afterward without him featuring in its pages, again according to Pais (1982). Such was the suddenness of his fame that when he first visited New York two years later, he was greeted by cheering crowds lining the

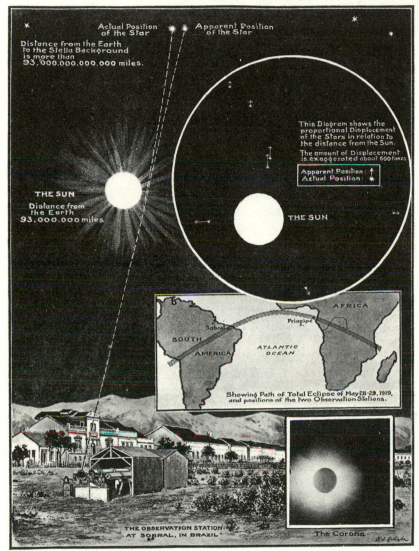

"STARLIGHT BENT BY THE SUN'S ATTRACTION": THE EINSTEIN THEORY.

FIGURE 19. The *London Illustrated News* explains the 1919 eclipse to its readers. The eclipse expedition attracted a great deal of attention in the press as the international cause of science caught the public's imagination in a war-weary world.

(Reproduced from the *London Illustrated News*, November 22, 1919. The illustration is by W. B. Robinson.)

streets. These enthusiastic people almost certainly had never heard of him before 1919.

The impact of the eclipse results extended well beyond physics and astronomy. This striking example of how science was conducted made an enormous impression on the philosopher Karl Popper, who strongly shaped the development of philosophy of science in the twentieth century. Popper was interested in the problem of demarcation. How can we identify what makes science, science? What distinguishes it from other human endeavors? Popper distrusted sociopolitical ideologies like Marxism and psychological theories like that of Sigmund Freud, which claimed to be scientific but were, in Popper's view, mere pseudoscience. He claimed that Marxism and Freudianism lacked falsifiability. Since they did not make quantitatively precise predictions, they were flexible enough to adapt to any events and could never be disproved. Any historical event could be made to seem like a prediction in hindsight by juggling the various tenets of Marxist theory. But Einstein had created a theory with none of this type of flexibility. It appeared to have no wiggle room at all, and Einstein made a point of telling people so. He was always quick to declare that the theory must stand or fall by its three tests. If it was shown to be wrong, it must be abandoned. One of its charms was its conceptual coherence. You could have the whole thing, in its elegant entirety, or you could drop it altogether.

Popper (1963) commented upon how the eclipse test in particular influenced him in a later article:

> There was a lot of popular nonsense talked about these theories [Marxism and psychological theories like Freud's], and especially about relativity (as still happens even today), but I was fortunate in those who introduced me to the study of this theory. We all—the small circle of students to which I belong—were thrilled with the result of Eddington's eclipse observations which in 1919 brought the first important confirmation of Einstein's theory of gravitation. It was a great experience for us, and one which had a lasting influence on my intellectual development.

This aspect of Popper's thinking about science has been very influential, especially among scientists. It is a damning indictment of a modern scientific theory to accuse it of being unfalsifiable. There is little doubt that the eclipse test played a key role in developing Popper's thinking. At the same time, other philosophers drew rather different conclusions. A philosopher named Ilse Rosenthal-Schneider, who was a student in Berlin in 1919, was a near witness to another dramatic moment in the eclipse story that took place quietly offstage. It was the moment when Einstein received his first news of the results, before the full announcement. It came by telegram from his Dutch colleague Hendrik Lorentz. Although the war was over, communication between the former enemies still largely proceeded through neutral Holland, as it had when de Sitter sent word of Einstein's theory over to Eddington in England. Now came the reply, again transmitted by the Dutch. Although the results were not final, the light deflection effect was established.

On the morning the telegram arrived, Ilse Schneider (as she was then) visited Einstein's home. We even have Einstein's earlier letter to her confirming their appointment (*CPAE*, vol. 9, doc. 104). Years later she recounted what transpired when they met. Einstein told her about the telegram and showed it to her. She was impressed by his air of sangfroid. What, she asked him, would he have done if the experiment had gone against his theory? "Then," he said, "I would have to be sorry for dear God. The theory is correct" (Rosenthal-Schneider 1980, 74).

Even senior colleagues of Einstein were somewhat in awe of his calm certainty in the face of possible experimental refutation. Planck and others commented upon it in letters to him at this time (*CPAE*, vol. 9, doc. 121). While others waited on tenterhooks for the news from England, Einstein seemed unruffled. Nevertheless, as soon as he got the telegram, he did, like a dutiful son, write to his mother with the good news (*CPAE*, vol. 9, doc. 113). The daughter of a close friend, Paul Ehrenfest, drew a scene of the whole affair, with Einstein sitting calmly and saying, as everyone rushes about him, "Oh well—I know it—I did calculate it" (*CPAE*, vol. 9, doc. 175).

While Eddington was notably nervous about the third test, Einstein was confident that others would come around to his way of thinking. One old colleague from his days in the Swiss patent office, Edouard Guillaume, had become an opponent—one of those who criticized relativity without really understanding it. Einstein patiently tried to explain to Guillaume where he was going wrong but to no avail. When Guillaume argued that the solar redshift test would doom the theory, given the lack of any experimental confirmation, Einstein calmly replied that "the line shift follows very strictly from the general theory of relativity and will soon be confirmed. In two years no one will doubt it anymore" (Einstein to Guillaume, February 9, 1920, *CPAE*, vol. 9, doc. 305). Amazingly enough, three years later even the previously implacable St. John had reversed his position on the Einstein effect.

It is as though Einstein invented the modern image of the cocksure theorist. But this image poses a problem. The corollary of Popper's falsifiability is that an Einstein can afford to let his theory be tested because he *knows* it is right. And if the theorist knows the theory is right, is there really any need for the experiment at all? Perhaps this is how science works—not with the elimination of theories one by one but with a triumphant march of pure thought, largely unimpeded by the piddling efforts of pedantic astronomers to restrain it. Eddington also played to the gallery, as many theoretical physicists have since, by affecting a calm certainty in the reliability of theory. This encourages a view of science which is "realist." If general relativity is the actual way that the universe behaves, then of course its predictions will be vindicated. Einstein's theory is not to be understood as a human effort to comprehend the universe about us that will one day, in its turn, be replaced. Rather, it is an example of a human mind hitting upon the true mathematical laws that govern reality. Once that has been done, experimental confirmation is a mere formality. But recall that he had been quite confident in 1914 also, when he had the wrong prediction made from the wrong theory!

No doubt, it is the maddening refusal by theoretical physicists to countenance being wrong that has made people skeptical of the whole eclipse story. But a key point to keep in mind is that however they may have liked to tease their students and the public, both Einstein and Eddington devoted a lot of time and effort to making sure this test would take place. This effort was not a charade. But exactly what it was surely goes to the heart of the question of what science is all about.

In Popper's day it was acceptable for a philosopher to draw attention to a very well-known event in the history of science and assume that his readership knew enough of the details to follow the main thrust of his argument. An anecdotal approach to the history and philosophy of science was quite common. But later in the twentieth century, Thomas Kuhn took issue with some of Popper's ideas. For him something larger was going on in certain episodes of scientific history than simply slotting a new theory into the place an earlier theory had previously occupied in the larger structure of scientific thought. He pointed out that sometimes new theories force changes in that overall framework, which in turn set off revolutionary changes in science, generally. He coined the term *paradigm shift* to describe what happens when scientists are forced to completely transform much of the way they do science in order to accommodate revolution-ary changes in their thinking. His use of the term *paradigm*, where Popper might have referred to a theory, was part of his strategy to portray science not as a body of knowledge but as a process. In order to understand that process, one had to carefully study it in action.

The modern approach to science studies is illustrated by Earman and Glymour, whose paper on the eclipse expedition is just as much history as it is philosophy. In contrast to Popper, they are interested in more than just a quick summary of what transpired. They want to take you with them as they follow Freundlich to Crimea and report what happened there. They have spent time with the literature, poring over the writings of the principal scientists and critiquing their calculations. They have philosophical points to make, but those points must be supported by detailed historical evidence. An even better example of the modern approach is the work of Peter Galison, especially in his book *How Experiments End*. Galison offers a perspective on how bias can lead to an incorrect experimental result without any desire to commit fraud. Experimentation, in Galison's view, is an open-ended process with no well-defined exit strategy.

One wishes to stop the experiment when the right answer has been found. But in research one does not know the right answer in advance. How are experimenters supposed to assure themselves that all reasonable steps to eliminate possible sources of error have been taken? Obviously, it is tempting to stop the experiment when the answer one always expected was right has been achieved. Thus, Eddington seemed happy enough with his pro-Einstein result until he learned of the pro-Newton result from the Sobral astrographic. Then, he seems to have carefully gone over his whole data reduction. He was relieved to learn of the four-inch results *"not only because of theory"* but because he claimed he could not reconcile his own data with that from the Sobral astrographic. But suppose the four-inch data also went against him. Would he have kept searching and tinkering with his results until he did achieve a reconciliation? Popper warned us against theories that are too flexible. Galison also reminds us that there is a certain flexibility in experimental work. Up to a point, you can keep tinkering with the data until it fits your expectations. If both sides of the theory-testing equation are flexible, then is it possible to come up with any result at all, given our prejudices? Perhaps instead of stumbling upon a truth about nature, as the realistic view of science insists, Einstein merely came up with a theory that was so convincing that scientists constructed the experimental evidence to confirm it in some kind of autosuggestive process.

Another pioneer of the detailed microstudy of scientific life is the English sociologist Harry Collins, who has made use of Earman and Glymour's work in his book *The Golem*, coauthored with Trevor Pinch. It is Collins who coined a term for the problem caused when you cannot decide whether your experiment worked correctly because the only test of its proper performance is whether it obtained the right result. This is the "experimenters' regress," and it goes to the heart of the eclipse debate. Did Dyson reject the data from the Sobral astrographic because the instrument performed incorrectly or because it got the wrong answer? Collins is also interested, among other things, in the collective nature of the decision to accept a new scientific result. While Galison focuses on the struggle of the individual scientist to interpret results, Collins studies how communities of scientists, viewed as social entities, decide between competing experimental claims. For him the interesting thing about Earman and Glymour's account of the three tests of general relativity is how the eclipse test influenced the third test. Before the eclipse, Evershed and St. John were skeptical of general relativity. Afterward they began to incorporate it into their interpretations of their results. Is this another way in which scientific results are socially constructed? Are humans just finding in nature what they expect to find?

Notice something important about the debate over whether science is socially constructed, as sociologists of science like Collins and Pinch would argue. One thing almost everyone who has conducted a close study of modern science agrees upon is that expertise is important. When I say that Eddington was the only scientist observing the eclipse on Principe, I am saying something critical to the whole enterprise. Only Eddington had the expertise to perform the experiment, reduce the data, and interpret the results. To be sure, Cottingham had important and relevant expertise to contribute, but he was not in a position to actually do the experiment. His role was merely to assist Eddington. Only someone with the appropriate knowledge and training can judge the verdict of nature in a case like this, and only someone who has actually performed the experiment can have a legitimate opinion on the verdict of nature regarding whether the Sun bends light. But for the theory to be accepted, a whole world of people, including the vast majority of physicists, must decide that the experimental confirmation is correct. How can they, since they never witnessed the experiment? Collins and Pinch would argue that part of what they must decide is whether they accept Dyson's and Eddington's expertise as credible. To do so they must themselves have some relevant expertise. In other words, Dyson's and Eddington's fellow physicists listen to the report, doing so in light of their level of trust in the reliability of the people involved, and decide whether to accept their word on it. One of my points is that the modern focus on Eddington is misplaced because it was Dyson who had the most relevant expertise, and it was Dyson who made the key decisions. The decision by many astronomers to accept the eclipse results was, in great part, an endorsement of Dyson's judgment.

The following quote from the minutes of the Royal Astronomical Society (RAS) meeting following the famous joint meeting suggests that they did endorse his judgment. The words are spoken by Dyson's colleague on the Joint Permanent Eclipse Committee (JPEC), Turner, who had made his own contributions to the eclipse preparations. He struck a tone of national pride in the whole eclipse endeavor and recalled a painful episode in British astronomical history. In that painful episode, a predecessor of Eddington's, James Challis, Plumian Professor of Astronomy and director of the Cambridge Observatory, and previous Greenwich director and Astronomer Royal George Airy allegedly failed to act upon John Couch Adams' prediction of the location of the planet Neptune. In so doing they let the discovery of the planet fall to continental astronomers.

> I would like to make some remarks of a purely domestic nature. About three-quarters of a century ago—to be exact, 72 years ago yesterday—the Society met to hear about a great discovery. The Astronomer Royal of that time read a report on the discovery of Neptune, and the Society learnt that the Astronomer Royal and the Plumian Professor had lost a great opportunity. I am not attaching blame to anyone. The occurrence has been compared to a dropped catch at cricket. To-day we have met to hear how the Astronomer Royal and the Plumian Professor have made use of an opportunity. In the former case the task was assigned to the Plumian Professor and he was favoured with good weather; in this case the Plumian Professor had no luck with the weather, but he did all he could. The Astronomer Royal in the old case had no instruments, but this time the Astronomer Royal took pains to use his best instrument—Mr. Davidson. Finally, I should say that in this case the Astronomer Royal made the opportunity by drawing attention to it. Even while the war was still going on, he never lost heart. And now we have to celebrate one of the greatest successes of the Society. (Jones 1919e, 427)

To this, the president of the society, Alfred Fowler, responded, "We all share in these remarks."

The Critics

Of course, not everyone was prepared to accept Dyson and Eddington's verdict, or to regard it as a triumph for British science. Following Turner's remarks, another astronomer rose to give the case for the prosecution. Hugh Newall was the director of the Solar Physics Observatory at Cambridge, which was situated close to the university observatory (of which Eddington was director) on the Madingley road. After World War II, the two observatories merged and now form part of the university's Institute of Astronomy. The Solar Physics Observatory was the successor of—and inherited much of the equipment and the staff of—the former South Kensington Solar Physics Observatory of Norman Lockyer. As such it historically played the leading role, along with Greenwich, in organizing British eclipse expeditions for a century beginning in the mid-1800s. It is striking, therefore, that it played no role at all in 1919, which would partly have been because Newall's assistant and eventual successor, Chubby Stratton, was serving with distinction in the armed forces during the war. Stratton was the leading figure in British eclipse expeditions of the twenties and thirties and the secretary of JPEC from 1923 (Chadwick 1961). Newall and Stratton were both in Crimea in 1914, but Stratton did not stay for the eclipse, returning to England and his military duties at the outbreak of war. It made little difference, as the weather was poor. Frederick Stratton was, in some ways, Eddington's direct opposite as a Cambridge astronomer. He was a decorated war hero who served as a very effective signals officer in the war.[1] He loved eclipse expeditions and approached them as military operations. He had been Third Wrangler in the year Eddington was Senior Wrangler. In recognition of Eddington's achievement, being only in his second year at the time, it was said that Stratton would have come second "in any normal year."[2]

Stratton was a widely loved character in British astronomy. The one thing he lacked was Dyson's luck. He was clouded out at almost every eclipse he went to (Chadwick 1961), so the English expeditions never really had the chance to benefit from the experience gained in 1919. It was left to others to build on what they had begun. In fairness,

Stratton was not much interested in the Einstein test, but it hardly mattered, since he was clouded out at least five times at subsequent eclipses.

Now, at the RAS meeting, Newall began his remarks with an urgent plea:

> We have already had an opportunity to congratulate the observers on their splendid results—splendid in whatever way they are interpreted. But I think it is only natural for one who has devoted his time to the consideration of the surroundings of the Sun to cry "Pause" in the interpretation of the observations. And surely we have a screaming warning not to overlook the effects of a possibly widely extended atmosphere in the enormous prominence seen on the limb of the Sun on the very day of the eclipse. (Jones 1919e, 428)

Newall was one of those who believed the light-bending effect was simply due to refraction. Knowing that the Sun has an outer atmosphere, the corona, which extends far out into space, Newall proposed that this must be much denser than commonly thought. He advanced the prominence as evidence for a dense atmosphere of this type. However, other astronomers believed, correctly, we now know, that this outer atmosphere is very hot and, accordingly, also very rarified—much too rarified to contribute a significant amount of refraction to the light-bending effect. But of course in 1919, this question was still up for debate, and many opponents of relativity theory seized upon it as a counterargument to Eddington and Dyson. Notably, this counterargument accepted the reported measurements but disputed their interpretation. Scientists like Newall would have joined with Earman and Glymour in decrying the theory-centric trichotomy of Dyson and Eddington. They would have insisted that proving the Sun has an effect on the positions of nearby stars was not necessarily making a statement about the gravitational field of the Sun unless other interpretations were first discounted. This is one of the chief problems with Popper's falsification idea. Popper had pointed out, very correctly, that one can never definitively prove a theory right; some aspect of it could always remain untested. But he tended to overlook that it is not particularly easy to definitively prove a theory wrong, either. Any experiment or observation contains so many contending influences that there is endless room to dispute the interpretation of a result, even if the result itself is accepted.

Frederick Lindemann then rose to rebut Newall's remarks. The two had sparred in a preliminary way at the joint meeting and had therefore had some time to marshal their arguments.[3] He argued, based on both the luminosity of the corona and the behavior of comets passing close by the Sun, that the corona could not have near the density required for such a strong refraction. Lindemann and Newall had something in common. Both were the sons of wealthy amateur astronomers who had made their money laying undersea telegraph cables. Lindemann's father being German perhaps goes a little way to explaining his relative enthusiasm for Einstein's theory, though in his later role as Churchill's most trusted advisor on scientific and technical matters, he had an enormous influence on Britain's prosecution of the next war against Germany. Of course, we have seen that he and his father had already attempted to do the light deflection experiment without an eclipse. At the joint meeting, Fowler, the RAS president, noted that John Evershed's attempt to use the Lindemanns' technique of using dark-red filters to image Regulus during the conjunction of August 21, 1917, had been clouded out. He still held out hopes (fruitless, as it turned out) that the idea could be made to work. As he put it, "Good opportunities might not be very frequent, but they would be more so than favourable eclipses" (Jones 1919b, 395). In everyone's mind the slow pace of experimental verification of this particular theory was galling.

A most ingenious attempt at an explanation was offered in the pages of *Nature* by a professor in Galway in the west of Ireland, Alexander Anderson (1919). He proposed that the atmospheric effects of the eclipse umbra itself caused the light deflection and presented a calculation intended as a plausibility exercise. The shadow of the Moon sweeps across the surface of the Earth and is associated with various meteorological changes. Anderson proposed that significant density changes in the atmosphere of the umbra might cause shifts in the position of stars close to the Sun, as seen from the place on the

Earth currently in total eclipse. It proved impossible to take the idea any further than that since the pattern of star shifts was decidedly nonradial. This was a strikingly different prediction from Einstein's. Still, it is possible that Anderson's idea had some influence in one way. Many critics focused on the moderately nonradial nature of some of the star shifts on the Sobral plates. The stars, as measured, were not all observed to shift dead away from the Sun. But though the skeptics kept darkly hinting at some significance to this, because it did not match Einstein's prediction, it is notable that the observed shifts were mostly radial and much more closely matched Einstein's prediction than Anderson's.

Fans of the ether theory were certainly vocal in trying to impose their own interpretations on the results. The Dutch astronomer Robert Jonckheere proposed, even before the expedition departed, that one explanation might be a tendency for the ether to be denser close to the Sun, causing a refraction effect to take place. Eddington responded to this by saying:

> Presumably the suggestion is that the condensation is due to the presence of the massive Sun, and that its effect is to modify the velocity of light. But a modification of the velocity of light in the neighbourhood of a massive body is just the effect we are looking for; so the suggestion amounts to a hypothetical explanation or illustration of the Einstein effect, and is not to be regarded as an alternative to it. (Eddington 1919b, 122)

Einstein's earliest version of his relativistic gravity theory focused on the idea that gravity modified the speed of light and that, indeed, the speed of light was effectively the measure of the gravitational potential. Here is one instance in which Popper's aversion to explanation, rather than prediction, fits well with the history. It would never have occurred to Jonckheere to predict the light deflection effect on the basis of the ether theory, but once it had been predicted, he was perfectly prepared to modify the theory in a qualitative way in order to accommodate the effect. Obviously, it can be a tedious business falsifying such a theory because its defenders make use of interpretative flexibility to defeat all opposition. But the truth is that sometimes advancing scientific knowledge can be a tedious business.

Scrutiny of Dyson and Eddington's claims was not restricted to the arena of interpretation. Their data analysis was not accepted uncritically, either. As a rule, critical examination of a published paper is fairly limited. An editor and a referee may read the paper, looking for problems, but they will rarely ask to see the unpublished data sheets associated with the measurement. Readers of a paper generally will not do so, either. If they doubt the work, they are more likely to ignore it than attempt to replicate it. Replication, if it occurs, will take the form of a separate experiment, typically, with a different experimental design. But some results demand an unusual level of scrutiny. In this case the entire suite of calculations to produce the final light deflection result from the original plate measurements was independently replicated. This unusual exercise was carried out by a team working under the direction of Louis Agricola Bauer, head of the Carnegie Institute of Washington's Department of Terrestrial Magnetism. Bauer was an American geophysicist and a veteran of several eclipse expeditions. It was he who had witnessed the 1919 eclipse from the coast of Liberia, a site rejected by the British expedition planners. As he put it in his résumé on the eclipse, "Though it is not necessary for the detection of this magnetic effect to have a clear sky . . . it has been my good fortune now three times to have a clear sky when others whose work absolutely depended upon clear weather were not so fortunate" (Bauer 1920, 1).

There can be little doubt that Bauer was skeptical of Dyson's claims on behalf of Einstein. His résumé devotes considerable space to a discussion of Campbell and Curtis' 1918 data. He quoted Larmor, saying, "The view asserts itself that the very important astronomical determination is to be regarded as a guide towards future theory rather than as the verification of the particular theory which suggested it" (Bauer 1920, 18) and ultimately concluded, "Possibly the best attitude to take is that of open-mindedness and to let no opportunity pass by for further experimental tests" (Bauer 1920, 18). But he was frank in admitting that he found no problems with Dyson's work, and he even cleared up one often-raised point. He carefully studied the nonradial aspects of the four-inch star deflections and concluded that they were small and consistent with the known measurement error in the data (Bauer to Dyson, July 17, 1920, RGO Archive 8, fol. 123, 3).

Scrutiny of the eclipse measurement was minute, indeed. Nowhere is this better illustrated than in a paper given by Davidson in 1923 with the title "The Amount of the Displacement in Gelatine Films Shown by Precise Measurements of Stellar Photographs." The problem here is that a bright object appearing on a photographic plate may draw in emulsion during the process of development, especially if it is overexposed. The question is whether some movement of this type could affect the positions of nearby stars on the plate. Although Davidson's paper is presented in the context of his ongoing work with Dyson on stellar parallaxes, it almost certainly was in response to Silberstein's raising this issue of what he called the "Ross effect" as a possible explanation of the eclipse data in a paper in 1920 (Silberstein 1920). Silberstein, in fact, moved to the United States and took up work with the Eastman Kodak Company around this time. Even this unlikely hypothesis to explain the stellar shifts in the 1919 plates seems to have raised hopes among relativity skeptics. Indeed, the audience followed up on Silberstein's point, asking, "Was [this] effect taken into account in the famous measurements which substantiated Einstein's prediction?" Davidson's reply was that "the point has been carefully considered. On account of the diffuseness and small density of the corona and the relatively large distances of the stars [from each other and the Sun] it seems unlikely that their positions can be affected. In any event the Ross effect is in the opposite direction to the Einstein displacement" and thus would not help those, like Silberstein, who favored Newton's theory.

All in all, most contemporary skeptics preferred to await further eclipse tests. For the moment the four-inch data from Sobral stood as an insurmountable obstacle to any frontal attack on Einstein's theory. The quality of the images, the prestige of Dyson and his assistants as observers, and the effectiveness of Dyson and Eddington's public presentation of their results precluded any serious effort to discredit this data. Bauer's reanalysis of the four-inch data reduction was evidence of the unusually close scrutiny of Dyson's work. No one seems to have felt that there was an opening to contest Dyson's finding in favor of Einstein's theory, given the acknowledged expertise of both Dyson and Eddington (Almassi 2008). But this was just one instrument. There was only limited support for the result from Eddington's Principe data, and no support from the Sobral astrographic. If the next eclipse produced any results falsifying general relativity, then the four-inch data would be left as an isolated result that might, in the end, turn out to be an aberration. The person who was most determined to perform this replication was Campbell. Later, in reviewing the situation as it stood in 1919, he had this to say in dismissing Eddington's contribution: "Professor Eddington was inclined to assign considerable weight to the African [i.e., Principe] determination, but, as the few images on his small number of astrographic plates were not so good as those on the astrographic plates secured in Brazil, and the results from the latter were given almost negligible weight, the logic of the situation does not seem entirely clear" (Campbell 1923, 19). The Greenwich team and their superior four-inch data were very much in the crosshairs as preparations began for the 1922 eclipse.

The 1978 Reanalysis

The debate about the interpretation of the Sobral astrographic plates has focused on quibbles over the decisions made at the time. These quibbles are based on little more than the written accounts left by the original eclipse team. Although not myself an astronomer, I am married to one, and my wife, Julia Kennefick, pointed out that the original plates were still preserved at the Royal Greenwich Observatory (RGO). She observed that with modern plate-measuring machines and astrometric data-reduction software, surely more could be done than had been possible in the days of merely human computers like Davidson and Furner?

In June 2003 I made a trip to Cambridge, primarily to look over Eddington's papers at Trinity College and the RGO's manuscript archive, now housed at the Cambridge University Library. I also had the idea of making inquiries about the original plates. To my surprise, I learned from Adam Perkins, the curator of the RGO archives, that a modern reanalysis of the data had already been done. This had occurred in 1978, to commemorate the Einstein centenary of 1979. He gave me the name of the man, Andrew Murray, who had been in charge of the RGO's astrometry team at that time and had set the project in motion.[4] It turned out that Murray's name was not on the published paper, but happily, he had occasion to write a letter to the *Observatory* on the topic ten years later, which I found (Murray and Wayman 1989), and this led me to the original paper (Harvey 1979). Thanks to Perkins and to Donald Lynden-Bell, former director of the Cambridge Institute of Astronomy (successor to the Cambridge Observatory), I eventually got in touch with Murray himself, who very kindly corresponded with me in some detail about the process of the 1978 data reanalysis. It seems clear that it is to Murray himself that we owe thanks for the inspired idea to subject the plates to a modern astrometric analysis.

In 1978 most of the Sobral plates still survived intact (in contrast to the Principe plates). One eclipse plate and one comparison plate taken with the four-inch lens were missing, and one of its eclipse plates was broken. A few of the astrographic plates were discolored. In general nothing stood in the way of the project. The plate measurements were made by E. D. Clements (known as Clem) on the Zeiss Ascorecard at the RGO, by then relocated to Herstmonceux Castle in Sussex. This plate-measuring machine permitted the operator to measure accurately and record the positions of every star on each plate, both eclipse and comparison. The data produced in this way could then be entered into an electronic computer. The astrometric data-reduction software was written by Murray, and the process of reduction was carried out by Geoffrey M. Harvey, author of the published paper. Both Clements and Harvey were members of Murray's staff at the RGO. The first item of note is that the modern methods had no particular difficulty with the astrographic data, although the images were sufficiently noncircular, on many of the plates from both instruments, to prevent the use of a completely automated plate-measuring machine, such as the RGO's GALAXY. The fact that the astrographic plates were measured in both declination and right ascension during the reanalysis does not even call for particular comment in Harvey's paper. Murray remarks, in an e-mail to me, written on November 22, 2003:

> In 1978 the plates were re-measured individually on the Ascore-card machine at Herstmonceux; each image was centered in a square graticule and the (X,Y) co-ordinates were recorded by means of moiré fringe gratings. The reasonable results obtained, particularly for the "inferior" Sobral astrographic images, would seem to indicate that the problem with the 1919 measurements was not so much in the quality of the images, but rather in the reduction method, which relied very heavily on the experimental determination of the scale constant e. The Herstmonceux plate reductions of course included both co-ordinates on each plate, giving a much better separation of the plate scale from the deflection on the eclipse plates.

Harvey (1979) compares the 1919 results with those he recovered using modern techniques:

TABLE 1. Gravitational Displacement at the Sun's Limb in Seconds of Arc

Determination	Displacement
Predicted from Einstein's theory	1.75
Four-inch plates reduced by Dyson, Eddington, and Davidson (1920)	1.98 ± 0.18
Four-inch plates measured on the Zeiss	1.90 ± 0.11
Astrographic plates reduced by Dyson, Eddington, and Davidson (1920)	0.93
Astrographic plates measured on the Zeiss	1.55 ± 0.34

Harvey (1979, 198) comments:

> For the 4-inch plates there is no great difference between the value obtained by Dyson et al. and that from the new measurements, but the error has been considerably reduced. For the Astrographic plates, however, a significant improvement has been achieved by the new measurements. Where the previous reduction yielded a value of 0".93 with an unspecified, large error, the

new determination is 1".55 ± 0".34. This is still a weak result, but does provide support for that from the 4-inch plates. Combining the two fresh determinations, weighted according to their standard errors, gives 1".87 ± 0".13, a result which is just within one standard error of the predicted value.

It is remarkable that the alternative value for the Sobral astrographic from 1919 happens to have an almost identical value (1."52 compared to 1."55) to that obtained by the modern reanalysis. Is this mere coincidence? Does it suggest that the argument claiming there was a loss of focus but no change in scale was correct? Certainly, the modern value casts grave doubt on the 0.93" value obtained by the original team. Murray, in the same e-mail to me, comments:

> We have to remember that in those days [i.e., 1919], the labour of computation was a problem, so short-cuts had to be taken. In particular, the Greenwich astrographic plates were only measured in one co-ordinate (declination). The general philosophy, both at Greenwich and Cambridge, seems to have been to determine the relative plate scale and individual orientations by some means or other, and then apply these to the measured displacements of individual stars to derive the deflection obtained from each of them.
>
> There can always be a problem trying to independently determine too many plate constants simultaneously in one solution because of correlations between them due to the actual geometrical distribution of the stars in the field. Presumably, there must be some such effect that has affected the scale derived in table IX [i.e., the derived scale that apparently gives rise to the erroneous result from the astrographic plates in Dyson, Eddington, and Davidson (1920)]. We should note that, although the deflection is greater in declination, there is a lot of information on the plates scale in the right ascension direction which has been completely ignored.

He adds, in a later letter (November 27, 2003):

> I can only infer that the inclusion [by Harvey in 1978] of the right ascension measures on the astrographic plates (which were

ignored by Dyson et al.) has considerably improved the solution for the deflection, in spite of its smaller effect.

There is a wealth of extra information in the modern analysis. Rather than being obliged to simply compare pairs of images plate by plate, the astrometric software permitted the construction of a database of positions from the comparison plates, against which the positions of the stars on the eclipse plates could be compared. Thus, the displacement of each star could be compared to the position of every other star on every comparison plate, not just to its own position on one or two comparison plates. In addition, the plate-measuring machine was able to provide reliable measurements of position in both coordinates for the astrographic plates. This wealth of extra data meant that there was no difficulty in calculating the scale change.

Since it now seems likely that the low "Newtonian" value was due to errors in reducing the data, it seems plausible that the problem lay in an incorrect value for the scale. Therefore, the assumption that the scale did not change much between eclipse and comparison plates was the correct one, though, obviously, in 1919 Dyson had no way of knowing this for certain. In hindsight, we can argue that the Greenwich team was correct to conclude that the prediction of general relativity was confirmed over the "Newtonian" prediction.

It seems, as one might expect, that the teams who took and handled the data knew best after all. But it is hard to stop a good story once it gets going, and the story of Eddington's bias is a case in point. Stephen Hawking (1988) included the following passage in his famous book *A Brief History of Time*.

It is normally very difficult to see this [light-bending] effect, because the light from the sun makes it impossible to observe stars that appear near to the sun in the sky. However, it is possible to do so during an eclipse of the sun, when the sun's light is blocked out by the moon. Einstein's prediction of light deflection could not be tested immediately in 1915, because the First World War was in progress, and it was not until 1919 that a British expedition, observing an eclipse from West Africa, showed that light was indeed deflected by the sun, just as predicted by the theory. This proof of a German theory by British scientists was hailed as a great act of reconciliation between the two countries after the war. It is ironic, therefore, that later examination of the photographs taken on that expedition showed the errors were as great as the effect they were trying to measure. Their measurement had been sheer luck, or a case of knowing the result they wanted to get, not an uncommon occurrence in science. The light deflection has, however, been accurately confirmed by a number of later observations. (32)

It appears that Hawking was aware that widespread suspicions concerning the original data reduction existed. He also apparently remembered that a reanalysis of the plates had been done. This in itself makes him nearly unique, since I have not found even one paper that cites their publication. The Science Citation Index lists only the Murray and Wayman letter to be discussed. It seems the RGO reanalysis team's results were reported at a meeting of the Royal Society in 1979, which might be where Hawking became aware of their efforts.

Knowing that some reanalysis had been undertaken and recalling the many stories of a dubious data analysis by the original team, he may naturally have jumped to the conclusion, when writing his book nearly a decade later, that the reanalysis gave birth to the story he remembered. Nothing, of course, could be further from the truth, as far as the reanalysis goes, and it was for this reason that Murray and P. A. Wayman of Dunsink Observatory in Dublin (possessors of the four-inch instrumentation used at Sobral) wrote to the *Observatory* in 1989 to object:

> The result from the 4-inch plates were thus confirmed, with a smaller standard error, and even the very low-weight 13-inch plates give a significant result.
>
> The last attempt to measure the deflection by optical observations [as opposed to observations in the radio band] at an eclipse was in 1973, by a team from the University of Texas. The result from that expedition was: Deflection (arcseconds) = 1.66 ± 0.19.

Conditions were very far from ideal on that occasion, but by comparison, the results from the 1919 eclipse were very respectable. It is completely unjustifiable to dismiss them as Hawking has done. A recent assessment by Will (Was Einstein Right?, p. 78) that ". . . these expeditions were triumphs for observational astronomy and produced a victory for general relativity . . ." would seem to be much fairer.

So strongly was the current running against poor old Eddington by that point that even good news was turned on its head in the book that millions of people actually read, while no one ever seems to have read the paper that contained the good news!

Dennis Sciama's (1969) book *The Physical Foundations of General Relativity* shows how skepticism of the 1919 results slowly increased from about 1970 onward. Sciama (1969) states:

> Eddington himself later referred to it as "the most exciting event I recall in my own connection with astronomy." Ironically enough, we shall see that Einstein's prediction has not been verified as decisively as was once believed. Between 1919 and 1966 there have been fewer than thirty [total solar] eclipses, giving altogether a total observing time of not more than about two hours. (The longest possible duration of a total eclipse is about 7½ minutes, and such an occasion occurs very seldom). . . . In fact results have been published for only six eclipses. (69–70)

Looking over the results from all these eclipses (including that of 1919), Sciama concludes:

> One might suspect that if the observers did not know what value they were "supposed" to obtain, their published results might vary over a greater range than they actually do; there are several cases in astronomy where knowing the "right" answer has led to observed results later shown to be beyond the power of the apparatus to detect. (Sciama 1969, 70)

To be sure, Sciama comes quite close here to charging that one might achieve any result from an eclipse measurement of light bending. But note that his real point is that the tests do not vindicate general relativity in particular. He does not claim that the data do not support Eddington and Dyson's main contention, which is to say that the results falsify the "Newtonian" result. I contend that much of the change in narrative concerning the eclipse is due to this shift from a contemporary focus excited by a test, the first ever to do so, that falsified Newton while leaving a rival theory standing to the view, since the 1960s, that Einstein's theory is undefeated and fending off rivals and that is struck by how threadbare seem the emperor's old suit of clothes. Fortunately for relativity, it has much less revealing clothes to wear nowadays.

All of these stories contain elements of truth, but in my opinion, the closest we can come to the truth is to say the following. First, the 1919 eclipse expeditions established clearly that light-bending in a gravitational field was a real effect. This was, after all, their principal goal. As Eddington put it, their mission was to "weigh light." Second, they showed that of the two predictions made by Einstein, the evidence strongly supported the higher general relativistic value and appeared to falsify the lower, which Eddington labeled the Newtonian result. This conclusion was supported by subsequent eclipse expeditions, by radio observation of occulting quasars from the 1970s on, and by the data reanalysis of the Sobral plates in 1978. Nevertheless, to paraphrase Eddington, myths have weight. They are not easily suppressed or replaced, and mere written accounts cannot necessarily hope to halt the spread of a good story. Historical nuance is generally alien to any story passed on orally. If my account is lucky enough to ever be passed on orally, who knows what will be made out of it?

Dyson's Decision

The Sobral data analysis was recorded on a large sheaf of pages still held in the RGO archive. These pages contain tabulations of measurements from the plates taken by both the astrographic and the four-inch telescopes used at Sobral. They also contain intensive calculations deriving the light deflection and the plate constants from those measurements. There is one particularly intriguing calculation on pages 208–10. The text on page 210 seems to be written in Dyson's hand. This calculation is of obvious importance because it closes with the words, "The results from the Astrographic telescope support those from the 'four-inch' telescope in showing an outward deflection. For the reasons given above they are not of the same value, and the figures obtained from the 4-inch telescope are to be accepted as the result of the expedition to Sobral."

This is very similar phrasing to what appears at the end of the discussion on the Sobral plates in the final published report (Dyson, Eddington, and Davidson 1920). The fact that it appears here certainly suggests that the calculation that immediately precedes it is significant. Presumably, this calculation caused Dyson to reach this conclusion firmly. What was that calculation? At first glance it is tempting to think that the preceding calculation relates to the four-inch telescope, since the line before this quoted passage reads: "The mean of these quantities is +r0.51 or 1″.98 at the limb."

A deflection of 1″.98 at the limb of the Sun is the value quoted in the report for the four-inch telescope. But a study of pages 208–10 quickly reveals that the quoted result is a mean of measurements taken by the astrographic instrument. How can the astrographic data be made to agree with the result from the four-inch? The answer is that the value of 1″.98 probably represents the top of the error bar associated with the alternative result of 1″.52 achieved by assuming a minimal change of scale. In effect, Dyson and Davidson show that if the scale change was small, then the light deflection on the astrographic plates was 1″.52 ± 0″.46. The significance of 1″.98 is that 1″.56 + 0″.46 = 1″.98, where the measurement has an associated error of some 30 percent. Thus, the error bars of the results from the two instruments actually overlap, if it is assumed that the scale did not change between the eclipse and comparison plates taken with the astrographic lens. Suddenly, the two instruments are not discordant, thus lending some more confidence to the four-inch results. It is worth also comparing this with the result from the 1979 reanalysis. They gave 1″.55 ± 0″.34 for the same plates (see the appendix for a more detailed explanation of Dyson and Davidson's calculation).

Dyson himself expressed all this clearly in his brief article for *Nature* on the eclipse results. In discussing Eddington's results from Principe, he noted, "There is no reason to suppose any change of scale." In discussing the Sobral astrographic results, he said, "If it is assumed that the scale has changed, then the Einstein deflection from the series of plates is 0.90"; if it is assumed that no real change of focus occurred, but merely a blurring of the images, the result is 1.56"" (Dyson 1921b, 787).

Up until now it has been thought that this dilemma, the possibility of a large change of scale in the Sobral astrographic plates, was a unique occurrence. Since the Greenwich observatory never obtained any more of their own plates at subsequent eclipses, historians had believed there was no way to know if Dyson and Davidson were consistent in deciding to throw out the results from the astrographic plates in 1919. But this turns out not to be true at all. Although the Greenwich team were themselves clouded out on Christmas Island in 1922, Davidson was engaged in measuring plates taken at that eclipse. An expedition with a station in South Australia, near the end of the track of totality, obtained plates they were unable to measure with sufficient accuracy. Accordingly, they sent these plates to Greenwich, where Davidson measured them and reduced the data. In his section of the published report on this expedition's results, Davidson had this to say about one of the eclipse plates: "It was at once evident that for some reason the scale of [plate] III was different from that of Nos. I and II, and as no Comparison field had been impressed on this plate, no further use could be made of it" (Dodwell and Davidson 1924, 158).

This is an identical situation to that obtained for the Sobral astrographic plates. The comparison field referred to here is the same concept as Eddington's check field, a different field from the eclipse field used to provide an independent determination of scale. At Cordillo Downs in 1922, the Australian expedition took these comparison plates during the eclipse on the same plate as the eclipse field, but this was not done for the plate III discussed here. Lacking a means of independently determining the scale and finding internal evidence that the scale change was great, Davidson simply dropped this plate from consideration. The argument about the data analysis in 1919 centers on the following question: Were the Sobral astrographic plates disregarded because they agreed with Newton or because they suffered from a possibly large change of scale? I have argued that it was because of the possible change of scale. We can at least conclude from Davidson's handling of the data in 1922 that he behaved consistently. Faced with a clear change of scale, he declined to use the data from plate III. He and Dyson made the same decision in 1919 after noting that the only way out, to assume no change of scale, would have ruined the agreement with Newton. A half century later, they would be amply vindicated by the work of Harvey, Murray, a plate-measuring machine, and automated data-reduction software.

13

Theories and Experiments

In the real world, and in real time, it is rarely crystal clear how the rules of theory testing are to be interpreted. Let us take Freundlich as an example. In 1929, after a decade and a half of trying, he finally obtained data at an eclipse to test general relativity. By that time, after the work of Eddington, Dyson, Campbell, and Trumpler, it was widely accepted that the eclipse measurements confirmed Einstein's theory and falsified Newton. In this context, Freundlich's results from the 1929 eclipse in Sumatra are easy to interpret. He obtained a deflection at the limb of the Sun even greater than that obtained in the two earlier expeditions, fully 2.24 seconds of arc, even larger than Dyson's value of 1.98" (Hentschel 1997, 144). Obviously, this still falsifies Newton's theory, but that is now old news. What struck Freundlich is that it does not agree well with Einstein's theory, either. In the new situation, Einstein's theory is the only one available to be falsified; and given his own results, Freundlich was in a position to reinterpret Crommelin and Dyson's result from 1919 as being too high for proper agreement with general relativity. He now argued that his results had falsified Einstein's theory.

This is one potential problem with falsificationism. What do you do if you have no theory left unfalsified? Theories are tools, and science without tools is not very effective. How do you do calculations with no theory to guide them? It is not surprising, therefore, that Freundlich did not entirely throw out general relativity. He argued that the theory failed to account for interactions between photons. Particles of light, he claimed, affected each other's motions, so they simply did not follow the paths predicted for them by general relativity but deviated from those geodesic paths by amounts depending on the number of other photons nearby. He thought that his theory could explain general relativity's failure (as he saw it) to pass either the light deflection or the solar redshift tests. But it turned out that other physicists and astronomers did not much like Freundlich's theory. Neither did they pay much attention to the disagreement between his 1929 measurements and Einstein's theory. It is a matter of interpretation how scientists react to experimental results. They chose to interpret Freundlich's results as confirming a large light deflection, as Einstein predicted, rather than a small one, as Newtonian theory predicted. Even Freundlich accepted general relativity, since one can argue that his photon-photon interaction is merely a new physical effect he introduced in addition to the gravitational interaction predicted by general relativity.

You can argue that Freundlich's experience offers a vindication of Eddington and Dyson's decision to frame their results in terms of the "false" trichotomy modern critics so denigrate. Here is Freundlich doing exactly what he is supposed to do, fearlessly putting forward a numerical result that lacks any theoretical support, and his contribution is roundly rejected, or even ignored. Crying "a plague o' both your houses" is an unattractive option when there are only two houses available in which to pass the night. There are reasons to think that the theory-testing process demands some amount of framing in theoretical terms, especially when the level of experimental precision is not high.

Theory testing can be highly situational. In 1929 there was only one theory left standing, so the appetite for falsifying it was obviously not great. It was clear to everyone that Freundlich's results were even worse for Newton than they were for Einstein. It was not until viable rival theories emerged much later that interest in falsifying Einstein really took hold. This does not mean that no one was interested in doing the measurements; it is that the current standing of the rival theories affects the interpretation of the results. This is especially true if there is a lack of rival theories to test.

Individual scientists also have views that alter with time. When Einstein and Freundlich first met, Einstein was anxious to have relativity tested, and Freundlich was flattered to be chosen by the great man to lead the charge. But in the 1920s, their relationship cooled until it became almost frigid (Hentschel 1997). They initially argued over a manuscript of Einstein's that Freundlich tried to sell to raise money for his new solar observatory near Berlin, the famous Einstein Tower in Potsdam. Einstein was outraged, claiming he had only loaned the manuscript to Freundlich. After this incident Einstein increasingly came to sympathize with Freundlich's many enemies in the astronomy community. He reacted to later news of Freundlich's falsification of his theory in the following words:

> Verification of the theory is unfortunately much too complicated for me. We're all just poor blighters, you know! But Freundlich doesn't impress me one bit. If absolutely no light deflection, no perihelion motion, and no line shifts were known, the gravitation equations would still be convincing, because they avoid the inertial system (this ghost, which influences everything, but to which things in turn do not react). It really is odd that people are usually deaf to the strongest arguments, while they are constantly inclined to overrate precision in measurement. (Letter to Born, 1952, translated and quoted in Hentschel 1997, 142)

How hard it is to recognize this Einstein as the same man who had urged Freundlich onward in his efforts four decades previously! Certainly, Einstein's own views had changed in that time, and his work on unified field theories increasingly led him into areas of physics where pure math was more of a guide than experimental results. And he had always been skeptical of experimental results that violated the principles on which he constructed his theories. Does this mean that Einstein was happy to be vindicated by experiment but unwilling to be falsified? Was Popper wrong in thinking that Einstein alone did science the way it should be done? Not entirely. In this case it is clear that he did not trust Freundlich's work. Einstein was certainly not the only one who suspected that Freundlich did not get his measurements right in 1929. Falsifying experiments may turn out to be mistaken, and scientists usually do discount work from sources they do not trust. Plenty of evidence indicates that Einstein continued to look for experimental guidance in his search for new theories throughout his career. But where he had once believed in the reliability of Freundlich's work, he stopped doing so later in life. We learn from this that the personal and the scientific are not always easy to disentangle. Scientists must balance the alleged experimental facts against their own scientific instincts (Einstein's contempt for the ghost of Newton's absolute space) and their judgment of other scientists' abilities in determining their reaction to tests of theories. Experts in the field were certainly prepared to countenance a light deflection result greater than that predicted by Einstein. For instance, Dyson, who obtained such a result in 1919, wrote to Bauer in 1920, saying, "If a larger value than 1.75" comes out in the next Eclipse then I shall be inclined to regard the difference from Einstein's law as real" (letter from April 6, RGO Archive 8, fol. 123). It may be that Freundlich's measurements came too late, after Campbell's work, which agreed so precisely with Einstein's prediction, had solidified acceptance of general relativity in everyone's minds.

Einstein's remark here shows he is no follower of Popper's in another important sense. Popper complained about the tendency of Marxists to value explanations of events. Marxist theory was lauded for its ability to explain historical events. Yet Einstein argued that it was his theory's ability to explain how the ghost of the inertial system can be avoided that made it so valuable. This is in stark contrast to Popper, who argued that explanations are mostly worthless in science compared to verifiable predictions. Yet predictions are slippery too, and though Popper might have denied it, they probably derive most of their cachet from a claimed numerical preciseness. But Einstein expressed himself skeptically here about this precision, and the reason is almost certainly his experience that precise measurements are sometimes false! You cannot always presume that a fidelity to precision measurement will guide you in the right direction scientifically. Sometimes numbers can mislead.

In considering the extent to which the scientists in our story were influenced by theory, I think it is interesting to look at Campbell's change of heart in the summer of 1919.[1] As Campbell set off for Europe, he made a point of cabling Curtis to give him plenty of ammunition for a strong statement to his English audience. Curtis readily replied, and Campbell did indeed tell the Royal Astronomical Society (RAS) that he thought Curtis had ruled out the general relativity prediction, though not Einstein's original prediction, which was Eddington's "Newtonian" value (Jones 1919c). Note the language. The null result is the only result that is not in danger of being ruled out at this juncture.

Once in Europe, Campbell's feet seemed to get colder and colder. He cabled Curtis shortly after the RAS meeting, urging him to proceed cautiously since his errors were large. In response, Curtis asked a journal editor to withhold his Pasadena paper from publication. Later, from Brussels, Campbell cabled more definitively, "Delay publishing Einstein results, Campbell." (Crelinsten 2006, 141). Now, it is possible to imagine Campbell's change of heart being entirely due to anxieties about the quality of Curtis' data. But if so, why did this anxiety come to the fore when he was in Europe and unable to examine that data closely? He seemed quite bullish before he left the United States. What was different about his environment in Europe?

Although mere speculation, my hunch is that he began to realize that no important European theorist really believed in the null result anymore. As far as the people he met in Europe were concerned, it was a choice between the lower light deflection result and the general relativity prediction. This may have made him very nervous about the trend of Curtis' thinking. Curtis believed the test was between the old etheric theory and the old relativity theory. Like any scientist, Campbell was prepared to be wrong at least occasionally, but he certainly did not want to be a laughingstock. At least, he wanted to be seen to be testing the right theories. A man who had left California thinking that Curtis was on the right track and confirming the reliable old theory, not of Newton but of the ether, now saw that modern theorists were speaking in radically different terms. Eddington was holding forth on what a shock it would be to find a null result. Campbell and Curtis thought they were coming down in favor of the safe bet. How unexpected was it to find they were daringly putting down a marker for the rank outsider! That would certainly require a second look at the data and explain why Campbell suddenly wanted Curtis to hold off on publication. In the end, the results were never published, and only a manuscript of Curtis' paper exists in the Lick Observatory archive, hunted up there by Crelinsten.

One reason why this hunch appeals to me is because of the well-known difference between American and European physics at this time. America boasted of its talented experimentalists, especially in the realm of high-precision experiments by men like Albert Michelson and Robert Millikan, two Nobel laureates. But Europe was considered the center of theoretical physics, and America had produced few notable theorists at this time. Certainly, discussions on relativity in America focused on Michelson's experiments and the mystery of the failure to measure the ether wind. Many American physicists and astronomers still believed in the ether and were not terribly up to date on the subtleties of the relativity theory replacing it in Europe. I strongly suspect that Campbell underwent something of an education in these matters upon his visit to Europe and discovered that coming out in favor of the null result, as Curtis was inclined to do, would invite serious scrutiny of his experiment. He was not at all sure it could stand up to such scrutiny. Certainly, we know he concluded that further work was needed and ended up conducting the measurements himself for several years (Crelinsten 2006, 185–89). He never did publish anything from Goldendale to do with the Einstein test. It is noteworthy, from the perspective of my hunch, that when he went to observe the eclipse of 1922, determined to finally have his say, he took with him an assistant who was trained in Europe, the Swiss-American astronomer Robert Trumpler.

What should we make of Curtis' approach to the problem? He had few stars near the Sun and poor-quality images. He even had complaints about the quality of his plate-measuring equipment

(Crelinsten 2006, 133–38). He divided the stars on his plates into an inner and an outer group in an effort to make the most of what he had. The idea here is that the outer group will indicate the size of the scale change (since they have a small deflection, being far from the Sun) that permits measuring the deflection for the inner group. But the deflections he measured were small, not least because the stars were not close to the Sun and had large associated errors. One of his two plates actually showed a deflection in line with the general relativity description, but after averaging this with results from the other plate, he ended up with the value he reported, which was lower than the half-deflection prediction. Jeffrey Crelinsten, the historian who has studied Curtis' papers, noted approvingly that this was the "proper scientific procedure" in contrast to "Eddington's" decision to exclude the Sobral astrographic results (Crelinsten 2006, 136, 144). But one must be careful here. As we have seen, Dyson made that decision, not Eddington, though Eddington concurred with it.

But here is the crux of the whole matter. What is the correct scientific procedure? The answer is that there are two main kinds of experimental error. The first is random error, also known as Gaussian error, or normally distributed error. Such errors are likely due to a complex of factors, impossible to disentangle in a given case, that should all average out over enough trials. In other words, truly random error will sometimes cause the experiment to underreport a value, sometimes to overreport a value, but the average will, presumably, remain close to the "true" value. Of course, taking the mean from only two plates is not enough to guarantee this, but it is still good practice to do the averaging over whatever results one has.

The other kind of error is systematic. Such an error may have only one cause and likely will not be random in nature. It is liable to make every run either under- or overreport the true value. Averaging does no good here. Indeed, if one believes that a particular run or a particular instrument is subject to a systematic error, then one should not average results from that run in with other results that may not be subject to that error. It is good scientific practice not to include in one's average results from a malfunctioning instrument. Thus, it may be that both Curtis and Dyson acted according to good scientific practice.

No foolproof set of rules exists for following good scientific practice. It is the judgment of the experimenter that must decide a particular case. There are guidelines, but they must be interpreted anew in each situation that arises during an experiment. If Dyson believed, as he clearly did, that the Sobral astrographic was unreliable, then averaging its results with another instrument that was working well would have been the wrong thing to do. Let us now bring the oft-criticized Eddington back in for his view on correct scientific procedure.

We know that Dyson was skeptical of general relativity before the expedition departed. Of course, in any scientific endeavor there comes the moment at which the scientist becomes convinced of the correct result and subsequently tends to become a partisan for that result, whatever their original bias. So it is amusing that in his first draft of the report from Sobral, Dyson endeavored to average the results from the two instruments there, having noticed their average was very close to the Einstein prediction of 1."75. In his draft of the paper, he stated, "The mean with these weights is 1."83 and is very close to the value required by Einstein's theory" (Manuscript of report, RGO Archive 8, fol. 150). So Dyson did try to average the results from his two instruments, including the Sobral astrographic. Doing so would have actually given him a result almost spot on with Einstein's prediction! Eddington objected:

> I do not like the combination of the astrographic with the other Sobral results—particularly because it makes the mean come so near the truth. I do not think it can be justified; the probable errors of both are I think below 0".1 so they are manifestly discordant. If the results are accepted with the weights assigned, the probable error of the mean (judged from their accordance) is about ± 0".20, which certainly does not seem to do justice to the results obtained. I would like to omit the last 5 lines of p.4. It seems arbitrary to combine a result which definitely disagrees with a result which agrees and so obtain still better agreement (Eddington to Dyson, October 21, 1919, RGO Archives 8, fol. 150).

I agree with Eddington on the definition of correct scientific procedure. Averaging the results from two different instruments when one was suspected to have suffered from a serious systematic error would have been a mistake. This is appropriate only where systematic errors have been eliminated, and only random errors remain. It was correct to publish all the data in an open and forthright way, and it was a reasonable judgment call to prefer the results from the instrument that manifestly performed better and to ignore those from the other one.

It may rankle some people that Eddington here referred to the theoretical prediction as "the truth." But I would argue that whatever you think of his terminology, he was being honest in acknowledging the experimenter's temptation to look to theory as a guide when there are no prior experiments to compare against. I believe that Curtis and Campbell behaved in very similar ways. The truth of the matter is that theory must guide experiment anyway, since one only performs experiments with some prospect of success, based, in this case, on Einstein's predictions.

Here we are confronted once again by Harry Collins' experimenters' regress. In many technological applications, it is obvious after constructing a device whether it works or not. A gun either shoots or it does not. But in pure research in disciplines like astronomy, a working experiment is one that gets the right answer. Yet you cannot use that to determine whether the experiment works because finding out whether the answer really is "right" is the point of the experiment. Suppose two experiments of this type disagree, as happened, in this case, when Freundlich disagreed with Campbell and Trumpler. How do you know which result is correct? As we shall see, the two sides began to critique each other's work and argued that the other side would agree if they had performed the experiment or data analysis properly. Because there are so many points upon which to disagree, such a controversy can last almost indefinitely. Certainly, Freundlich and Trumpler never did reconcile over this disagreement. The regress sets in because the best test of an experimental apparatus is to use it again and see if it works, but "working" is actually part of the issue that is up for debate. Furthermore, in the case of the eclipse, the regress is particularly pernicious because there may never be an opportunity to do the experiment again, especially if you have Freundlich's luck with weather and war.

Certainly, it is a known problem that a theoretical or other prior expectation may influence a scientist to find that predicted result even if it is wrong. Peter Galison has shown how there is always something more to try in order to improve a complex experimental apparatus. But experimenters must end their experiments at some point in order to report a result. How do you know when you have found the right result? Programmers are familiar with the dictum that there is always one more bug in a code. You stop when the remaining bugs are relatively small and benign. Similarly, no experimental apparatus works perfectly. The judgment of the operator must decide when it is working well enough. Then one stops the experiment and announces a result. Galison discusses a famous case of getting the wrong answer, and it involves another character in our story, Einstein himself. In 1915 Einstein conducted an experiment with a young Dutch physicist, Wander De Haas, to measure the gyromagnetic ratio of the electron. On theoretical grounds, Einstein expected the quantity g to be one. That is the result they found. But in subsequent years, some other experiments reported a higher value, and eventually, it transpired that electrons have an unusual spin of one-half, which means their g factor is equal to two. This seems to be a typical case of letting a theorist into the laboratory, even a competent one like Einstein. He goes and gets the answer he expected to find! One can see why many people expect that Eddington did the same thing. For that reason, it is important to remember that Dyson actually played a bigger role in this experiment than Eddington, and Dyson was no theorist.

One critic of Eddington is the modern physicist and historian Francis Everitt, who is forthright in rejecting his theory-based approach. Everitt was the director of one of the key gravitational physics experiments of recent decades, Gravity Probe B, which tested one of the most dramatic and difficult to verify predictions of general relativity. This prediction, known as frame dragging, claims that the Earth drags spacetime around as it rotates, making nearby orbiting bodies spin with it. The rotational effect is transmitted through the gravitational field itself. Everitt arranged for the analysis of the Gravity Probe B data to be done in the blind, with a key ancillary part of the experiment conducted by a separate group who were to keep their results secret until after Everitt's team presented their own measurements. Here is another example of good scientific procedure. Since Everitt did not know the final piece of the puzzle required to get the frame dragging, he could not be tempted to nudge closer to Einstein's prediction. But note that the data analysis of Gravity Probe B turned out to be a very difficult and protracted affair because of unexpected problems in the experiment. It took years for Everitt and his team to feel confident enough to announce a result. It takes money and other resources to do this kind of careful approach—money that NASA has to spend. In addition, Gravity Probe B was so expensive that there is no likelihood of anyone repeating the experiment, so special care was warranted. Dyson and Campbell, on the other hand, planned to repeat the experiment at the next eclipse, with more time to prepare. Although some physicists value a non-theory-centric approach, it seems that such elaborate precautions as Everitt's are not necessarily the norm.

It is easy to forget how much time data analysis can take. Einstein himself seems to have been taken aback at the length of time, a full half a year, that elapsed between the eclipse and the announcement. In August (*CPAE*, vol. 19, doc. 93), he was already surmising that the delay in making the results public was because the British astronomers "may be waiting about half a year in order to take comparison exposures of the relevant sky region with the same instrument." He must have been on tenterhooks waiting until November! In fact, Eddington did not wait six months for comparison plates, but it did take months of work to complete the data analysis and prepare everything for publication. Einstein knew about the British plans to test his theory at the eclipse from late 1917, when reports of the expedition began to appear in German journals. His own estimate of the time needed for analysis was much too short. Writing to his mother in June, he expected the final results in six weeks (*CPAE*, vol. 9, doc. 61)!

In was not until September that there was any news at all. Eddington and Cottingham gave a presentation on the eclipse to a meeting of the British Association for the Advancement of Science in the middle of the month, and a Dutch physicist attending the meeting heard Eddington claim that his own result from Principe fell between the two results predicted by Einstein (the half- and full deflection). He passed this on to Lorentz, who quickly sent a telegram to Einstein on September 22 (Einstein 2004, 35; *CPAE*, vol. 9, doc. 110). This was the telegram that Einstein showed Ilse Rosenthal-Schneider. In a follow-up letter, Lorentz told Einstein that though Eddington had spoken tentatively, since the analysis was not final, "the reality of the phenomenon [of light deflection] was established beyond doubt" (*CPAE*, vol. 9, doc. 127). On the basis of the communication from Lorentz, Einstein published a short note on the result in the leading German science journal, *Die Naturwissenschaften* (Einstein 1919). It is worth pointing out that the discovery that light has weight, which was all Eddington claimed at this point, was taken by Einstein as a strong personal vindication. As he commented to his mother about this time, he was "very frequently asked about the result, both in conversation and in writing" (*CPAE*, vol. 9, doc. 99). Clearly, everyone in Einstein's circle awaited the news with bated breath.

Two Master Science Popularizers: Eddington and Dyson

We have seen how Eddington drew fire from some philosophers for his trichotomy, his framing device for the eclipse results. If labeling the half-deflection as the Newtonian result was misleading, does this mean he should have abandoned theory testing altogether? Some philosophers argue that an acceptable theory was at hand to test against Einstein's, that of the Finnish physicist Gunnar Nordström. Einstein had heavily contributed to Nordström's theory so that it counted as a legitimate metric theory of gravity with none of Newtonian gravity's pathologies in the age of relativity. It was easy to ignore this theory since it was not very good at passing tests anyway. It failed the eclipse test because it predicted no light deflection, and it had earlier failed to pass the Mercury perihelion test as well. But it did form a contrast with Einstein's theory for the eclipse test since Nordström's theory predicted no light deflection at all. So is this what Eddington should have done? Should he have billed his test as a showdown between Einstein and Nordström and explained to the audience that Newton had already been slain offstage? As a philosophically correct course of action, this may have had its virtues, but as compelling narrative, it is obviously lousy. It is like telling the story of "Little Red Riding Hood" but leaving the wolf out of the final scene at Granny's house. Perhaps, the wolf has sensibly fled with Granny, leaving Red Riding Hood standing there alone when the woodman walks in, chats for a bit before realizing he has no purpose in the story anymore, and wanders out of the scene. It is simply not narratively satisfying to kill off the major villain in a voiceover and allow a minor character to mosey around before being inevitably removed from the scene. That would have been a good way for Eddington and Dyson to lose their audience. It is hard to imagine Whitehead comparing such a presentation to a Greek drama!

The narrative qualities of the trichotomy must have helped spread the word about the eclipse results much more effectively than mere numbers. Unless Eddington or Dyson composed a song, it is difficult to think of a more suitable framework for effective oral transmission. Three possible results feel right, as in the story of "The Three Little Pigs" or the "Three Billy Goats Gruff." You have one little pig who gets knocked off right at the outset, another little pig who almost makes the cut, and one little pig who emerges unscathed. It is true that Eddington could have tried to attach Nordström's name to the first result in the trichotomy, assuming that he even knew about this result, which was published not by Nordström but by Einstein and a young colleague during the war when Eddington had no access to German journals. In fact, this probably was the case since before his departure, he discussed how surprising it would be if the null result were correct. This would have been a natural place to at least mention Nordström's theory, had he actually been aware of it.

In fact, all three results were due to Einstein, when you think about it. As we have seen, the so-called Newtonian result was just as much his as the general relativity result. It is true that the null result was the accepted one, but only by default. Einstein was so brilliant that he forced people to think about something that had never occurred to them, and once they thought about it, they realized they had no idea what would happen. The old ether theory, which might have been consistent with the null result, was dead, or at least terminally ill, by 1919. The only coherent theory that actually predicted the null result was Einstein's formulation of Nordström's theory, so the null result really belonged to him, just as the other two results did. So in the context of 1919, Einstein had not only put two of the three elements of the trichotomy on the table but had also played a major role in the only effort to make the null result respectable. It is a rare moment in the history of science when we have a test between three predictions all made by the same man!

Einstein being brilliant enough to think about things in a way no one had dreamed of before makes it all the more extraordinary that the news that he had overthrown Newton met with such an enthusiastic and favorable reception. It is tempting to give the credit here to Eddington and Dyson, two men who were well known in their day as gifted public expositors of science. Alistair Sponsel has written about the effectiveness of their campaign to bring their results before the scientific community and public. Some modern commentators have spoken about the publicity surrounding the eclipse results as if something were faintly disreputable about it. Surely, if Dyson and Eddington were so carefully presenting their results to get the desired response from their audience, this implies that they were being underhanded? Sponsel disagrees, and I do so even more vehemently. There is a tendency to presume that the truth needs no advocate; that only someone unsure of the scientific basis of an argument would trouble to engage with the media and the public so vigorously. I feel that an idea or opinion that is not defended is one that will go unheard and ignored. It is silly to pretend that, somehow, members of the public, or even many scientists, would seek out Dyson and Eddington's report and carefully study it. Instead, the news had to be brought to them in a way that made it comprehensible. Of course, this would involve framing it in particular ways, which Dyson and Eddington hoped would give people a sense of why Einstein was probably right. Does this mean that the other viewpoint went unheard? No. Many opponents of relativity theory had their say. Dyson and Eddington made some effort not to be overly slanted in their presentation, but it is silly to imagine they were somehow obliged to mute their own rhetoric in order to let people make up their own minds.

Even before the expedition, a notion that there was something important about this eclipse seems to have gotten around. The edition of *Punch* for May 14, 1919, contained the following item, part of a series of humorous news snippets culled from fictional newspapers: "Sun Eclipse in May, Wireless Operators' Help Asked—Daily Paper 'We Ought All to Put Our Shoulders to the Wheel and Make This Victory Eclipse a Big Thing.'" So it is perhaps not surprising that by November the London *Times* was anxious to cover the announcement of the results. Some modern writers try to give the impression that Dyson and Eddington were masterly manipulators of the media. I am not completely convinced of this. They certainly proved to be quite savvy media operators, as Sponsel shows, but the media probably took their cue from a wider sampling of expert opinion. Sponsel's most interesting revelation concerns the caution with which Dyson and Eddington approached expert opinion. It turns out that one presentation of the results actually preceded the joint meeting. That was one made by Eddington before the $\nabla^2 V$ club in Cambridge on October 22. This club existed to facilitate the discussion of mathematical physics among a limited number of Cambridge scholars. This was its first meeting since mid-1916, according to Sponsel (2002, 456), and it came just the day before Dyson moved to the Royal Society that the joint meeting be held on November 6.

Sponsel gives the minutes of the club describing the meeting, which took place in the rooms of Ebenezer Cunningham, one of Cambridge's experts on relativity theory:

> The President [this was Cunningham] called on Professor Eddington to read his paper on "The Weight of Light." The paper dealt with the attempt to verify experimentally Einstein's Theory of Gravitation based on the Principle of Relativity . . . A general discussion followed; the President remarked that the 83rd meeting was historic and that the results announced there would probably

involve the changing of the name of the club from $\nabla^2 V$ to something more barbaric. (Sponsel 2002, 456)

This refers to the fact that the club's name was the Laplace operator that features in a particularly compact expression of Newton's gravitational law. Therefore, the club's very name was in homage to Newton's theory and would now need to be altered with the advent of general relativity—presumably, to something like the Einstein tensor. So, as Sponsel points out, Eddington could now be assured that even in Cambridge, home of Newton, theorists were prepared to accept the importance of the results and acknowledge that they could lead to the overthrow of Newton's theory. This certainly must have fortified him and Dyson, who certainly would have feared some conservative backlash against their report to the joint meeting. We know from Eddington's letter to Dyson dated October 21 that he and Dyson planned to meet in London on the twenty-third, so he may have reported to Dyson the favorable reception he had received from the younger generation of Cambridge theorists (RGO Archive 8). We also know from correspondence between Dyson and Fowler, the secretary of the RAS (October 10), that they had previously settled on October 23 as the date upon which Dyson would move to the societies that the joint meeting take place shortly after (RGO Archive 8).

However emboldened Eddington may have been by the positive reaction from the $\nabla^2 V$ club, Dyson, who was known for his conviviality on social occasions, must have been most pleased by the enthusiasm shown by members of the RAS Club after the joint meeting. This dining club is as old as the RAS itself and meets regularly for dinner after RAS meetings. Dyson told the club that "Newton wanted 0.87 seconds of arc, Einstein 1.74, but Cottingham wanted to double this amount." Cottingham, in response, protested he had never doubted Professor Eddington's sanity but "all the same, I must confess I was very pleased when Eddington said to me one morning, after making a few plates measurements from those we developed on the island—'Cottingham, you won't have to go home alone'" (Wilson 1951, 193–94). The dinner closed with a drinking song proposed by the Joint Permanent Eclipse Committee's (JPEC) H. H. Turner. It will be noted that he deftly turned things around and put the old-guard defenders of the ether in the dock as the ones opposing Newton's theories.

> The idea that light has mass, we got
> From Newton, its bequeather,
> They "waved" aside his views as rot,
> And filled all space with ether.
> But once more comes a change of scene,
> The ether's swept away, Sir,
> And space is emptied now as clean
> As the bottle of yesterday, Sir.
>
> The mysteries of time and space
> Demand investigation.
> That space durates and Time has place
> Is now the explanation.
> But *we've* long known that obvious link,
> I hardly need to say, Sir,
> 'Tis ruled that whensoe'er we drink,
> A bottle shall space a day, Sir.
>
> We cheered the Eclipse Observers' start,
> We welcome their return, Sir,
> Right manfully they played their part,
> And much from them we've learned, Sir,
> No toil or pains they thought too great,
> Nor left Einstein unturned, Sir,
> Right cordially we asseverate
> Their bottle a day they've earned, Sir.

(WILSON 1951, 194)

Cambridge theorists like Cunningham could be expected to understand the theoretical issues that so inclined Eddington toward Einstein's theory. The harder sell was really to Dyson's fellow astronomers. And so we find that at one RAS meeting shortly after the joint meeting, amid presentations by the theorists, including Eddington, explaining what relativity was all about, Dyson took the trouble to bring along a slide showing an enlargement of one of the eclipse photographs. He remarked, "I wish to remove misconceptions which some Fellows may have with reference to the observations. (Showing on the screen an enlargement from one of the eclipse photographs of the stars κ_1 and κ_2 Tauri, he remarked that the diameters of the images were 4″ and 3″ respectively, while the displacements were quantities of 1″ and less.) In linear measurement the quantities dealt with are of the order of 1/100th of a millimetre. These displacements are small, but Astronomers familiar with stellar photographs will, I am sure, support the statement that quantities of this order are readily measured, if only good photographs are secured" (Jones 1919a, 106).

So Dyson emphasized his own confidence in his ability to make the call on the falsification of Newton's great theory. And this level of confidence from an astronomer with impeccable credentials did have an effect—an effect that was compounded with each passing month as he took the trouble to further engage with his fellow astronomers, sending out eclipse photos, demonstrating measurement techniques, and very publicly making plans to repeat the test at the next eclipse. The question was, given the skepticism of so many of his fellow astronomers, exactly how many of them he could convert. Each conversion that did occur made others more likely as the discourse became ever more friendly to Einstein's ideas. To achieve acceptance, a theory has to cease being something that is tested and become instead something that is used—a tool for research.

The idea of a theory as a tool, which is one aspect of Kuhn's term, *paradigm,* is really illustrated by what has been called "the conversion of St. John" (Hentschel 1993). In 1919 Charles St. John was still claiming that he could not see how to reconcile Einstein's solar redshift prediction with the observations. He was coming close to saying that general relativity had been falsified, and many people, like Guillaume and Silberstein, couched his results in these terms. Einstein himself admitted that the theory could not survive such a falsifying test. Yet a few years later, St. John completely reversed course and offered up an interpretation of the solar redshift in which Einstein's theory played an integral part. Is this the bandwagon effect? Was

St. John now convinced that general relativity was the right answer, and he had to make sure his work produced the right result? That is certainly the impression that some commentators have given. There is a flavor of it in Earman and Glymour's influential paper on the solar redshift test of general relativity (Earman and Glymour 1980), to which their eclipse paper is a companion. In fact, it is arguably this paper, rather than their eclipse paper, that had such an impact on the thinking of authors like Collins and Pinch. After all, surely it tells us something important about the way science is done if the work of Dyson and Eddington could play such an influential role in transforming the way solar astrophysicists like St. John and Evershed approached their efforts to test general relativity.

Of course, this kind of conversion bias can play a role in science. They say there is no zealotry like that of the convert. But actually, something else is going on here also. The solar redshift measurements of St. John, Evershed, and others were never primarily concerned with theory testing. They predated Einstein's prediction of the gravitational redshift. Here is an example of what Kuhn would call normal science in action. In normal science one is not concerned with trying to test theories or overthrow existing paradigms. One is problem solving. The solar redshift was a long-standing problem in solar astrophysics. The aim of this research was to explain the origins of the observed shifts in the absorption lines in the Sun's spectrum. These shifts were usually redshifts, but the size of the shift varied from line to line and from location to location on the Sun's disk. Only at the limb of the Sun was it always a redshift. In the center of the Sun's disk, it was actually sometimes a blueshift. Such a blueward shift was impossible to explain using Einstein's theory; neither did Einstein predict the variability between lines and locations. It was this qualitative difference between Einstein's predictions and the observations that motivated St. John and others to argue, around 1919, that they did not see much evidence that Einstein was right.

But after 1919 there was independent evidence that general relativity was right. After 1922 that evidence greatly strengthened. Now St. John was obliged to consider more seriously the claims of Einstein's theory. He was in no greater position than before to provide a precise test of the theory, but he was in a position to do something more interesting. He could use the theory in a model explaining the solar redshift as the superposition of two different effects. It had always seemed likely that the line shifts in the solar spectrum were due to Doppler shifting. If columns of gas were rising and falling in the solar atmosphere in convection currents, then this would produce redshifts for the falling gas (moving away from Earth) and blueshifts for the rising gas (moving toward Earth). What was difficult to explain was why most of the shifts were redward. Surely, there ought to be as much rising gas as falling gas. But if one assumed Einstein's effect was always there, then the sum of the gravitational redshift and the Doppler shift due to convection would be clearly redward and only occasionally blueward. Near the limb of the Sun, where one views the columns of gas from the side, the effect would be mostly gravitational, and one should see a more consistent redshift of about the amount predicted by Einstein. From 1923 this is the picture St. John presented, after a monumental study of many solar lines. The real puzzle is why this explanation took so long to emerge. Evershed had been struggling with the "limb effect" for years. He should certainly have adopted Einstein's theory much earlier. That he did not is undoubtedly because of the outright skepticism most astronomers expressed about Einstein's theory. It is important to keep in mind that the mental barriers to accepting relativity theory were high. The focus on Eddington's predisposition in favor of the theory has clouded modern commentators to the reality of the situation in 1919. The mental bias at that time was heavily the other way. Only after Dyson came out in favor of general relativity did the weight of scientific opinion begin to shift and make it even possible to start thinking about general relativity in an unbiased way. Another aspect of the problem is that interest in general relativity focused attention on it in a way that demanded a response from observers like St. John. His "conversion" of 1923, as Hentschel has shown, was the result of intensive and painstaking measurement by someone acknowledged as a master of the field. The difficulty of measuring such tiny spectral line shifts over so many different parts of the spectrum placed the problem, as with the eclipse experiment itself, close to the limits of what could be achieved with the instrumentation of the day.

By 1923 St. John was no longer testing relativity—he was making use of it. When a theory has been tested and adopted, then one uses it for its explanatory power. There had always been puzzling aspects of the solar redshift that were difficult to explain. Looked at from the post-1922 vantage point, St. John and others realized that accepting general relativity was right meant that they could provide, for the first time, a qualitatively convincing explanation of the solar redshift. This still has aspects of a test. If a theory is proving useful and is making measurements more comprehensible, this certainly is some evidence of its essential rightness. But it is not a conventional test of the theory. We have passed into a new stage of research in which the correctness of the theory is assumed, and it is used to interpret the results at hand. The solar redshift never did properly test general relativity, though, happily, improved experimental techniques meant the experiment was done very accurately in the laboratory in the second half of the twentieth century.[2] An important point to notice here is that Popper's interest in demarcation may sometimes get in the way of inquiring about actual scientific practice. Popper would like us to learn the moral that scientists are less impressed by explanatory power and more interested in precise quantification than other people. But this may not always be the case, even when considering such precise and painstaking work as St. John's solar redshift measurements. In this case general relativity was valued for its explanatory power, in addition to its precise predictions. A satisfying but partly qualitative understanding of the solar spectrum emerged in the early twenties. Because the velocities of the rising and falling columns of gas were unknown, this explanation could not be accompanied by a precise numerical test. But it was clearly valued, nonetheless. Scientists are as adaptable to the realities of what is possible as anyone else. Sometimes explanatory power is what one has, and it can form the basis for further measurements leading, hopefully, toward increasing precision.

The RAS Gold Medal Affair

Hard on the heels of the excitement of the joint meeting came a startling development. The Royal Astronomical Society voted to award Einstein, a German, its annual Gold Medal award. This is one of the most distinguished honors that an astronomer can receive. Many people found it astonishing that it would be given to a German such a short time after the war. German science was still broadly boycotted at this time, and it seemed, for a moment, that the proponents of international scientific reconciliation had swept all before them. Eddington was already conscious of the positive effects in this way when he wrote his first letter to Einstein not long after the announcement. He began:

> Dear Professor Einstein, It was a great pleasure to receive your letter from Holland, and to be in personal communication with you. I was sorry not to be able to come over to meet you. Our results were announced on Nov. 6; and you probably know that since then all England has been talking about your theory. It has made a tremendous sensation; and although the popular interest will die down, there is no mistaking the genuine enthusiasm in scientific circles and perhaps more particularly in this university. It is the best possible thing that could have happened for scientific relations between England and Germany. I do not anticipate rapid progress towards official reunion, but there is a big advance towards a more reasonable frame of mind among scientific men, and that is even more important than the renewal of formal associations. (*CPAE*, vol. 9, doc. 186)

He chivalrously added a note regarding the ambitions of German men of science to establish this German theory.

> Although it seems unfair that Dr. Freundlich, who was first in the field, should not have the satisfaction of accomplishing the experimental test of your theory, one feels that things have turned out very fortunately in giving this object-lesson of the solidarity of German and British science even in time of war.

This letter was written on December 1, 1919. That same month came evidence that Eddington had been too pessimistic in imagining that official moves toward reconciliation were some way off. In December the RAS voted to award Einstein the Gold Medal. Eddington was so excited that he made an unfortunate error. A Quaker acquaintance of his was departing for Berlin, and Eddington asked him to take Einstein news of this remarkable event. As a consequence, he had to write a difficult letter the following month:

> Dear Professor Einstein, I have just heard from Mr. Ludlam that he has seen you. I am sorry to say an unexpected thing has happened and at the meeting on Jan. 9 the Council of the R.A.S. rejected the award, which had been carried by quite a large majority at the previous meeting. The facts (which are confidential) are that three names were proposed for the Medal. You were selected by an overwhelming majority in December. Meanwhile the "irreconcilables" took alarm, mustered up their full forces in January, and managed to defeat the confirmation of the award in January. So for the first time for about 30 years no Gold Medal will be awarded this year! I confess I was very much surprised when the motion was proposed and carried originally (it was proposed by two men who during the war have been violently "patriotic"); but until a day or two before the January meeting we all regarded the confirmation as a matter of course. I did not write before as I was doubtful if Ludlam would see you. I am sure that your disappointment will not be in any way personal; and that you will share with me the regret that this promising opening of a better international spirit has had a rebuff from reaction. Nevertheless I am sure the better spirit is making progress. Mr. Ludlam mentioned the possibility that you might come over about May. We should be delighted to have you and your wife to stay with us at the Observatory; and you would get a most cordial welcome in Cambridge, and especially at my own college, Trinity (the college of Newton). You will see that there would be some awkwardness in visiting the Royal Astr. Society after what has happened; although many astronomers would be delighted to meet you.

If you should be unable to come on to England, I would make every effort to meet you in Holland. I am sorry that I sent the message to you and troubled you prematurely; but Mr. Ludlam's visit seemed such an excellent opportunity. Yours very sincerely, A.S. Eddington. (*CPAE*, vol. 9, doc. 271)

English Quakers were at the fore in providing relief to those going hungry in postwar Germany, which was still recovering from the effects of the British blockade. The blockade had continued until the Treaty of Versailles was signed in mid-1919, and the German economy was naturally slow to recover. Ernest Ludlam was engaged in distributing food to the needy and had visited Eddington before departing for Berlin in December.

As for Eddington's irreconcilables, here is what one of those had to say, writing to Dyson to advise him of his intention to vote down the confirmation of the Gold Medal award. This confirmation required a three-fourths majority, whereas the original vote required only a majority. The writer was none other than Eddington's Cambridge neighbor Newall, director of the solar observatory, and he was clearly engaged in a two-front war against German science. He was looking, on the one hand, to overthrow the experimental verdict in favor of Einstein and, on the other hand, the popular verdict, as expressed in the original Gold Medal vote.

My Dear Dyson, I ought to have written you long ago to thank you for so kindly sending me the details of method of reduction of the eclipse plates but I have been so immersed in atmosphere calculations [most likely calculations to show that the light deflection could be due to an extensive solar atmosphere], that I have let the days go by. Anyhow now I send you hearty thanks for your letters.

I have been thinking a great deal about the medal + the more I think of it, the less I like the situation.—I very much regret that we did not choose R.[3] for it seems to me that his contributions are of a deep import in astronomy. However the choice fell otherwise.

Much as I hate the idea of opposing confirmation of a selection made in due form by a majority of the council, yet I hold that this award is premature scientifically and in a political sense forestalls the expression of what may be desired by the nation in the resumption of relations with the Central Powers. And I hold this view so strongly that I must vote against the confirmation. I don't want to do so without letting you know.

Ordinarily we would wish to give a personal welcome to the medalist. In this case we not even know by what channel we can send the award. I am writing to Turner and to Fowler. (Newell to Dyson, January 4, 1920, RGO Archive 8)

Turner was, in fact, one of those who had nominated Einstein, the other being Jeans. Recall that Turner had, during the war, been vocal in proposing that German scientists be ostracized after the war, and he and Eddington argued about this in public. Now, it seemed, he had had a change of heart, or at least regarded Einstein as an acceptable German. After three nominations were put forward at the November meeting, held in the wake of the famous joint meeting, a majority had favored Einstein at the December meeting. But his candidacy failed unexpectedly at the final hurdle in January. News of the affair leaked to the German press, where it was assumed that the British had initially mistaken Einstein for a Swiss, later denying him the honor when Einstein declared he was German (*CPAE*, vol. 7, 370). In fact, this confusion about Einstein's nationality may be what Newall referred to in his remark about "channels." This version of events was quite untrue but does reflect Einstein's famous statement to the *Times* of London in November:

By an application of the theory of relativity to the taste of readers, today in Germany I am called a German man of science, and in England I am represented as a Swiss Jew. If I come to be represented as a bête noire, the descriptions will be reversed and I shall become a Swiss Jew for the Germans and a German man of science for the English. (*CPAE*, vol. 7, doc. 26)

The *Times*' rejoinder was to note that "in accordance with the general tenor of his theory, Dr. Einstein does not supply any absolute description of himself" (*CPAE*, vol. 7, 215).

In legal terms, Einstein, though born in Germany, had renounced his citizenship as a teenager, probably in order to avoid the draft. He had subsequently taken out Swiss citizenship and only acknowledged German citizenship after coming under considerable pressure to do so upon receiving a much bigger award, the Nobel Prize, a few years later. He resigned that citizenship for a second time in 1933 after the Nazis' rise to power.

The boycott of German scientists did indeed turn out to be a serious and long-term matter. In 1921 Einstein was the only German scientist invited to attend the prestigious Solvay conference. The ambiguity surrounding his citizenship seems to have played a role then, as the "only German invited is Einstein, who is considered for this purpose to be international" (*CPAE*, vol. 12, intro, 30). So it is quite plausible that men like Turner simply viewed Einstein as not really German. The fact that he had renounced German citizenship in protest at the country's war machine and did in fact travel on a Swiss passport seems to have made a real difference in the years of bitter feelings after the Great War.

As it turned out, Einstein did not attend the 1921 Solvay conference. He went to America instead to support the cause of establishing the Hebrew University of Jerusalem. During the war Einstein had become aware of the horrible strength of European nationalism, and seeing the treatment of Jewish refugees from Eastern Europe in Berlin made him conscious of the plight of those he viewed as his own people. He became a Zionist and agreed to visit the United States at the invitation of leaders of the Zionist movement. In America he received a tumultuous reception, along with important Zionist leaders, with crowds lining the street to catch a glimpse of him on his arrival in New York. He seems to have been a bigger draw than the statesmen accompanying him, for Chaim Weizmann, the Zionist leader, was asked on arrival by the press for his opinion about Einstein's theories. He supposedly replied, "During the crossing Einstein explained his theory to me every day and by the time we arrived I was fully convinced that he really understands it."[4] For a man who would have been unknown to most Americans only two years earlier, it was a sudden and dramatic ascent to fame.

On his way back from America, he decided to finally pay a visit to England and left New York aboard the White Star Line's steamship *Celtic* in June 1921 (*Cork Examiner*, June 9, 1921). When I saw the identity of his ship, I was quite excited, as it regularly called at my hometown of Cork in southern Ireland on its route to Liverpool.[5] Had Einstein perhaps visited the harbor of Cork on this voyage? It turns out that he could not have. The War of Independence was raging in Ireland, nowhere more fiercely than in Cork, and the British government had forbidden ships to land in the harbor (*New York Times*, May 30, 1921). Einstein passed on to England, where he received a very warm welcome. A few years later, he returned again to finally receive the Gold Medal. Ironically, by that time he may actually have traveled on a German passport! For a while Einstein's became a familiar face in England. In the early 1930s, he spent time in Oxford, at the invitation of F. A. Lindemann. This was part of his effort to stay out of Germany during the Nazi rise to power. His fame had, by 1921, grown to prodigious proportions. The man who had done so much to make him so famous, Eddington, noted the difficulty in getting to spend time with him when he wrote to Lindemann in 1932 inquiring about Einstein's plans: "I remember the first time he came to England I had invitations to dinner to meet him in London and Manchester on the same night. This is doubtless explicable by the Principle of Indeterminancy; still I hope on this occasion Ψ will have a more constrained distribution."[6]

Of course, it was Einstein himself who developed the notion that the motion of particles is governed by a probability function (Ψ, here) in the form of a wave, which he called the *ghost field*. It accompanies particles as they move and tells observers where, statistically speaking, the particle is likely to be. At this time, a similar indeterminancy hovered over the question of whether Einstein was really German or Swiss, or partook of a dual nature. Such is the absurdity of the debate over Einstein's nationality (and he would have certainly called it absurd) that we must leave the final word to *Punch:*

A patriot fiddler-composer of Luton
Wrote a funeral march which he played with the mute on,

To record, as he said, that a Jewish-Swiss-Teuton
Had partially scrapped the Principia of Newton.
    (*Punch*, November 19, 1919, 422)

Later Eclipses

Why have many modern physicists, philosophers, and other commentators sounded so critical of Eddington's work on the 1919 eclipse? One reason is that they want to counter the impression that this one experiment confirmed general relativity as the correct theory of gravity. The idea of the *experimentum crucis*, the crucial experiment that convinces everyone of the truth about nature, is a romantic ideal. In one movie account of the 1919 eclipse data analysis, Eddington conducts it for the first time at the joint meeting of the two societies in front of the audience! The truth is that many people were unhappy with and skeptical of Eddington and Dyson's results. Even those who accepted the results wished to see them confirmed by further observations. Replication is a cornerstone of science, and so a single experiment can probably never be the final arbiter in theory testing. The fact that no one was going to be in a position to repeat the experiment for several years is what set the eclipse results apart. In this sense Eddington and Dyson's work is unusual. There was a considerable period of time in which they were the only experimenters who had falsified the old gravitational theory, and yet it is true that many people did accept, during those few years, that Einstein's theory was likely to replace the old theory. But there can be little doubt that it was a provisional state of affairs. By late 1919, astronomers were already making their plans for the next available eclipse.

Unfortunately, the next four eclipses after May 29, 1919, were either partial or annular and therefore unsuitable for the light deflection measurement. The eclipse of October 1, 1921, was total, but the track of totality ran across the Southern Ocean and Antarctica, making an expedition to observe it impossible. The first eclipse of 1922 was annular, leaving September 21, 1922, as the first opportunity to confirm, or disconfirm, the result of 1919. As Curtis put it in 1920, "During the next 10 years we are going to have available, with luck, only 18

minutes for further tests of the deflection effect, and . . . the Australian eclipse . . . is one of the best of the lot" (Crelinsten 2006, 196).[7]

By then the war was long over, and eclipse expeditions could
be mounted from several countries, as we learn from the definitive
account given in Jeffrey Crelinsten's book *Einstein's Jury* (Crelinsten
2006). But to the fore were the Lick and Greenwich observatories,
who were determined to put their best foot forward after the handi-
caps imposed on them by the war in their planning for the eclipses
of 1918 and 1919. Campbell, in particular, was stung by the failure
of his 1918 effort. He felt he had committed the "mistake of having reported, though guardedly, on Curtis' results at the meeting
of the Royal Astronomical Society" in July 1919 (Crelinsten 2006,
201). He spent a great deal of his time continuing to work on the
Goldendale data. Curtis left Lick for a new position during 1920
while the Goldendale data reduction remained unfinished. Curtis
always felt that the results did not confirm Einstein's theory, but now
Campbell himself took over the measurements. As he worked on it
more and more, he was less than impressed with Curtis' efforts. He
wrote, "Computing is not your strong point . . . the sheets contained
so many errors that we were led to regard your final results as fairly
representative of your original measures, because the computational
errors were so numerous, as to be themselves subject to the law of
accidental errors!" (Crelinsten 2006, 193). At the risk of sounding
like a supercilious physicist, it is worth mentioning that Curtis began
his academic career as a professor of classical languages before starting
all over again as an astronomer (Campbell 1920).

Campbell now had a stroke of good luck. He had a young astronomer visiting Lick on a fellowship and quickly offered him a position
as an assistant when he learned Curtis was leaving. Robert Trumpler
had two advantages over Curtis in working on the light deflection
problem. The first was that he was familiar with relativity theory.
He was Swiss and had been educated, like Freundlich, at Göttingen in Germany and had the mathematical training to understand
relativity theory very well. Perhaps more importantly, his expertise
was in astrometry, having worked, like Dyson, primarily on parallaxes and proper motion. Curtis, by contrast, seemed to have had no

particular experience in this area of astronomy. His research focused on the spiral nebulae, and he was among those who showed that these are actually galaxies beyond our own. He was one of the two figures in the "Great Debate," which helped establish our modern understanding that the universe contains vast numbers of galaxies like our own. But his skill set was not suited to the demands of the Einstein test. In Trumpler, Campbell was fortunate enough to have an astronomer arrive on his doorstep with Freundlich's knowledge of the theoretical stakes and Dyson's measurements skills (if not quite his vast experience). He certainly helped himself, in the intervening years, by assembling the best possible equipment for the task.

Meanwhile, the British were to suffer a total systems failure at the 1922 eclipse. The missteps began at the earliest planning stage, with the dependable Hinks failing to think outside the box. The track of totality in the 1922 eclipse crossed the Indian Ocean south of Indonesia and then cut straight across Australia. It made landfall on a desolate stretch of the western coast of Australia known as Eighty Mile Beach.[8] Conditions there were ideal for the eclipse in terms of duration of totality, since it had the longest totality of any spot on land, and climate, since it rarely rained there. But there were no port facilities or railroads at either the coast or inland along the track's path through the desert of Western Australia. In Eastern Australia the eclipse would take place late in the day, with the Sun low in the sky.[9] For these reasons Hinks rejected Australia altogether when choosing eclipse sites. Conventional wisdom dictated that the first step in choosing an eclipse site was to draw the track of totality and then circle places on the track with port or rail infrastructure. Eighty Mile Beach, with nothing whatsoever of this sort, seemed too risky to Hicks. But in making this choice, he condemned the Greenwich expedition to one of the few specks of land available outside Australia, Christmas Island in the Indian Ocean, for the eclipse. Campbell, by contrast, was to end up being rewarded for his typically American willingness to hit the beach. He and his wife, Elizabeth (née Thompson), were expert eclipse planners, and this was to be their crowning achievement.[10] It seems we must also give some credit here to Heber Curtis. As chair of the Eclipse Committee of the American

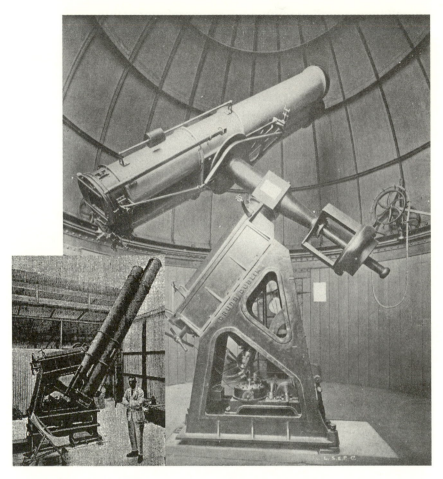

FIGURE 20. The astrographic telescope inside its dome at Greenwich. This is the same dome whose exterior is depicted in figure 2. There are actually two telescopes on the same mount. The nearer tube is the actual astrographic telescope. Notice the flat end of the tube where the photographic plates are inserted. The other tube (mostly hidden in this view) is used as a guide scope, since the astrographic itself has no eyepiece for an observer to look through and is not designed to properly focus light for the human eye. It was designed to focus light only for photographic plates, which were primarily sensitive to blue light. Only the object lens from this telescope was taken to Sobral in 1919. But after the bad experience with the coelostat mirror in 1919, Greenwich shipped the entire telescope to Christmas Island for the eclipse of 1922. The inset shows the instrument on-site in a photograph printed in the *Times* on the day of the eclipse, with the two tubes clearly visible. The person next to the telescope could be the expedition's leader, Harold Spencer Jones, chief assistant at Greenwich. The expedition was completely clouded out in 1922 and obtained no data.

(Main picture courtesy of Graham Dolan; inset is reproduced from the *Times* [London], September 21, 1922, 14.)

Astronomical Society (AAS), he had already reported to Eddington, in a letter dated December 23, 1920, that "from information I have rec'd I do not believe that Wollal, on the ninety-mile beach, is more inaccessible than many stations which have been occupied in the past. In length of totality, and in favorable weather conditions, it seems to me to be better than all other localities. Were I going, I should have chosen that spot" (Crelinsten 2006). Presumably he passed on this preference to his successor as chair of the AAS Eclipse Committee, Campbell.

Campbell persuaded the Australian government to provide him with a schooner, called the *Gwendolen,* to land him directly onto the beach at Wallal Downs, a remote sheep station close to the site of a telegraph post. Campbell's meticulous research had uncovered the fact that schooners were habitually used to deliver supplies and pick up wool from the sheep stations along the beach. In addition, they provided a steamer, the *Governor Musgrave,* to tow the sailing vessel down the coast from the town of Broome to the vicinity of the landing site. Having achieved this planning success, he chivalrously offered places on the ship to his rivals. But by then Greenwich was committed to Christmas Island. Spencer Jones, who led the expedition, found on arrival that the weather was completely unsuitable. The sky was rarely clear, and they were clouded out on eclipse day. Perhaps the habitually unlucky Freundlich's presence on the island was the decisive factor. However, John Evershed, director of the Kodaikanal Observatory, took up Campbell's offer, having been foiled in his initial hopes of observing from the Maldive Islands directly south of his own observatory in southern India. He had not been able to find a suitable vessel to land there. But his equipment failed completely, in particular the sixteen-inch coelostat loaned to him by Dyson. Its driving mechanism did not work properly, and no useful data was obtained. Could this have been the very one that Davidson had used in 1919, about which there had been so much worry? In his report on the expedition, Evershed was scathing about the instruments supplied by British manufacturers, having made trials with several unsatisfactory coelostats (including one made by Grubb). He contrasted this failure with the excellent equipment brought by Campbell, who did not make use of co-elostats, preferring to bring complete mountings for his telescopes. Campbell had experienced a disaster of his own in 1905 when a clock drive broke down minutes before totality (Pang 1996, 31). Landing tons of this equipment onto a beach and hauling it by donkey train a mile inland to the campsite was a mammoth undertaking, but somehow it was pulled off.

Campbell continued his high-stakes approach in designing his observing program. He decided to image as many stars as possible by taking only two exposures during the eclipse, each fully two minutes in length. This left little margin for error, especially since long exposures demand excellent tracking. Clearly, Campbell had total confidence in his equipment, and he had also devoted a lot of time to practice runs and rehearsals in the years leading up to the eclipse. In particular, he made trials of long-exposure photographs of stars near the full Moon, which has a similar brightness to the solar corona. This allowed him to check that he would not fog his plates through overexposing the corona. With meticulous planning, he surmounted all difficulties. The main problem with Wallal turned out to be the impracticality of leaving his equipment in place for months on that isolated shoreline in order to take comparison plates. So he ended up having Trumpler stop off on Tahiti en route to Australia to take the comparison plates. This required them to take check plates from which they would measure the change in scale between the eclipse and comparison plates taken at different locations. Thus, Campbell and Trumpler ended up adopting an identical scheme for obtaining the scale change as that used by the much-maligned Eddington in 1919! Of course, they would have far more stars to work with, thanks to better weather and superior equipment, not to mention exceptional seeing conditions in the drier climate. In the end they took advantage of the fact that they had so many stars far from the Sun on their eclipse plates that they could assume the most distant ones were unaffected by light deflection and therefore useful for determining the scale change.

Several teams had been clouded out on Christmas Island, and Evershed's equipment had failed him at Wallal. That left only three teams with data: the Australian team already discussed, who sent their data to Greenwich for analysis; Campbell's Lick team; and a team from the University of Toronto, who had also landed from the schooner at Wallal. In fact, the Toronto team published first and pronounced in favor of Einstein. But not all experimental tests are equal. Their results showed a lot of scatter, with few stars imaged close to the Sun, and no one seems to have ever taken their data as constituting a decisive test of the theory. Everyone was waiting to hear from Campbell. Like Eddington in 1919, Campbell had no wish to keep his audience waiting. He hoped to analyze the data before he left Australia, but in the end this proved overly ambitious. Nevertheless, the parallels with Eddington continue, according to Crelinsten, because the initial reductions they did produce before returning home were indecisive, falling midway between the half and full deflections (Crelinsten 2006, table 205). After a trip home to Switzerland, Trumpler got down to business with Campbell, and they measured everything again. As in 1919 the data analysis took months, and this time even more people were waiting with bated breath.

As the moment of truth arrived, and with the decision now taken out of his hands, Dyson began to worry that he might have backed the wrong horse in 1919. According to Crelinsten, a colleague wrote to Campbell, telling him, "When I saw Dyson last he said that he would not be in the least surprised if the 1922 photographs did not confirm the Einstein effect. He thought that possibly they in England had stressed Einstein a little too much" (Crelinsten 2006, 205). Dyson went so far as to write to Campbell himself, fishing for news:

The results of the 1919 eclipse were in accordance with Einstein, but it may be that your results of 1922 will not confirm this. If so, will you be kind enough to let me know as early as you can. My reason, is of course, that if there is disagreement we must regard the point as unsettled and every endeavour must be made to test the matter again next September.[11] In that case it will be desirable for us to send an expedition but it will hardly be necessary if you confirm the 1919 results in which case the question may be regarded as settled. (Crelinsten, 205–6)

Dyson was as good as his word when Campbell made his announcement, stating:

> I don't think there is "any possible shadow of doubt" about the correctness of Einstein's prediction of the deflection of light, whatever difficulties may be found with the rest of his theory. It is hardly likely that anyone will be coming from this side for the eclipse in California. (Crelinsten, 213)

Turner, writing in the *Observatory*, definitely felt that the English astronomers, as a result of their verdict in 1919, had tied their reputations to Einstein's prediction. Now they could rest on their laurels.

> The English observers were determined to go on even if the American verdict had been against them; there are no Courts of Appeal or Houses of Lords in Astronomy for getting verdicts reversed—the only possible recourse is to hammer on as before with renewed vigilance for possible flaws in the evidence but for this they were quite prepared. It is, however, a considerable relief to find that the necessity for this further campaign is now removed, and if any English observers are able to visit America this year it will no doubt be merely as a return to their old love, the Corona. (Quoted in Crelinsten 2006, 213)

In the end Campbell made good use of his data once he had the leisure to review things at home with Trumpler. His success in obtaining images of over seventy stars on his plates, which had a very wide field of view, now paid off. The outermost stars on his plates were so far from the Sun that they could be assumed to have only a small Einstein deflection. Accordingly, any shifts in position for them could be assumed to be due to the change in scale. So in addition to the check plates, Campbell and Trumpler could control for any scale changes on the eclipse and comparison plates. For reasons like these, there are few quibbles with the results they finally announced. Their average for the measurements from their plates—most plates measured independently by both men—came out to 1".74, compared to Einstein's prediction of 1".75. Is there any suggestion of a bias here? Again, Campbell seems to have been a skeptic all along. Asked about what his hopes had been regarding the test of general relativity, he replied, "I hoped it would not be true" (Douglas 1957, 44).

Although neither Dyson nor Campbell returned to the Einstein test, Freundlich was still determined to press ahead. He only obtained data in 1929. He had already had his falling-out with Einstein not long before the 1922 eclipse. At the end of 1921, Einstein had resigned from the board of the foundation set up to fund Freundlich's experimental tests of his theory, citing "an irreparable discord" with Freundlich (*CPAE*, vol. 12, doc. 340). Although Freundlich interpreted his results as in conflict with Einstein, the fact that his measurements were even larger than the general relativity prediction meant that it was largely read as a further falsification of Newtonian gravity. Eddington, as usual, saw the humor in the situation. When Freundlich described his result of a larger than Einsteinian light deflection to the RAS in 1932, Eddington repeated Dyson's remark to Cottingham—that Eddington shall go mad in the event that the result is larger than Einstein predicts—to the delight of the audience (Chandrasekhar 1987, 117).

14

The Unbearable Heaviness of Light

What happens in science when scientists disagree about the results of their measurements? If we all agree that experiment is the ultimate arbiter, then there is a problem because not everyone can perform the experiment. Most people, including most scientists, must decide whom to believe without the benefit of their own experience performing precisely that experiment. One major factor is the reputation of the scientists making the claims. If I believe professor X is a reliable and expert experimenter, I will take his word over that of professor Y, whom I know to be inept and untrustworthy. But my colleague professor Z is a friend of professor Y and inclined to take his part. Now what? Well, if Z now does his own experiment, which supports the work of Y, then it may be that the number of scientists who trust either Y or Z is greater than the number who trust X. Soon, I may find myself in a minority. Take the case of Freundlich, who, a decade after the original light deflection experiments, finally gave his own verdict. That he was late on the scene was not his fault, but it did pose a problem. He did not enjoy the kind of reputation that would allow him to singlehandedly overthrow what seemed to have been settled by several earlier experiments. Freundlich wanted to upset the consensus in favor of Einstein. But his results appealed little to the Newton faithful and even less to the true believers in the ether. He thus lacked a constituency, and he had no one else who claimed to have obtained a similar result to the one he had found.

It was obvious to Freundlich that he would need some supporting results if he were to win his case for a large value of the light deflection greater than two arc seconds at the limb of the Sun. He felt that the four-inch measurement from Sobral, which was almost that big, counted for something. But other results seemed to go against him. He did what any good scientist would do; he redid the analysis of his rival's data and discovered that they actually agreed with him after all! In his paper Freundlich claimed that, as reanalyzed by him and his collaborators, all previous eclipse measures actually returned a value of better than 2" at the limb. In particular, he claimed that this was true of the Lick 1922 results. Obviously, this called for a response from Trumpler. He returned the favor, not only rebutting Freundlich's claims about the Lick 1922 data reduction but also redoing Freundlich's analysis of his own data from the 1929 eclipse and finding that these results actually agreed nicely both with Trumpler's own results and with Einstein's prediction of 1."75 at the limb.

So what value can we place on the dependability of science when two professional astronomers can so cavalierly dispute each other's analysis of their own data? Trumpler's complaint about Freundlich's analysis of the Lick 1922 data was that Freundlich did not respect the weighting of the different stars used by Campbell and Trumpler. The Lick team gave some stars less weight than others in a complex way that Freundlich simplified, for instance, by simply throwing out stars with very low weight. In one way you could argue, with Freundlich, that fine choices of weighting were simply a way to sift the data to get to the desired result. On the other hand, Trumpler could respond that only the scientists who actually handled the instruments and the data are in a position to make such fine judgments. Reasons for weighting stars differently in 1922 included underweighting stars that did not appear on every plate, which seems quite reasonable. The details are complex, but at any rate we may note that it was not just Dyson and Eddington who failed to make "the logic of the situation . . . entirely clear" in their data analysis, to throw Campbell's own charge back at him. The logic of the situation is often not entirely clear in science, at least to those looking in from the outside. In this sense, science, especially experimental science, is a bit like a bad joke. Sometimes, you just had to be there to get it.

Trumpler's complaint about Freundlich's own 1929 data was that he had made use of a unique method of obtaining the scale change. Freundlich superimposed a reseau, a grid-like pattern, on his photograph plates using an actual grid positioned within the instrument's optics, with the aim of providing an absolute definition of scale. But Trumpler pointed out that the grid might have expanded with temperature and therefore changed size during the eclipse compared to at night when taking the comparison plates. He redid the analysis ignoring the reseau, deriving the scale change from the plates themselves as he had done with Campbell in 1922 and as Dyson had done in 1919. He claimed to get 1."75, precisely on Einstein's prediction. Obviously, in order to do this, Trumpler had to have access to Freundlich's plates, which he was granted, in a true scientific spirit. In this game of claim and counterclaim, it is difficult to know who exactly to credit, and the nonscientist will perhaps be dismayed to see how little certainty there is to rely on in such controversies. Nevertheless, Trumpler made two points that are telling. His argument that Freundlich's reassessment of someone else's data is dubious does carry some weight. And his claim that the reseau might have changed size because of a physical expansion of the grid is also plausible. It is certainly worrying that he got such a different result by trying the conventional method of data analysis on Freundlich's plates.

Now, it has been claimed that Campbell was unbiased, in contrast to Eddington. This, of course, is untrue. Campbell was just as biased as Eddington, but in the opposite direction. But just as with those who ignore Dyson, those who fixate on Campbell ignore a crucial protagonist, Trumpler. The argument can be made that Trumpler was the important figure in the 1922 data analysis. He understood relativity better than Campbell and probably had more experience in astrometry. I do not want to minimize Campbell's role, but

Trumpler's contribution was significant, and there are valid reasons for suspecting that Trumpler was favorable to relativity. Certainly, it is suspicious that Trumpler's results always managed to come out very close to Einstein's, even when handling someone else's plates. Did Trumpler come so close to the "truth," closer than the supposedly biased Eddington, because he was a brilliant measurer or because he was determined to get close to it? "Who knows" is the answer to that. Keep in mind that, given the size of his error bars, random error should have sometimes prevented him from getting the "true" result no matter how capable he was. Personally, I am inclined to respect his competence, but he could be both competent and susceptible to unconscious bias at the same time. I am sometimes inclined to think that the eclipse observer least apt to give in to unconscious bias was actually the much-maligned Eddington! Trumpler, Dyson, and Campbell all seem to me to have been anxious to get close to 1."75 once they decided that the data favored Einstein. Eddington stuck to his guns and, as we have seen, refused to be swayed by arguments that suitable averaging or weighting could shade the result closer to that magic value.

Finding Our Place in the Galaxy

Although the drama of Einstein supplanting Newton as the arbiter of gravity had seized the public imagination, the antirelativists were still in love with their old luminiferous ether. Somehow, they imagined that the old certainties of physics would all come back, and the mathematical and geometrical subtleties of the new physics would all go away with the reemergence of the ether. Thus, the light deflection results were never all that strongly contested on empirical grounds. Alternative explanations were common enough, but these could hardly suffice to restore faith in the ether. So it is perhaps not surprising that the main experimental challenge to relativity in the 1920s came from a replication of the old Michelson-Morley experiment. This experiment was still regarded, especially in America with its empirical bent, as the only really major obstacle to belief in the ether. It was a landmark in the development of modern precision optics and essentially involved using an interferometer to measure the speed of light in two perpendicular directions to see if it was different along the direction of the Earth's motion through space. The original experiments concluded that light traveled at the same speed in all directions and that interferometers of that type could not be used to detect the Earth's motion. The man who laid down the new challenge to relativity was Dayton C. Miller, Michelson's successor at the Case School of Applied Science in Cleveland, which is now part of Case Western Reserve University.

Einstein first heard of Miller's ether drift experiments while visiting Princeton University during his first visit to the United States in May 1921. He and Miller met in person later than month at the Case School,[1] where Albert Michelson had earlier conducted his famous experiments on ether drift with Edward Morley. Beginning in 1900, Miller took up the collaboration with Morley (who taught chemistry at the adjacent Western Reserve University), and they together published important papers confirming the null result of the Michelson-Morley experiment (Morley and Miller 1905). However, like Michelson, Morley and Miller were firm believers in the ether theory of light. They became convinced that the null result was due to ether drag, blaming the "heavy stone walls of the building within which the apparatus was mounted" (the lab was in a basement) for the null result (Morley and Miller 1907). The Michelson-Morley experiment set out to measure the speed of the ether wind created by the Earth's motion through the ether. But what if the Earth dragged the ether along with it in its motion? Then ether wind experiments conducted on Earth would naturally fail. For this reason, Miller became convinced that the original experiments failed because they were, in effect, conducted underground.

Miller, accordingly, set up the apparatus "on high ground near Cleveland, and covered in such a manner that there is nothing but glass in the direction of the expected drift" (Morley and Miller 1907). Einstein could not examine the apparatus in 1921 because it had by then been moved to Mount Wilson in California, at the suggestion of the observatory's director, George Ellery Hale (Swenson 1972, 192). Nevertheless, he was suspicious that Miller's subsequent results were due to a failure to adequately control temperature in the vicinity of the apparatus. Miller's desire that his apparatus should be, as far as possible, open to the ether wind, tended to render it also unusually open to the elements, especially sunlight. This can be highly detrimental to the precision of optical experiments because differences in temperature, as we have seen, play havoc with sensitive optical equipment. Modern experts who have reanalyzed his data have concluded that his results are consistent with variations due to temperature changes (Shankland et al. 1955; Lalli 2012).

During their encounter in 1921, Einstein and Miller discussed a feature of the ether drag theory that would play a central role in the 1925 debate between their respective supporters. We know this from a sketch in Einstein's own hand recording their discussion (Illy 2006, 303). If the solid matter of the Earth dragged the ether along with it completely so that no ether wind, or ether drift, as it was often called, could be detected at the surface of the Earth, still it must be true that in space far from the Earth, the ether must be unaffected by the Earth's motion. Accordingly, it was natural to suppose that a gradient must exist so that at sufficient altitude above the Earth, some ether drift must be measurable. When the two men met in 1921, Miller had already begun (in April) a new series of experiments (with substantially the same apparatus) in another lightly constructed building at the Mount Wilson Observatory, over five thousand feet up in the San Gabriel Mountains of Southern California. There he claimed to detect a positive ether drift, in contrast to the earlier null result in the Cleveland basement. He always claimed, however, that there had been a similar but smaller positive effect in the slightly elevated location in Cleveland just mentioned. This result from Cleveland encouraged him to conduct the experiment at a much higher elevation, though the idea to conduct the experiment at altitude seems to have been put to him by Ludwik Silberstein, after the latter's move to the United States.[2] Silberstein played, for decades, an extraordinary dual role as one of the three men in the world who understood relativity (according to himself) and yet was the most persistent and dangerous of its many detractors. He was in fact an early and authoritative expert on the theory, yet the glee with which he repeatedly greeted falsifying results manages to give the impression that he only entered the precincts of relativity theory in order to act as a double agent. After the April 1925 National Academy of Sciences meeting at which Miller announced his results, Silberstein crowed, "I had . . . the honor of killing officially Einstein's Relativity of 1905." Curtis, who was present, reported that "Silberstein acted as chief mourner" as Miller shot relativity "full of holes" (Crelinsten 2006, 277). Silberstein seems to have acted throughout as Miller's champion, publicizing his results and giving them a veneer of theoretical coherence.

I do not mean by this that Miller was a fool, but there is little doubt that he did not talk the language of theoretical physics with easy familiarity, especially relativity theory. Silberstein could articulate his results with a sophistication they otherwise might have lacked, given Miller's isolation from current theoretical thinking. Different communities of physicists may speak in different languages, and Silberstein acted as a translator for Miller. Silberstein himself makes a counterpoint with Eddington. Both men understood very well the importance of Einstein's relativity as a tool or Kuhnian exemplar. The usefulness of the tool seduced Eddington into favoring the Einsteinian paradigm. Silberstein obviously took another tack, seeking a way back to the comforting paradigm of his youth, wielding the newfangled tool he had mastered. In the meantime, the results of Miller's 1921 experiments were published the next year (Miller 1922) and reached Einstein through his close friend, the German physicist Max Born. Born was shocked by the news: "The Michelson experiment belongs to things that are 'practically' a priori; I believe not a single word of the rumor" (*CPAE*, vol. 13, doc. 320). Einstein took a similar view.

Before 1925 Miller published only brief reports on these experiments. Finally convinced that he was consistently seeing a small ether drift on Mount Wilson, he came forward confidently in the middle of that year, presenting his work at a meeting of the National Academy of Sciences. His work was featured in the magazine *Science*, and Einstein was asked for a response by the editor (Miller 1925). Einstein answered cautiously: experiment being the supreme judge, he awaited more complete details (*CPAE*, vol. 15, doc. 13). The same reserve can be seen in his letter to Robert Andrews Millikan, the great American physicist. He "supposed that Miller's experiments rest on error sources. Otherwise the entire theory of relativity collapses like a house of cards" (*CPAE*, vol. 15, doc. 20). So far, so very Popperian. Most physicists expressed serious doubts, first among them Millikan and his staff at the California Institute of Technology (Caltech), who were in a position to see the apparatus on nearby Mount Wilson. A year later, one of them reported to Einstein that "the repetition of the Michelson experiment by [Roy] Kennedy at Mount Wilson gave a completely negative result, even though the sensitivity of the apparatus was four to five times higher than that of Miller's" (*CPAE*, vol. 15, doc. 372). One eyewitness of Miller's apparatus, Max Born, "was horrified by the mess of the experimental arrangement," as his wife, Hedwig Born, wrote to Einstein (*CPAE*, vol. 15, abstract 408). Astronomers also doubted Miller's claims that the ether behaved differently at altitude. Eddington demanded to know why observatories at altitude, such as the one at Mount Wilson, did not report any difference in stellar aberration from that observed at sea level.

A clear implication of Miller's results was that his instruments upon Mount Wilson were actually measuring the solar system's motion through the luminiferous ether. In fact, several of Einstein's correspondents were suspicious of Miller's failure to make a bold statement regarding the velocity and direction of the Earth through space, including not only the Earth's rotational and orbital motion within the solar system but also the solar system's motion through interstellar space. The last line of Miller's 1925 papers merely stated that he was working on calculating such a quantity. Some scientists suspected Miller of being unwilling to let his own theoretical model undergo a possibly falsifying test. It must be admitted, though, that some confusion existed because astronomers' understanding of the motion of the solar system was undergoing a major transformation at this time. Up until then, they believed that the solar system was moving within the system of nearby stars at a velocity of twenty kilometers per second in the direction of the constellation Hercules. The

Sun was apparently moving at this speed toward the point within Hercules called the solar apex, discovered in the eighteenth century by William Herschel. Observations with the new generation of big telescopes in the mid-1920s resulted in a revised estimate of the solar system velocity upward by an order of magnitude.

Coincidentally, a major contribution to this revolution was presented at the same session of the National Academy of Sciences at which Miller presented his results in 1925. The author of the new results on the motion of the solar system, Gustaf Strömberg, an astronomer at the Mount Wilson Observatory, presented results showing that the solar system moves at a velocity of some three hundred kilometers per second relative to certain old stars (now called *halo stars*), the globular clusters, and the spiral nebulae (other galaxies than our own). According to Crelinsten, Strömberg regarded his own results with a certain amount of confusion. It was not apparent to him why the rapidly moving stars (the halo stars) should be moving at the same speed as the spiral nebulae. He wondered if some property of the ether restricted them to motions below a certain value and, apparently, entertained hopes that his work would ultimately disprove relativity theory. Sharing a certain anti-Einsteinian outlook with Miller and working in close proximity on Mount Wilson, he helped Miller shape his own results, since the idea naturally arose that they were both essentially measuring the motion of the solar system through the ether.

Of course, the best interpretation of Strömberg's results was, and is, that the solar system orbits the center of the galaxy at the velocity he measured for the halo stars. The reason that we see the halo stars and nearby galaxies moving with the same large velocity is that both these kinds of object are essentially standing still as we whiz along at high speed through the galaxy. The earlier observed velocity of roughly twenty kilometers per second is merely our velocity with respect to nearby stars that share our orbital motion about the galaxy's center. Strömberg's results contributed strongly to our current understanding of our motion and position through the Milky Way galaxy, first correctly interpreted by the Dutch astronomer Jan Oort (1927) based on suggestions from the Swedish astronomer Bertil

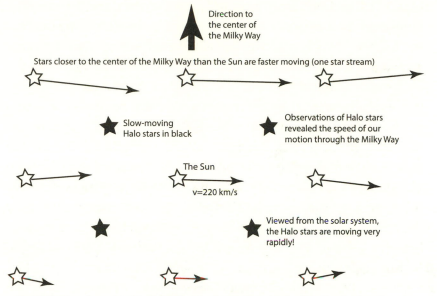

Direction to
the center of
the Milky Way

Stars closer to the center of the Milky Way than the Sun are faster moving (one star stream)

Slow-moving
Halo stars in black

Observations of Halo stars
revealed the speed of our
motion through the Milky Way

The Sun

v=220 km/s

Viewed from the solar system,
the Halo stars are moving very
rapidly!

Stars farther from the center of the Galaxy than the Sun are slower moving (the other star stream)

FIGURE 21. The Sun moves around the center of the galaxy at 220 kilometers per second. Stars a bit closer to the center move faster, and stars a little farther away move slower, which explains the star streams seen by Kapteyn and confirmed by Dyson and Eddington. Some stars, known as halo stars, are moving slowly, and they enabled Strömberg to measure our true velocity around the center, which is some 220 kilometers per second. This is the velocity that Miller should have been able to measure if the ether theory of light is correct. In order to explain how the galaxy could hold us in orbit at this enormous velocity, astronomers were forced to acknowledge that the Milky Way must have over a hundred billion stars and be far larger than thought at the time. We now know, from measuring the cosmic microwave background, that our galaxy itself is moving through the universe at a velocity of some six hundred kilometers per second. (Reprinted with the permission of the author.)

Lindblad. It might be wondered why there was initial confusion in interpreting the rapid motion of Stromberg's fast-moving stars and nebulae as simply our own solar system's motion. The reason is because it was difficult to see how such a rapid motion could be part of any settled system in equilibrium. Where were we going at such a speed? Lindblad proposed the complex system of linked rotations, which we now believe in, and Oort then deduced from this the direction and distance to the center of the galaxy and also the enormous mass of the galaxy, which is necessary to keep us in orbit at the incredible speed we are moving. Thus, in a few short years, astronomers learned that we are part of a galaxy with a total mass of a hundred billion stars and are more than twenty thousand light-years from its center, when the previous best estimate, by Jacobus Kapteyn, had us only two thousand light-years from the center of a system of stars numbering in the millions, not billions.

Strömberg eventually helped Miller come up with a specific claim of the solar system's motion through space: two hundred kilometers per second in a direction about twenty-three degrees away from the solar motion in the galaxy found by Strömberg (1926). This was close enough to Strömberg's conclusions to put new heat into the reception of Miller's work. It seemed that he had succeeded in measuring this dramatic new motion of the Earth in the galaxy. This even caught the imagination of many physicists, as is shown by correspondence between two famous European physicists, Wilhelm Wien and Erwin Schrödinger. Their discussion of Miller's results kicked off on September 16, 1925, when Wien asked Schrödinger, "What do you think about the positive result of the Michelson experiment?" He continued, "I have just obtained new information from America, according to which one can hardly doubt the correctness of the result. The observations on the shift of the interference patterns reproduce the motion of the solar system in the universe, in agreement with the latest observations of Strömberg!! It is the most astounding result in physics." Wien and Schrödinger sprang into action, and their correspondence was, for a time, dominated by their efforts to organize German and Swiss experiments designed to test or replicate Miller's results. One even detects a certain degree of satisfaction in their anticipation of the forthcoming counterrevolution in physics. Wien's previous quote concluded: "If the observations are substantiated—as can hardly be doubted any more—relativity theory, the special as much as the general one, is finished, and we must go back again to our old ideas of 25 years ago." In his reply Schrödinger commented upon "the hardly noticeable 'counter-propaganda' in the Jewish circle of physicists" trying to play down the result, which he decried as unfair to Miller (translated and quoted in Mehra and Rechtenberg 1987, 453).

It was probably this triumph of Miller's that set in motion the most important of the replicating experiments inspired by his work. At the end of 1926, Hale wrote to Einstein informing him that Michelson himself would perform an interferometer experiment atop Mount Wilson (*CPAE*, vol. 15, doc. 425). It was several years before results were announced, but Michelson firmly ruled out any inertial motion of the Earth even a tenth as great as astronomers, whose conclusions were by then firmly established, demanded (Michelson, Pease, and Pearson 1929).

The question might reasonably be asked as to how Miller, who was making measurements indicative of a motion of the Earth in the plane of the interferometer of roughly ten kilometers per second at most (Miller 1933, table 3), could end up predicting a velocity through the ether for the entire solar system of over two hundred kilometers per second, roughly in line with Strömberg's discoveries. The answer is that, in doing the calculation so many physicists and astronomers had demanded of him, he of course had to determine the true apex of the solar system's motion through the galaxy. That is, he had to decide which point on the sky we on the Earth are moving toward. If this "Miller apex" turned out to be nearly perpendicular to the plane of his interferometer, then of course ten kilometers per second would turn out to be only a small component of a far greater motion. This is what Miller claimed, at first only orally and in short communications and eventually, in more detail, in his 1933 paper. It is worth noting that the apex he claimed for the Earth's motion is nowhere close to the actual apex known today, which is not all that far from the old solar apex in Hercules, while the Miller apex is far in the Southern Hemisphere, in the direction of the large Magellanic cloud. It seems plausible that in deriving his best estimate of the Miller apex, he naturally enjoyed considerable latitude within his quite noisy data and exploited this, consciously or unconsciously, to find an apex that would give him a solar system velocity compatible with Strömberg's results.

In discussing these three kinds of motion of the Earth, it is worth mentioning that a particularly difficult point for Miller is the fact that the rotation of the Earth is not inertial. It therefore follows that the noninertial aspects of it can be measured in a laboratory without violating the principle of relativity. Thus, interference fringes produced by the rotation can be measured in a ring interferometer, in which light is passed in opposite directions around a closed circuit before the two beams are rejoined. This experiment was first performed by the French physicist George Sagnac in 1913. After him it is still known as the Sagnac effect. A large-scale version of the Sagnac experiment was performed in 1925 by Michelson himself, together with Henry Gale, using a nearly two-kilometer-long ring interferometer large enough to correctly measure the rotational velocity of the Earth. Sagnac merely showed that his device could measure rotation by permitting its support to rotate. Michelson, by contrast, measured the rotation of the Earth itself, carrying his instrument around with it as it turned. As Michelson pointed out, his experiment invalidated the hypothesis of complete ether drag (Michelson and Gale 1925). If the ether is dragged around by the Earth to such an extent as to make it impossible to measure the ether wind at ground level (or basement level) using a Michelson interferometer, then it should also preclude the Sagnac effect. Only special relativity, with its distinction between inertial and noninertial frames, predicts the observed null result for the Michelson-Morley result combined with a positive result for the Michelson-Gale experiment.

One other astronomical result, which Eddington pointed out, played a role in the controversy. At the same remarkable session of the National Academy of Sciences at which Miller and Strömberg presented their results, Hale reported on work by Walter Adams, an astronomer at the Mount Wilson Observatory who demonstrated the existence of gravitational redshift in the spectrum of the white dwarf star Sirius B (Adams 1925). This result was in agreement with earlier calculations by Eddington (1924). Gravitational redshift was still a favorite topic of relativity skeptics (Einstein 2004, 37–40). Eddington drew attention to this major success for Einstein's general theory, providing an underpinning for the weakest leg of the three tests that upheld it. Since, as Adams remarked in his paper, the result is the first direct evidence in favor of Eddington's claim that white dwarfs have enormous densities, the result also suggests that some astrophysical systems could have far stronger gravitational fields than are found in the solar system. Thus, the relevance of general relativity to astronomy might eventually prove to be far greater than it at first appeared. Miller's result, on Eddington's reading, thus runs counter to the current of scientific history.

It must be admitted, however, that Adams' result has been questioned by recent critics, who observe that his measurements do not agree with modern work on Sirius B and were probably wrong by a factor of four. This went unnoticed at the time because he had managed to match Eddington's theoretical prediction, also out by a factor of four (Holberg 2010). It may be that this is a case in which one of the results that influenced contemporaries in favor of general relativity does not really stand up to closer scrutiny today, as far as precision measurement is concerned (Peebles 2017). Nevertheless, the result is important because astronomers came to realize that white dwarfs are almost certainly small, highly dense stars. As such, their surface gravity is extraordinary. Einstein's theory demanded that they exhibit an unusually large redshift, and Adams showed, at least qualitatively, that one did. Others followed with measurements of other white dwarfs. That Einstein's theory provided a convincing explanation of the odd fact that small, hyperdense objects have large redshifts is important evidence in its favor. After all, there is no logical reason why all white dwarfs should be moving away from us at high speed, the usual explanation for redshifts. We see once again, contra Popper, that explanation can play a role in confirming a theory in scientists' minds and that it is not really all about precision and falsification. Sometimes, measurements are not quite as precise as they are claimed to be, but the qualitative picture still holds true.

If, as Einstein believed, the source of Miller's results was not an ether drift, the difficulty was to explain what was causing it. Einstein suspected small temperature differences between the arms of the interferometer were to blame. "Temperature differences in the air between the two beams, of an order of magnitude of $1/10°$, would suffice to cause the whole fuss," as he wrote to his close friend Paul Ehrenfest (*CPAE*, vol. 9, doc. 71). Einstein's honest opinion, expressed to his confidant Ehrenfest, was that "I have a low opinion of the Miller experiment at the bottom of my black soul, only I can't say it out loud" (*CPAE*, vol. 15, doc. 49). He told Millikan privately that he did not trust Miller's result but had no right to say so in public (*CPAE*, vol. 15, doc. 58). To Millikan he expressed his most instinctive opinion: "Apart from any special theory, it seems to me almost impossible that the small difference in the locations of Cleveland and Mount Wilson would entail so big a difference in the ether wind" (*CPAE*, vol. 15, doc. 58). Characteristically, he invoked the creator to Ehrenfest: "The difference between Cleveland and Mount Wilson cannot be so significant, considering the grand scale on which the old one created the world" (*CPAE*, vol. 15, doc. 49), crediting God "with more elegance and intelligence" than that in another letter (*CPAE*, vol. 15, doc. 26). When he had first learned of Miller's experiments in 1921, he expressed his distrust in a similar vein: "Subtle is the Lord but malicious He is not" (*CPAE*, vol. 12, 53).[3] Now we see what exactly he had in mind. God may be slick, but it seems that he ain't mean after all.

As always, scientists put their faith in replication. But even here Einstein had doubts, desperately noting the expense involved: "What can one do now to bring some order to this epidemic? It would be a pity to spend too much money on this shady matter" (*CPAE*, vol. 15, doc. 86). Many of those engaged in such endeavors communicated their intentions to Einstein, some seeking help in raising funds. While Kennedy's experiment at Caltech was certainly influential in turning opinion against Miller, the balloon-based experiment carried out by Auguste Piccard turned out to be of the greatest significance for the history of aeronautics. A Swiss physicist based in Brussels, Piccard informed Einstein of his intention, as Einstein put it, to "Miller" in a balloon (*CPAE*, vol. 15, doc. 85). Although balloons had been used in science previously, and although Piccard himself was an experienced aeronaut, his success in helping rebut Miller encouraged him to combine his physics career with his passion for ballooning. Significant funding was needed for manned scientific ballooning, and Piccard saw the Miller affair as a perfect opportunity to interest potential funders, especially with Einstein's "moral support" (*CPAE*, vol. 15, docs. 74, 87). It seems clear that his success in conducting Miller's experiment in a balloon encouraged him to go higher. His twin brother (and fellow aeronaut), Jean, recalled that just after the Miller work, Auguste first spoke to him about trying to set an altitude record. "My twin brother, Auguste, first discussed this proposed flight with me in 1926. He wanted to go to a greater height, not to establish an altitude record, but to determine, if possible, the action of cosmic rays and their quality and intensity at different altitudes" (Ziegler 1989, 960). Piccard hoped to contribute to the debate on the origin of cosmic rays, which were widely thought to be of extraterrestrial origin. Some, such as Millikan, had argued they were terrestrial. Since Millikan's unmanned balloon flights had provided unexpected evidence for the noncosmic origin, Piccard argued that a manned balloon flown to high altitudes would enable a more reliable experiment. Accordingly, he invented a sealed capsule that enabled him to set an altitude record in 1931 by becoming the first person to ascend to the stratosphere, nearly sixteen thousand meters up. The resulting fame saw him immortalized in the character of Professor Calculus by his fellow Brussels resident Georges Remi (better known by his pen name of Hergé) in the pages of *The Adventures of Tintin*.

If Einstein felt unable at the outset of the controversy to publicly air his skepticism of Miller's findings, he was gradually encouraged to express his doubts more freely. He was even quoted in the press advising the public not to place any bets on the confirmation of Miller's findings (*CPAE*, vol. 15, doc. 161). At the end of 1926, Einstein took a public stance on the matter. In this short paper in a popular scientific journal, he first summarized the improvements Kennedy and Piccard had made on their apparatuses compared to Michelson's original instrument; then he concluded that even though they could not completely eliminate the disturbing effect of environmental temperature, their results disproved Miller's main statement— namely, that there is a drag of the ether by the Earth that changes with altitude. Einstein concluded with a chivalrous funeral oration for the initiator of the debate: "No doubt, it was Prof. Miller's outstanding merit that he initiated a meticulous re-examination of Michelson's important experiment" (*CPAE*, vol. 15, doc. 478).

The 2017 Eclipse and the End of Eclipse Expeditions

The eclipse of August 21, 2017, was the first total solar eclipse visible across a broad swath of the United States since the eclipse of 1918 that Curtis observed from Goldendale, Washington.[4] By a quirk of history, nature conspired to produce highly visible supernovae in our galaxy in 1572 and 1604, but there have been none since Galileo pioneered the use of the telescope in astronomy in 1609. In the same way, it kept useful eclipses away from the continental United States over the period when Americans played a critical role in testing gravitational theory. One could argue, as Crelinsten does in his book, that the United States played a dominant role in experimental tests of general relativity, through the work of astronomers like Campbell, St. John, and Walter Adams and precision experimentalists like Albert Michelson, Robert Dicke, and many others in the period after World War II. The 1918 eclipse came at an inconvenient moment, when American astronomers were handicapped by wartime conditions. The 2017 eclipse came too late. Astronomers no longer needed eclipses to test relativity. Why was this? What had happened in the meantime?

Part of the history of eclipse expeditions has been a process whereby the study of eclipse phenomena permits scientists to develop ways to observe those phenomena during the daytime. Early eclipse expeditions in the mid-nineteenth century focused their attention on the chromosphere, until Norman Lockyer and Jules Janssen discovered how to observe this without an eclipse. After that the corona was the main object of attention at eclipses, until methods for observing it were developed in the 1930s, making such research less dependent on travel to eclipses. Similarly, radio astronomers in the 1970s discovered radio-bright quasars that are occulted every year by the Sun. Since the Sun is not bright in the radio, these astronomers could use their antennae to follow the quasars up to the limb of the Sun, measuring the light deflection the whole way, right up to its maximum value. They could do this without moving their equipment and repeat the experiment every year. Naturally, the precision of their measurements within a decade was superior to anything ever achieved at an eclipse, and astronomers stopped going to eclipses to perform the light deflection measurement. The 1973 University of Texas expedition to Mauretania in West Africa is often cited as the last professional expedition to perform the Einstein test. This was still in the days before photographic plates were superseded by charge-coupled device (CCD) cameras, so in this respect the 2017 eclipse presented a great opportunity. By then the quality of equipment available to amateur astronomers surpassed anything from the pre-CCD days of astronomy. With very little travel needed, any experienced amateur astronomer could conceivably outdo Dyson and Eddington and replicate the Einstein test with a new level of accuracy.

Interestingly, the 2017 eclipse occurred on an important centenary. Evershed's was the first attempt to test general relativity by measuring light deflection during a conjunction of the Sun with Regulus, as recommended by the Lindemanns. This conjunction took place on August 21, 1917. The attempt was fruitless and was, apparently, handicapped by cloud. Exactly a century later, Regulus was once again in conjunction with the Sun, but this time, by chance, the Sun was in total eclipse across a great swath of the most technologically advanced country on Earth. Yet oddly, few attempts were made to take advantage of this. I myself was involved in two grant proposals to raise money for a large number of citizen scientists to replicate Dyson and Eddington's experiment. Both were turned down. The reason is that far more accurate measurements have been made by radio astronomers. Even in the regime of visible light (what astronomers call the *optical*) space-based telescopes, like the Hipparcos satellite, have greatly outdone what is possible with eclipse measurements. Admittedly, a large number of observers might be able to do much better than anything previously done at an eclipse, but the scientific gain would be small for the money. The proposals were pitched more as science outreach events than research efforts.

As we have seen, the light deflection test was performed quite a few times up until 1929. There were few expeditions after that date. This was partly due to the Great Depression and the war. It also happens that interest in general relativity declined after the initial excitement generated by the 1919 eclipse. Since Einstein had only identified three tests, there was little else for astronomers to do, and the dramatic discoveries of quantum mechanics and nuclear physics dominated physics in the mid-twentieth century. Mathematicians, not physicists, performed much of the work on general relativity during this period. This work was obviously theoretical. Little work was done on experimental gravity. But from the mid-1950s on, interest in the subject revived. One reason was the rise of radio astronomy, which led to the discovery first of quasars and then of pulsars, objects that, it quickly became apparent, were characterized by exceptionally strong gravitational fields. Only Einstein's theory could correctly explain their behavior. Work like that of Chandrasekhar on gravitational collapse now came to the fore, and the science of black holes developed. The term *black hole* was coined during the sixties. It was not just in astronomy that technology developed in a way that led to new breakthroughs. New experimental techniques permitted physicists to test Einstein's theory without the need for astronomy. The Pound-Rebka experiment used sensitive new techniques developed in nuclear physics to test the gravitational redshift in purely terrestrial experiments. Since astronomical attempts to perform this test had always been somewhat inconclusive, this was the first really definite confirmation of this important test of the equivalence principle that helped give birth to a new field of precision tests of gravity. But the real pioneer in experimental gravity was a Princeton physicist named Robert Dicke. He was obliged to point out that the existing tests did not have the required precision to really give general relativity a stress test. At the first international meeting devoted to general relativity, held in Chapel Hill, North Carolina, in 1957, Dicke had this to say regarding the need for new experiments:

> It is unfortunate to note that the situation with respect to the experimental checks of general relativity theory is not much better than it was a few years after the theory was discovered—say in 1920. This is in striking contrast to the situation with respect to quantum theory, where we have literally thousands of experimental checks . . . Professor [John Archibald] Wheeler has already

discussed the three famous checks of general relativity; this is really very flimsy evidence on which to hang a theory. (DeWitt 1957, 5; Peebles 2017, 183)

In putting the case so well for experimental gravity in 1957, he may have unwittingly gotten the ball rolling that led to the many recent accusations of bias or false claims of accuracy against Eddington and Dyson. But Dicke himself accepted the 1919 results as far as they went. He just wished to go further. It is also worth noting that he did not regard precision testing of general relativity as a pro forma exercise. He was confident that general relativity would be falsified by his efforts. He even had a wager with John Wheeler, the most noted relativity theorist of the day and his colleague at Princeton, that general relativity would ultimately fail the light-bending test when it became more precise. He specifically bet Wheeler that the true light deflection value would turn out to be less than 96 percent of Einstein's predicted value (Peebles 2017). So much for Freundlich's high value. Dicke thought he had evidence to support his claim that general relativity was wrong from his efforts to measure the non-spherical oblateness of the Sun, so he was quite confident about this bet. Wheeler, by contrast, was relying on his conviction that the theory's beautiful formal simplicity must be in accord with reality. In fact, it seems that settling this bet was one of the aims of the last eclipse expedition, the Texas-Mauretania expedition of 1973 (Matzner 1975). Remarkably, Wheeler has been proved right and Dicke wrong by the quasar observations (the Texas results were not precise enough to discriminate between them). Einstein developed his theory from a very narrow empirical base, relying primarily on his extraordinary scientific imagination. Yet it has passed every subsequent test. How he could do this is something of a mystery. The power of the human imagination to guess the results of an experiment ahead of time seems almost inexplicable. Furthermore, if humans possess this amazing faculty, then why was it only Einstein who was capable of discovering general relativity, since no one else was really close to doing so?

Naturally, the level of precision of Dicke's experiments, as well as what was achieved by his students and others he inspired, put the efforts of eclipse astronomers in a new light. Previously, they had been the leaders in the field of testing general relativity. The perihelion test was impressive, but it had a potential Achilles' heel, which came to the fore after Dicke and his student Carl Brans came up with a new relativistic theory to rival general relativity in the 1950s. This inspired Dicke to point out something that the antirelativist Charles Lane Poor had originally argued in the 1920s. Poor believed that the correct explanation of at least part of Mercury's anomalous perihelion advance was that the Sun is not a sphere but is oblate, with an equatorial bulge like the Earth's, due to rotation. Such an equatorial bulge on the Sun would affect Mercury's orbit and throw out the precise agreement between Einstein's theory and observation. Poor failed to get anyone interested in performing these observations in the 1920s (Crelinsten 2006, 226), but Dicke took up the challenge in the 1960s. Since the Brans-Dicke theory has an adjustable parameter, it could survive any such alteration in the unexplained part of Mercury's motion, but general relativity would not. So Dicke aimed to take advantage of the fact that his own theory was less falsifiable than Einstein's since it contained an adjustable parameter. Of course, looked at from Dicke's perspective, this was an excellent opportunity to constrain and measure that unknown parameter. In the end Dicke never did succeed in establishing that the Sun is significantly oblate.

Because the solar redshift test was so difficult, the light deflection test was originally the strongest pillar that supported the belief in general relativity. Now it became, inside a couple of decades, the weak link. What was particularly suspicious was the failure of its precision to improve with time. To see why this was so, let us look at the case of two astronomers, one Russian and one American, who took up the experiment in the mid-twentieth century.

The Yerkes Observatory was founded by George Ellery Hale while he was a professor at the University of Chicago. It played an important role in the rise of astrophysics, as it pioneered the integration of discoveries in the other sciences into astronomy. Hale continued this approach when he moved to California and founded the Mount Wilson Observatory. An astronomer at Yerkes, George van Biesbroeck, decided to use new techniques to redo the Einstein test at eclipses in

1947 and 1952. Like Trumpler, van Biesbroeck was a European immigrant to the United States, having been born and educated in Belgium before moving to Yerkes during the First World War. The motivation for his 1947 test came from the National Geographic Society. Lyman Briggs was organizing the society's expedition to Brazil to observe this eclipse and invited van Biesbroeck to accompany them to perform the Einstein test. Briggs is best known to history as the head of the Advisory Committee on Uranium, the predecessor to the Manhattan Project in the effort to develop nuclear fission technology in the United States. Reflecting the new financial resources available to science after World War II, the expedition to Brazil received funding from the U.S. Army Air Corps. Soon afterward, this became part of the new United States Department of the Air Force, which was a major funder of research in general relativity until the 1970s.

Van Biesbroeck knew that earlier efforts to measure light deflection were handicapped by the difficulty of measuring the scale change between the eclipse and the comparison plates. He decided to improve matters by using a semireflecting mirror to place two different images on the same plate. One would be of the eclipse field; the other would be of a check field visible in the sky at the same time. Since the images would be on the same plate and produced at the same time by the same optics, it could be assumed that the scale was the same. The check field would be visible on both eclipse and comparison plates. The stars in the check field would be used to measure the scale change. Since there would be no Sun near those stars, there could be no light deflection. This scale change would then be applied to the eclipse field, hopefully permitting a more accurate measurement of the light deflection. This would improve upon the technique used by Eddington in 1919. Although van Biesbroeck did not know it at the time, this technique had already been used by a Russian astronomer, Aleksandr Aleksandrovich Mikhailov, at an eclipse in 1936. This eclipse crossed the Soviet Union, and Mikhailov observed from Kouybyshevka in the Soviet Far East on the Trans-Siberian Railway.

Mikhailov encountered one serious problem, the Siberian climate. The correct time to take comparison plates was six months after the eclipse, which took place in June. This would be in the middle of the winter. Siberian winter conditions are not conducive to astronomy. He waited until late March 1937 before taking his comparison plates, but the temperature was still colder than twenty below, while the eclipse temperature was 23.6 Celsius. This enormous temperature difference seems to have caused a larger-than-usual scale change between the eclipse and the comparison plates. This problem was compounded by the fact that Mikhailov's split-plate technique failed in 1936, possibly due to the fact that the driving mechanism of the telescope did not work for both fields simultaneously. The driving mechanism was set for the eclipse field, and the star images on the check fields were elongated as a result. For whatever reason, Mikhailov's results, like Freundlich's, gave a light deflection result considerably larger than Einstein had predicted, but his effort to improve on Eddington's basic technique was unsuccessful. The experience left him somewhat doubtful of the whole enterprise: "Now I must confess that after obtaining the value of deflection from a complicated calculation one has no great conviction in the reality of an apparent repulsion of the stars from the Sun at all, as this is not very evident from the measurements themselves" (Mikhailov 1959, 603).

Mikhailov did not benefit from his experience at the 1936 eclipse because the next eclipse visible from inside the Soviet Union took place on September 21, 1941. Although he attempted to observe the eclipse at Almaty (then known to Mikhailov as Alma-Ata) in Kazakhstan and enjoyed favorable weather there, "war conditions frustrated observations" as he put it in 1959. On September 21 of that year, German troops were eliminating Soviet armies of some half a million men inside the *Kessel* around Kiev,[5] a catastrophe for the Red Army that very nearly made possible the Nazi capture of Moscow shortly afterward. He tried again after the war at several eclipses but was clouded out on each occasion, twice in the Soviet Union and once in Brazil, which is where he met van Biesbroeck and compared notes. There was plenty to talk about because van Biesbroeck's very similar scheme did not meet with success, either. In Brazil he was handicapped by the nature of the eclipse itself. As he put it, "The sun was, at the time of the eclipse, located squarely in front of one of the extended dark regions in Taurus," and he thus had no stars at all close to the Sun (van Biesbroeck 1950, 50). It is clear why Dyson had been so eager to get to the 1919 eclipse with its favorable star field. This made it all the more problematic that van Biesbroeck's check field had serious astigmatism. Mikhailov thought this might be due to the same driving problem he blamed for his difficulties in 1936. Van Biesbroeck disagreed, suspecting that heat from the Sun had affected the half-silvered mirror used to project the check field onto the same photographic plate as the eclipse field. This was a similar complaint to the one made by the British about their coelostat on the astrographic telescope at Sobral.

Van Biesbroeck was at least fortunate enough to have a chance to benefit from his experience in Brazil. He went to Khartoum, in the Sudan, in 1952 to have another go. This time he used a fan to ventilate his half-silvered mirror, but strong and gusty winds caused vibrations in the telescope, which he blamed for the fact that he obtained less than a dozen measurable stars on each plate. His method of measuring the scale change from the check field was actually used this time, fortunately, given the small number of star images acquired, and he calculated a value for the light deflection in close agreement with general relativity. Nevertheless, his dissatisfaction with the whole experience may be reflected in his paper, which Mikhailov describes as laconic because it is only two pages long. It is noteworthy that he seems to have made no further attempts. Nor, as far as I can tell, did Mikhailov, who as late as 1979 urged that the test be repeated at the 1981 eclipse. But as the best location was again in Siberia, halfway between Krasnoyarsk and Lake Baikal, it is perhaps not surprising that Mikhailov, by then past ninety years of age, seems not to have contemplated an expedition. Although the eclipse itself was in July, the best time to take comparison plates would presumably have again fallen in the Siberian winter, as in 1936.

The last hurrah for eclipse expeditions in the old style was the University of Texas expedition to Mauretania in 1973. This effort had something in common with the British expedition of 1919 in that theorists who were experts in general relativity were, as with

Eddington in 1919, closely involved. Cécile DeWitt-Morette and her husband, Bryce DeWitt, worked with a large team of astronomers under McDonald Observatory director Harlan Smith. The expedition benefited from the relative ease of modern travel and logistics. Whereas Mr. Hicks had chosen sites in 1919 by poring over the atlas and writing letters, Cécile DeWitt-Morette traveled personally across West Africa, taking advantage of her native French to identify the best observation site. Mauretania was chosen, and the expedition went so far as to construct an air-conditioned observatory and telescope on-site to eliminate the problems with temperature and other environmental factors experienced at previous eclipses.

Very little had been left to chance, since the desert site in Mauretania was usually cloud-free. But on the day of the eclipse, strong winds created a great deal of dust in the atmosphere, which scattered so much light that the sky remained quite bright during totality, leaving only Venus visible to the naked eye. This reduced the number of stars the team expected to record on their eclipse plates from one thousand to close to one hundred. This still represented a major improvement on earlier eclipses, but another problem was discovered when it became apparent that the screws holding the plates in place had not performed adequately, resulting in different comparison plates having different scale changes with respect to eclipse plates, compromising the data reduction. Nevertheless, the team was confident that their results represented a better test of Einstein's theory—a test the theory passed—than any previous eclipse measurements. It is worth mentioning that the Texas team did not make use of the split-plate technique but instead took check plates during the eclipse by moving the telescope to a different star field away from the Sun. They could afford to do this since their plate technology had improved to the point where long exposures were no longer required to obtain a deep image with faint stars on the photographs.

The experiences of van Biesbroeck, Mikhailov, and the Texas team illustrate the difficulties that eclipse conditions impose on observers. Unable to repeat their experiments, scientists contend with multiple ways in which a given measurement might go wrong. In a laboratory or observatory setting, such commonplace problems are overcome by simply repeating the experiment many times and refining the procedure, often minutely, with each repetition. It is often difficult to tell precisely how one should modify the apparatus, and much tinkering may be required. In Brazil in 1947, van Biesbroeck and Mikhailov disagreed about the reason for the astigmatism in the star images on van Biesbroeck's comparison plates. Van Biesbroeck had to wait years to find out if his chosen fix would work. Since he never performed the experiment after that, and since conditions still were not ideal at his second try, it is impossible to tell with certainty which man was right, especially since Mikhailov never got a proper second chance. It might seem that progress could have been made collectively since many astronomers did perform the experiment over the decades. But this overlooks the limitations in communicating via scientific papers or occasional meetings in the field. An experimental apparatus is far too complex to be described on paper, and studies of scientists in action (Collins 1974) tend to show that they learn how to make experiments work through actually working on the apparatus together. There is a considerable exchange of essentially nonverbal tacit knowledge, which is often required to achieve success in experimentation. The importance of tacit knowledge, which is the knowledge you do not know you possess, is emphasized by sociologists of science like Harry Collins. Just as you cannot learn to ride a bike from reading a book, Collins would argue that you cannot learn to perform an experiment from reading a scientific paper. Much scientific knowledge is quite tacit and is learned in the laboratory, or at the eclipse site, from watching and imitating. However, a given eclipse team is highly likely to get only one chance to take data. If they are left with a feeling that there were lessons learned, they may never have the chance to put those lessons into practice and may find it hard to articulate the most vital steps to others.

In the postwar period, astronomers and physicists themselves became increasingly skeptical of the claims made on behalf of the 1919 eclipse measurements. The 1973 eclipse team was trying to bring their precision more into line with the experimental standards of the second half of the twentieth century, and they discovered this could not be done under eclipse conditions. The fact that their error bars were no smaller than those claimed by astronomers working more than half a century before was particularly galling and deeply suspicious. Of course, they were frustrated to be unable to falsify either relativity or its Brans-Dicke rival, which had been their goal. But scientists, especially physicists, are used to precision improving consistently with time. Why did this not happen in the case of the light deflection measurement? Perhaps, they thought, the original precision had been overstated. One of those closely involved with the 1973 expedition, Cécile DeWitt-Morette, certainly made that claim to me when I spoke to her in 2004. I believe that views such as hers convinced many scientists that there was something dubious about Eddington's work. By the 1970s, Dyson's role had been largely forgotten, and his contributions were ignored. The fact that he was unbiased and had an excellent track record in astrometry therefore was not taken into account. It is true, of course, that he was almost unique in getting his measurements right the first time, but this was partly because of the unique advantages of the 1919 eclipse, which he had foreseen, and because he was sensible enough to use two instruments, one of which was tried and trusted in eclipse work. And of course, he was lucky, but he was always lucky at eclipses. He had all the skills necessary for the task, including the one that Napoleon allegedly regarded as most important in a general, luck.

I also think that physicists and astronomers are inclined to overlook the key reason why precision in the eclipse test failed to improve. Scientists regard continuous improvement in measurement as a hallmark of science that is unremarkable except where it is absent. If it is absent, it tells us nothing except that someone involved has behaved in a way that is unscientific or incompetent, or both. No one asks *how* this improvement takes place, except perhaps by some sort of osmosis. But in a typical research setting, one has students, postdocs, and professors moving from one lab to another, where experiments are performed on a daily basis; and learning occurs in a way that is all the more unnoticed for being tacit and unspoken. This did not happen with the eclipse experiment. One could not witness it being performed without going to an eclipse; and teams typically consisted of one, two, or three astronomers. Eddington, remember, was the only scientist on Principe during the eclipse of 1919. Of course, astronomers discussed their techniques and their equipment. They debated such things in person and in print. They even sometimes independently came upon the same idea, as with Mikhailov and van Biesbroeck. But it is one thing to have an idea and another thing to implement it. Implementation involves a hundred or a thousand factors that must be handled correctly for the technique to work. There are too many of these factors to discuss more than a few of them in any depth. It is the body of practice that, over time, leads to technical improvement; and the body of practice is transmitted, by and large, through social interaction in the laboratory or at the observatory. This was not possible in the eclipse experiment. Even real technological advances, such as improved photographic plates, could be offset by poor environmental conditions or other accidents, such as the Saharan dust and the faulty screws of Mauretania in 1973.

I will use the term *experimenters' progress* to describe scientists' expectations that the precision of a measurement will continue to improve with time until a once-contentious experiment becomes routine and uncontroversial. Scientists tend not to worry about the experimenters' regress, which we discussed earlier, because in their view it is countered by the experimenters' progress. A controversial measurement at the limit of precision today will be easy to replicate in the near future. Therefore, in time it will become obvious which scientist in a controversy was wrong.

What we see in the case of the eclipse experiment is that the experimenters' progress can, in some circumstances, be foiled by what we might call the *experimenters' malaise*. The malaise occurs when the precision of a measurement fails to improve with time. In this case a measurement can remain controversial, or even become more controversial with time. Thus, Eddington is accused today in ways that were not raised against him in his own day. The malaise is important because experimenters' progress is, according to scientists themselves, the demarcation between science and nonscience. Agreeing with Popper that precision is important, scientists regard science as characterized by measurable progress. A phenomenon is unreal if no improvement in precision or detection ever occurs. Failure to improve is evidence of pseudoscience. It is often viewed by scientists as a moral issue; hence, we can compare it to the sociological concept of anomie. *Anomie* refers to the antisocial behavior of individuals that is caused by weak social bonds. Just as with anomie, I believe insufficient socialization within the group created the experimenters' malaise. Just as people regard anomie as a sign of the bad character of the people in the social group, so scientists regard the experimenters' malaise as evidence of bad character in the scientists involved. So it is that Eddington ends up being accused of sharp practice or even fraud, despite a lack of evidence that I can see. I use the term *malaise* because scientists are uneasy about the lack of progress in the eclipse experiment but unable to put their fingers precisely upon the reasons for it.

The case of the planet Vulcan is an example that scientists would give of a spurious phenomenon unmasked by the absence of experimental progress. In the nineteenth century, several observers, some at eclipses, claimed they had seen Vulcan. It even appeared on some nineteenth-century charts of the solar system. But the planet did not become more routinely observable with time as knowledge of its orbital motion increased, and photography did not help in proving that it existed. If the precision of an experiment or the detectability of a phenomenon is failing to improve, this could be because the phenomenon is not real. But physicists do believe in the reality of light deflection, so other reasons must account for why a malaise occurred in this case. Of course, some physicists counter that the malaise did not occur in the eclipse test because they claim that Dyson and Eddington overstated their precision in 1919. They claim that the precision of the 1973 eclipse was actually much greater than that of the 1919 eclipse. I contend that this is untrue. The plate measurements of the 1973 eclipse were performed at Herstmonceux because the Texas team sent the plates to the astrometry group at the Royal Greenwich Observatory (RGO), for data reduction, just as the Australians sent their 1922 plates to Davidson at Greenwich. Thus, the Texas plates were reduced by machines and people similar to those who performed the 1978 reanalysis of the 1919 eclipse plates. They claimed similar precision for both sets of measurements (1."66 ± 0."19 for the Texas 1973 plates and 1."90 ± 0."11 for the reanalyzed 1919 plates). The Texas team sent their plates to the RGO for measurement because of the vast experience in astrometry the astronomers there possessed and the excellence of their plate-measuring machines. There are no serious grounds for believing that the 1973 results were more accurate than the measurements of 1919. So what happened?

The experimenters' progress can be the result of improved technology or experimental tools. So the malaise might occur if technology fails to improve. One can argue that this held back the redshift test for decades. Another scenario can occur in which scientists cannot manipulate the apparatus to improve measurement. This is often the case in astronomy, where scientists are observers, not experimenters. A replication of the Mercury perihelion advance would have been nice, but our solar system has only one planet close enough to the Sun to perform this test. Over time, the advent of space travel permitted more solar system tests of relativity involving satellites. Then the discovery of binary pulsars (a pulsar is effectively a planet with a highly accurate clock on board) permitted very precise measurements of the perihelion precession.

The experimenters' progress is often the result of improved experimental techniques by the experimenters themselves, considered as a group. This often depends on the transfer of tacit knowledge between experimenters working together in the research setting. The key point is that in this case the experimenters' malaise occurred because the eclipse measurement was not amenable to repetition, and this prevented researchers from gaining the practical experience necessary for the experimenters' progress to take place. Each scientist only got to perform the experiment once or twice, for a few minutes at a time. There was little scope for learning from one's own mistakes. Even more importantly, scientists could not learn from each other. Normally, students or other junior scientists would expect to be working alongside experienced scientists, learning by observing how the experiments are performed and tacitly absorbing essential techniques that are not frequently discussed in research publications. Thus, the problem with the screws in the plate holders that hurt the 1973 expedition would have been easily corrected if they had ever performed the experiment again. It is true that Dyson did discuss, in his report, his efforts to make sure the plate-holder screws worked well on his astrographic instrument in 1919. He retained Father Cortie's old tube for the four-inch because he knew that it had previously performed well at eclipses. The structure of JPEC helped the British because it encouraged astronomers from different institutions to cooperate and share information as best they could.

So in what circumstances can the experimenters' malaise occur? Obviously, if too few talented young people are coming into a field, this could be a problem. If ideological or other prejudices discourage experienced experimenters from performing an experiment, they may fail to pass on their expertise. But in this case, it was the lack of repeatability of the experiment that prevented younger scientists from learning the kind of tacit knowledge that is indispensable to performing an experiment well. Essentially, the experimental art failed to advance, and this led to a malaise. By the same token, sometimes the experimental art advances rapidly or with a sudden introduction of a new technique, and in this context a malaise can be overcome quite suddenly.

Just as there is an experimenter's art, so there is a theoretician's art. One can argue that the sudden progress in the theoreticians' art that relativity theory represented is what really caused problems for Newtonian gravity. A gratifying advance in theoretical technique encourages scientists to give a new theory serious consideration. Also, improvements in either experimental or theoretical art may cause problems for existing theories or paradigms that had previously seemed perfectly acceptable.

The issue in the eclipse experiment that challenged experimenters the most was the question of how to measure the plate scale change. Very quickly, most practitioners decided that Eddington's check-plate idea was the way to go. But they felt that it could be improved by taking the check plates during the eclipse rather than at night in order to preserve conditions as close as possible to those obtained while the eclipse plates were taken. The split-plate technique of Mikhailov and van Biesbroeck was ingenious but proved problematic. More popular was the idea of moving the telescope during the eclipse to take images of the check field. This costs precious eclipse time, but as plate technology improved, it was possible to accomplish this and take an exposure long enough to image many stars. Nevertheless, even in 1973 everything did not go entirely according to plan. Taking the check plates, along with the eclipse plates, under time pressure, with no chance of a do-over, definitely challenged the experimenter's art. If one could rely on a solar eclipse occurring in the same location every few years, it is easy to imagine a body of the requisite skills and experience being built up by astronomers at that location, but sadly, nature did not cooperate in this respect. One research group devoted itself to this problem over many years and did acquire considerable expertise, that of Freundlich at Potsdam. They also had their own ingenious method involving a double camera to photograph an eclipse field and a check field simultaneously without moving the telescope. Dogged by bad luck, they were ultimately broken up with the Nazis' rise to power just in the years when they might have really made use of their accumulated expertise. A later review article by one member of Freundlich's old group (who was by then in exile from Germany) gives a very considered discussion of the different techniques employed and their strengths and weaknesses (von Klueber 1960).

Ultimately, an improvement of technology happened in another area of astronomy, radio. In an obvious sign of frustration with the malaise, the eclipse test was dropped completely. No professional astronomers have attempted it since, unless they were acting as amateurs and purchasing their own equipment. As I myself know, it is very difficult to obtain any funding for performing this experiment today. This is one problem with discussions about how science should be done. In the real world, scientists have to balance the claims of "doing it properly by the book" with the need to conserve resources. Certainly, it would be nice to keep performing the eclipse experiment, but this costs money, and there are many other things that science ought to be doing. If the eclipse test is not going to improve its precision much, there is no point in continuing with it. Similarly, scientists did not continue to replicate Miller's experiments indefinitely and were ambivalent about the level of expense involved in doing as much replication as they did.

In a final note about the 2017 eclipse, the experience of astronomers performing the experiment at that eclipse suggests that CCDs and other modern equipment do seem to have overcome the problem of the malaise up to a point (Bruns 2018). So really big improvements in technology can overcome the malaise, but it may take a century. Yet it is striking that this, one of the biggest technological advances in twentieth-century astronomy, had no effect for years because by the time CCDs arrived, professional astronomers had given up on the eclipse test following the success of the quasar occultation technique. This is what makes 2017 so interesting. Now an individual can afford to do what previously required a major organizational effort for most of the twentieth century. A large-format CCD camera with a modern telescope, both quite portable pieces of equipment that can fit inside a large automobile, can be taken to an eclipse site and mounted on an inexpensive equatorial mounting. If the weather is bad in one location, the possibility of moving to a different site is quite open! Such a CCD is capable of imaging the thousands of stars hoped for by the Texas team in 1973. One could even consider doing without comparison plates altogether, since so many stars now have their sky positions measured to high precision that one can look up the comparison positions. Electronic computers can do all the computations for a vast number of stars, which was impossible in 1919. Another eclipse will cross the United States in 2024. It may be that the experience gained in 2017 can be applied then to obtain quantities of data large enough to put quite precise bounds on Einstein's theoretical prediction. It is still unlikely that the eclipse measurements will rival quasar occultations in precision, or even the light-bending measurement from Hipparcos, but this would still be a fitting way to commemorate the achievement of 1919. But a word of caution is still advisable: multiple observers from 2017 have reported, even with modern equipment, technical challenges that prevented them from obtaining publishable results.

General Relativity Today

One reason the eclipse test failed to improve is simply because not many people were working on it after 1929. Mikhailov and van Biesbroeck could have investigated together whether their split-plate technique's problems with astigmatism in the check field was due to driving or temperature issues, but this was impossible since they lived so far apart. Mikhailov's efforts were little known in the West because the Cold War imposed real limitations on interchange between scientists on either side of the iron curtain. Furthermore, the two men had other things to occupy their time. Mikhailov was director of the Pulkovo Observatory outside Leningrad from 1947 to 1964. This was the same observatory in which Dyson and Campbell's equipment resided during the First World War. During World War II, it was completely flattened by German forces during the three-year siege of Leningrad. Mikhailov was in charge of its reconstruction—obviously, a considerable undertaking.

The eclipse experiment of 1919 was one of the most important experiments of the twentieth century. It therefore seems odd that by 1947 Mikhailov and van Biesbroeck could have a quiet chat together in Brazil and by themselves encompass most of the people working on the problem at that time. But the fact is that interest in general relativity was at its lowest ebb by that time, and even when the renaissance began ten years later, attention had moved on to different tests and different techniques. That renaissance was to give the eclipse test of 1919 a new significance. As famous as it was in its own day, from 1955 onward, each passing year brought new reasons why its legacy had grown and its influence broadened.

A brief survey must suffice to give a flavor of the rapid development of general relativity after 1955. In cosmology the discovery of the cosmic microwave background in 1965 provided evidence that the big bang theory, one of the most remarkable theories to develop out of general relativity, was true. The discovery of quasars and pulsars in the fifties and sixties led to the development of the concept of the black hole and the refinement of thinking about collapsed stars such as neutron stars. Interest in a unified field theory or theory of quantum gravity concentrated attention on general relativity precisely because it had so successfully passed all experimental tests to date. The theory of black holes led to fascinating concepts like black hole thermodynamics, Hawking radiation, and the Penrose process. As recently as the 1980s, when I entered the field as a young student, one could still hear people claim that general relativity was more a part of mathematics than of physics. Astronomers then still spoke skeptically of the existence of black holes. Now, they are widely accepted. The detection of gravitational waves, which has taken place only in recent years, is one of the greatest triumphs of the theory. Finally, the light deflection effect itself plays a major role in modern astronomy via the phenomenon of gravitational lensing. Certainly, the 1919 eclipse expedition helped make all of this possible. But what if it had not found in favor of general relativity? After all, it certainly seems possible that they could have found against Einstein's theory. They might easily never have taken the four-inch with them. It was only the fortunate fact that Cortie was denied entry to Russia in 1914 that even made that telescope available. How much is science affected by mere chance? How would things have gone differently if Dyson had come out against general relativity?

Certainly, the role of random chance in science was acknowledged by the 1919 team from the outset. Before the expedition started, in an after-dinner speech, we are told that Crommelin proposed the following puzzle: "If C, C', D and E each speak the truth once in three times, independently, and C affirms that C' denies that D declares that E is a liar, what is the probability that E was speaking the truth?" (Chandrasekhar 1987, 118). Can we take this as Crommelin's prediction of what would happen on the expedition he had been asked to join at the last minute?

If so, it was a remarkably prescient summary of what actually transpired. First, Crommelin, in giving only one-third reliability to members of the eclipse team, forecast that only one of their three instruments would obtain really reliable data. Of course, that instrument was the one he had charge of! His foresight goes deeper, though. He correctly has Davidson (D), the operator of the Sobral astrographic, disagreeing with Eddington (E). Cottingham (C') disagrees with D—trivially,

FIGURE 22. Charles Davidson and Frank Dyson at Greenwich during preparations for the total solar eclipse at Giggleswick in Yorkshire in June 1927. This was the first total solar eclipse visible in England since the 1724 eclipse, whose track was drawn by Edmond Halley in figure 3. The coelostat mounted between them is the same one belonging to the Royal Irish Academy that was used in 1919. The light deflection experiment was not attempted at this eclipse because it was of such short duration (less than thirty seconds) and took place so early in the day in England. The Sun was too low in the sky. In any case, Dyson decided not to attempt any professional observations at this eclipse, commenting, "I've never really seen an eclipse myself: I've always been too busy taking photographs" (Wilson 1951, 224). Most observers in England were disappointed by heavy clouds and rain, but Dyson was lucky for the fifth (and penultimate) time.

(Reproduced from Wilson 1951, 224; the photograph first appeared in *Meccano* magazine, June 1927.)

since he did not have his own instrument and merely aided E with his measurements. Obviously, he must agree with E. Finally, Crommelin himself (C) agrees with C' and E and therefore disagrees with D.

Eddington later published this puzzle and gave as his solution that the probability that E (i.e., Eddington himself) is speaking the truth is twenty-five to seventy-one, or considerably less than 50 percent (Chandrasekhar 1987, 118). So he tacitly entertained the possibility that he might have gotten it wrong. It is possible that if the British had come down in favor of the half-deflection in 1919 that Campbell would have felt compelled to water down any evidence for a larger deflection in 1922. Arguments would no doubt have gone on for years, especially once Freundlich entered the fray, but the uncertainty might well have put theorists off and delayed the early development of the theory. How this might have altered the twentieth-century history of physics is difficult to say. However, even if we argue that we might still have ended up, more or less, with the same science we have today, this one experiment's influence would still have reverberated across several decades, which is remarkable.

As it was, the strangeness of the new science of gravity was noted by everyone; and Eddington, who himself had made some of the most important contributions to twentieth-century astrophysics, ended up being known primarily as the man who had proved Einstein correct. When he visited California in the twenties, one of his hosts, who was aware of Eddington's fondness for Lewis Carroll, even immortalized his role in verse:

> The time has come, said Eddington
> To speak of many things
> Of cubes and clocks and meter-sticks
> And why a pendulum swings
> And how far space is out of plumb,
> And whether time has wings.
>
> I learned at school the apple's fall
> To gravity was due.
> But now you tell me that the cause

Is merely $G_{\mu\nu}$[6]
I cannot bring myself to think
That this is really true.

You say that gravitation's force
Is clearly not a pull.
That space is mostly emptiness,
While time is nearly full;
And though I hate to doubt your word,
It sounds a bit like bull.

And space, it has dimensions four,
Instead of only three.
The square on the hypotenuse
Ain't what it used to be.
It grieves me sore, the things you've done
To plane geometry.

You told me that time is badly warped.
That even light is bent.
I think I get the idea there,
If this is what you meant.
The mail the postman brings today,
Tomorrow will be sent.

If I should go to Timbuctoo
With twice the speed of light.
And leave this afternoon at four,
I'd get back home last night.
You've got it now, the Einstein said,
That is precisely right

But if the planet Mercury
In going round the sun,
Never returns to where it was
Until its course is run,
The things we started out to do
Were better not begun.

And if, before the past is through,
The future intervenes,
Then what's the use of anything;
Of cabbages or queens?
Pray tell me what's the bally use
Of Presidents and Deans.

The shortest line, Einstein replied,
Is not the one that's straight;
It curves around upon itself,
Much like a figure eight,
And if you go too rapidly
You will arrive too late.

But Easter day is Christmas time
And far away is near.
And two and two is more than four
And over there is here
You may be right, said Eddington
It seems a trifle queer.

But thank you very, very much,
For troubling to explain;
I hope you will forgive my tears,
My head begins to pain;
I feel the symptoms coming on
Of softening of the brain.

(CHANDRASEKHAR 1987, 125-27; THE POEM WAS WRITTEN BY
PROF. W. H. WILLIAMS OF BERKELEY)

In spite of all the intricacies of Einstein's theory, Eddington always liked to emphasize that the main aim of the expedition had been to weigh light. The idea that light waves have mass and should be affected by gravity was original to Einstein and is inherent in his early work on relativity, which led to the famous equation $E = mc^2$. Eddington would point out to people that the early work on radioactivity, which showed that the emission of gamma rays (a form of light) was associated with a loss of mass by radioactive materials, also suggested this. With his gift for science outreach, he put the heaviness of light into very concrete terms when he said:

> If light has weight in proportion to its mass like matter, I can give you some idea of what that weight is. There is an appalling amount of light in an ounce; in fact, the cost of light supplied by gas and electric companies works out as something like £10,000,000 an ounce. This points the moral of Daylight Saving; the Sun showers down on us 160 tons of this valuable stuff every day; and yet we often neglect this free gift and prefer to pay £10,000,000 an ounce for a much inferior quality. (Quoted in Sponsel 2002, 449)

During the war, daylight saving time had been introduced in Britain for the first time as a wartime cost-saving measure. It was unpopular and was abandoned after the war (Sponsel 2002). Eddington and Einstein had always viewed the expedition's main goal to be to prove that light was heavy. It was an extra bonus for theory that Dyson felt able to go further and confirm that space was curved and not flat.

15

The Problem of Scientific Bias

Einstein scholars are regularly reminded of just how great his fame is. I am often asked (especially by journalists) whether he really first proposed the "Rule of 72" or if he was actually autistic or left-handed. Many things are believed about him that have no basis in fact or evidence. It is particularly common for well-known quotes to become attached to his name. In oral culture the most memorable phrases, or memes, quickly become credited to the people who are themselves most memorable. In our culture anything wise or science related becomes associated with Einstein, who is our preeminent modern sage. This enormous fame descended upon him as a result of the 1919 eclipse expedition, as Einstein himself admitted when he wrote, "The English expedition of 1919 is ultimately to blame for this whole misery, by which the general masses seized possession of me" (*CPAE*, vol. 13, doc. 1263). The great New Zealand physicist Ernest Rutherford concurred and later said to Eddington, "You are responsible for Einstein's fame" (Chandrasekhar 1987, 115).

According to Chandra, who was present in the senior combination room of Trinity College, Cambridge, in 1933 when Rutherford said these words to Eddington, the context was some British dissatisfaction that Einstein's fame exceeded that of Rutherford's, even though

Rutherford was the principal founder of nuclear physics. But such is the way of it. Rutherford himself, according to Chandra, attributed the drama of the eclipse expedition, with its message of postwar reconciliation, to Einstein's sudden rise to great fame. Chandra goes on to quote James Jeans, who divided credit for the eclipse expeditions equally between Eddington and Dyson, on the occasion of Dyson receiving the Gold Medal of the Royal Astronomical Society. Yet here too there was room for only one scientist's name to remain in the public consciousness. Eddington, the founder of stellar astrophysics, quickly became the only name, besides Einstein's, associated with the eclipse experiment. Interestingly, this tendency to ignore Dyson is even found among scientists and historians, who have wondered if Eddington's alleged bias in favor of Einstein's theory influenced the data analysis of the Sobral plates, which Eddington had, in all probability, nothing to do with.

If we do acknowledge that Eddington may have been biased (insofar as we can look into his heart), what does this mean? Can a biased person do good science? It sometimes seems that the mere allegation of bias is enough to invalidate a scientific result. Therefore, the claim that Eddington was biased has led to claims that the eclipse results were somehow suspect. Certainly, when one looks at Eddington's experimental procedures, one can criticize certain things he did. But this is true of all scientific work. It is just that most scientists' work is not scrutinized so closely. Of course, some experiments are performed more carefully and scrupulously than others. We might argue that Dyson and his team were more careful than Eddington, for instance. But no experiment is immune from scrutiny. No matter how careful one has been, once people start looking, there are elements of any scientific work that may be critiqued. It is one thing to allege that questionable things were done; it is another to say that they led to a false result being reported.

To better understand the role of bias in experimental science, let us compare Eddington with other eclipse experimenters. Bias, after all, may be a widespread phenomenon. An obvious counterpart to Eddington is Heber Curtis, who performed the 1918 Goldendale experiment. Curtis was biased the opposite way from Eddington, against Einstein and for nineteenth-century physics. Perhaps, as a result, he reported his measurements as favoring the result that light is entirely unaffected by gravity. Is this evidence that science is often merely an expensive means of confirming one's own prejudices? It is true that Campbell did not, in the end, feel that Curtis' data was good enough to publish and that later on, Campbell's own measurements of the 1922 eclipse favored Einstein. But this could be simply an example of the bandwagon effect. Undoubtedly, the existence of the British expeditions played a role in dissuading Campbell from publishing the suspect 1918 results, and working in 1922, he may have been concerned with replicating the result already made famous by previous expeditions. It is not just theory that can bias an experimenter. As anyone who has performed a laboratory experiment in school or college knows, one is expected to replicate the same result as others have done before. Recovering a different result may not be taken as evidence of an exciting new moment in science. It is more likely to be taken as evidence of an incompetent experimenter.

It is also worth noting that Campbell himself had a bias, and like most astronomers, it was against Einstein. As Curtis began the data reduction, Campbell wrote to the Mr. Hinks who had provided so much help with the site selection: "We hope that a week of intensive computing may give us at least a hint as to what the final results of his work will be. . . . I must confess that I am still a skeptic as to the reality of the Einstein effect in question, but I would not be willing to undertake a technical defense of my skepticism. I am quite ready to welcome a positive result, though I am looking for a negative one" (Crelinsten 2006, 133). In this respect Campbell's viewpoint was simply the reverse of Eddington's. His bias went in the opposite direction, but he claimed, like Eddington, to be open to other possibilities. How open Curtis was, given his later flirtation with the anti-Einstein company, is open to question. His students later claimed that he remained opposed to Einstein's theories all his life (Virginia Trimble, personal communication).

What scientists are trying to do, in performing an experiment, is get the right answer. In this respect it is like doing a crossword puzzle. Most of us know the right answer to a crossword puzzle will be published in the next day's newspaper and keep trying until we get it. Even if we do not have a copy of the relevant newspaper, we may ask a friend for the solution. We try to agree with the result that everyone else got. Similarly, engineers strive to make their devices perform exactly like all other devices of similar type. Research scientists do not have this luxury. If you are the first ever to perform an experiment, you do not know what the correct answer is. You are essentially trying to guess what answer others will get in the future and agreeing with that! One way to do this is to be careful, cautious, and methodical. But, as Peter Galison has pointed out, a time will come when your experiment must end. You must eventually stop being cautious and state a result. Undoubtedly, if no previous experimental results exist, there is a temptation to agree with theory. Indeed, this tendency is obvious in the 1919 team's presentation of their results. They repeatedly framed their experiment as being a choice between three theoretical possibilities. Logically, any result for the deflection of light might have been possible, which Eddington himself acknowledged when he wrote, "It is easy to calculate that the total deviation [due to gravity] of [a material body] on passing the Sun, if it grazed the surface, would be 0." 87, or half the Einstein deflection. It may happen that the ratio of weight to mass for light is not the same as matter. If so the deflection will be altered in the same proportion. The problem of the eclipse may, therefore, be described as that of *weighing light*" (Eddington 1919, 121). But the theoretical issues at stake were of such significance that it made sense to frame the experiment in such a way as to highlight its theory testing aspect.

Of course, theory often plays an essential role in science. It was theory that predicted the size of the effect. Had theory predicted a much larger gravitational deflection of light, Eddington and Dyson would have approached the experiment differently. Had theory predicted a much smaller deflection, they would never have embarked upon it at all. Theory must guide experiment because otherwise we would not know which experiments are interesting and achievable! In fact, had Einstein not pointed it out, most twentieth-century astronomers would never have believed that the Sun's deflection of light could even exist.

Another interesting comparison with Eddington is the case of Dayton Miller. Like Curtis, Miller was skeptical of Einstein's theory. Unlike Curtis, he published his results. I am suspicious that Miller's bias affected his results. His calculation of the direction of the Earth through the ether found a "Miller apex" that conveniently converted the small velocity (roughly ten kilometers per second) of the ether wind measured by his device into the large velocity (two hundred kilometers per second and above) found by Strömberg for the Earth's motion through the galaxy. Did Miller do this consciously and therefore fraudulently? It is possible. It is also impossible to prove, one way or the other. But it is highly plausible that, facing a calculation of his solar apex that was complex and had large error bars, he "kept at" the calculation until he got a result that agreed with the best evidence from astronomy. He had an advantage over other physicists because he worked at the world's leading observatory and had firsthand knowledge of those latest results from astronomy. He could easily appear to have found a result confirmed by the very latest, and very extraordinary, findings in astronomy. But notice how the advance of scientific knowledge forced Miller into making these bold predictions that rendered the theory he favored more falsifiable. In this way the experimenters' progress enables falsifying tests.

Finally, let us compare Eddington with Dyson. Dyson's case is different from Eddington and Miller, but not because he was neutral. That must be rare. Normally, the very fact that you are performing an experiment at all is because you expect a non-null result. No one went to the trouble of hauling equipment to an eclipse before Einstein came on the scene simply to prove that the stars do not change their positions because the Sun is nearby! Dyson was not neutral, but he appears to have changed his mind during the experimental process. He probably started out at least a little skeptical of relativity, like most astronomers, but he ended up confirming the theory. It is interesting that once he changed his mind, he exchanged one bias for another. For instance, he wanted to average the results from his two instruments to get an answer very close to Einstein's prediction.

Eddington had to persuade him that this was not kosher. Here we see the desire to let theory guide you to the right answer in its purest form. Dyson had no prior bias toward Einstein's theory, but once he decided in Einstein's favor, he was reasonably anxious to let the theory guide him. If you know that your error bars are large and that others will perform more exact experiments later, you may feel anxious for vindication in the future by coming as close as possible to the right result now. Theory is sometimes your only guide as to what that right result might be! And of course, once Dyson had nailed his colors to Einstein's mast, he knew his own reputation was bound up with Einstein's because Campbell's results would render a verdict not only on Einstein's theory but also on Dyson's previous experiment. So he breathed a sigh of relief when it seemed as if no "shadow of doubt" remained about Einstein's prediction, as he wrote to Campbell. Once that happened he canceled plans to repeat the test. Greenwich, in fact, never performed the Einstein test again, though they did try in 1929, when they were clouded out. Dyson might be surprised to find that while no shadow of doubt remains about Einstein's theory, there is, nevertheless, a distinct shadow of doubt hovering about his and Eddington's experiment!

Even realists, who believe that science is telling us how the world really is, must acknowledge that we do not have some inborn ability to comprehend the physical world. It takes great acquired expertise to perform scientific experiments. Unfortunately, it is of little use for these experts to do their work without telling the rest of us. By definition, the knowledge they gain about the world must then pass through society to become commonly accepted. If scientific ideas are memes, then we must accept that successful memes are not true—they are simply often repeated. Is it possible that we simply make the science that fits our preconceptions? It is not that simple, but we can say that we do not live in a world where we are born knowing about atoms but have trouble communicating with each other. Instead, we are born into a world with little correct knowledge about its workings but with excellent abilities to communicate with each other. Science is, by necessity, a social enterprise. Only the people who have performed the difficult experiments have empirical knowledge of the way the world really is. In 1920 only Eddington and a few others had personal knowledge of whether starlight is really affected by the

Sun's gravity. It follows that the rest of us must come to accept or reject Einstein's theory through social interaction with people who themselves have interacted with those who performed the measurements. For most of us, it is not observing photons move through spacetime that makes us trust Einstein's ideas. It is the way ideas move through society that makes us believe the "truth."

Ultimately, whether science is socially constructed or determined by the hard facts of reality is irrelevant. What we know is that the hard facts of reality are won with difficulty by people with unusual levels of expertise and skill. How those brave few convince the rest of us about the nature of reality is surely worthy of study, whether we are fooling ourselves about the laws of physics or whether we are on the right track. Indeed, if the social transfer of knowledge in our culture is such as to keep us on the right track, then it is all the more worthy of careful study! It is easy to assume that the study of reality must be straightforward, but it is not. Karl Popper alerted us to the difficulties of confirming a theory. While it is fine for Popper to say that we cannot prove a theory but we can falsify it, we must remember that in practice, falsifying theories is also problematic. If a theory is apparently falsified, it may be that the experiment was wrong or that our assumptions were incorrect. Sometimes, experimental results are just wrong. Just as we should be prepared to ditch theories when necessary, we should also be prepared to hang on to them when necessary. How do we know which of these contradictory bits of advice informs the right course of action in a given instance? We don't.

The fact that the light-bending experiment ceased to be performed after 1973 gives us a further clue about the way science is done. The dilemmas of research recede when experimental technology and technique mature. If an effect is well above the noise associated with measurement, there is less reason to worry that another competent researcher will disagree with the result your measurement has achieved. This is another way of saying that as research work becomes more routine, experimenters have confidence that another expert operating similar equipment will obtain similar results. Radio astronomers used their regular equipment, mounted in its usual location, to do their work. Thus, they were able to check against themselves, repeating the experiment year after year. They could collaborate with colleagues and observe their methods much more easily. In this way they worked steadily to reduce noise and had more faith that another observer would obtain similar results. They felt confident in the experimenters' progress and suffered no malaise.

The eclipse experimenters, however, suffered from a strong malaise. Their precision failed to improve, undoubtedly, because of the difficulties in repeating the experiment. Of course, some observers went to multiple eclipses, but the vagaries of weather and history mostly precluded them from obtaining more than one set of data. For instance, Freundlich traveled on at least six eclipse expeditions but only obtained data once. The 1973 team actually constructed a specialist observatory in the path of totality to try to overcome the problem of using transportable equipment. But they still fell afoul of the lack of repeatability when a technical problem that could have easily been fixed once discovered compromised their measurements. Elizabeth Campbell, Wallace's wife (like Eddington, Campbell went by his middle name with his intimates), stated it most clearly when she said that the difficulty of eclipse work was "two minutes, with no chance to correct mistakes, if there were any" (Pang 1996, 36). In spite of frequent rehearsals, tension before an eclipse was so great, she noted, that "there is never any sleep for the men in charge the night before an eclipse—so many little delicate things have to be done as nearly at the last moment as possible" (Pang 1996, 32). By contrast, occultations of quasars are observable anywhere in the world when they occur. Observers remain happily ensconced in their regular facilities, and every year a new set of observations provide a chance to improve upon the accuracy of the measurements previously taken.

In spite of it all, we have seen that science can progress even when scientists are handicapped by circumstances. But can it progress when they are biased? Progress in science is not guaranteed, and certainly, scientists sometimes change their minds, or have to backtrack. One can accuse many people in our story of bias, so why has Eddington attracted so much criticism over the decades? Primarily, it is because of his fame, of course, but also because he was perceived to have unscientific biases affecting this particular measurement. Some scientists are outraged, for instance, at the idea that he might have favored relativity because he sought reconciliation between English and German science after the war. But given how unpopular Eddington's antiwar views were, this accusation rings false to me. It is true that pacifism gained in popularity after the war as a reaction to its horrors, but this was not predictable in 1919.

There remains the charge that Eddington, like Einstein, his fellow theorist, simply believed in relativity so much, seduced by its theoretical charms, that he wanted it to succeed. Eddington himself cheerfully admitted that he hoped that relativity would be vindicated. When he wrote to Dyson in October 1919, he used the phrase that he was relieved "not only because of theory." He frankly admitted that he had hopes for the theory. When an experimenter performs a measurement and finds that the existing apparatus is the right tool for the job, she is relieved. This is the sort of relief Eddington felt when the eclipse experiment vindicated general relativity. We must acknowledge the bias without jumping to the automatic assumption that it influenced his analysis of his data. The mere existence of bias is not proof of scientific wrongdoing. Bias is normal. It is rare to be completely neutral in science or in life. Punishing Eddington for being honest in articulating his preferences is merely asking that scientists dishonestly suppress their views.

There is a sense that Eddington is in the dock with Einstein on charges of behavior inappropriate for a scientist. Both are theorists accused of being too guided by theory and insufficiently respectful of the role of experiment. When Einstein pitied the dear Lord, who must put nature, his humble creation, to the test against the certainty of Einstein's theory, he reinforced the image of the cocksure theorist who disdains the humdrum work of experimental confirmation. Eddington also, during his career, played to the gallery in this way. Yet we know that Einstein worked hard to encourage astronomers to test his theory. He discovered the possible tests and calculated his theory's predictions. He published papers and wrote letters to leading astronomers to publicize what would need to be done. He collaborated with Freundlich and others and helped raise funds for their efforts. He did everything practical that was required. Eddington did all this and participated in the observational work. It seems strange that all of this practical involvement in the effort to test the theory is ignored, and we are instead confronted with a playful remark clearly meant in jest. This does not mean that Einstein would immediately have capitulated if Curtis and Campbell had published their 1918 results vindicating Newton's theory. He would have insisted that the theory was correct and that their experiment was wrong, and he would have been justified in doing so. Experimenters are sometimes wrong! It was only over the course of many years that it became clear that relativity's prediction of the light deflection was completely correct. But that does not mean the public were wrong to lionize Eddington and Einstein in 1919. A new result is exciting, even if we acknowledge the possibility that it could later be overturned. When it comes with the dramatic overthrow of a famous theory, it is all the more exciting.

The celebration of Einstein's triumph over Newton in 1919 seems so sudden that it naturally raises suspicions of a rush to judgment. Why was only one experiment sufficient to overthrow such a well-established theory? Part of the answer is that Newton's theory had already been overthrown years before. The advent of special relativity, and the theoretical developments accompanying that theory, had rendered it inconsistent with important features of the new physics. Theorists were in the uncomfortable position of having to do a job with a tool they knew to be defective, at least in principle. Imagine working with such a tool, which your client imagines to be the very latest technology, while you know it may only be a matter of time before its shortcomings are apparent to everyone. Obviously, you are in the market for a replacement but conscious that your client will be suspicious if you are not seen working with the acknowledged tool of your trade. This was the situation that Eddington and Einstein found themselves in, and Eddington seized the opportunity to dramatize the changeover from the old theory to the new. It is true that he would have been disappointed if ambiguous experimental findings had ruined this great opening for him. But he would nevertheless have done his best to use the publicity to educate the public and his fellow scientists. I cannot prove that this kind of theoretical bias did not influence his handling of the data, but I am certain that it was his main source of bias and much more relevant than his alleged desire for postwar reconciliation, for instance. Nor am I willing to admit that it was an improper bias. Theorists have preferences for desirable outcomes, and in this they are no different from anyone else in science, or in society.

In this respect it was the fame of the 1919 eclipse that created the problem. Karl Popper was so impressed by Einstein's willingness to put his theory to the test that he developed his ideas on falsification. Popper's ideas have been highly influential, to the point where they now stigmatize a characteristic aspect of Einstein's approach to science. He was famous in his day for being willing, as a theorist, to challenge the validity of experimental results. He did so against early experiments that appeared to falsify special relativity and again against Miller's ether drift experiments. One lesson learned from modern science studies is that scientists fight hard for their beliefs. Science is not about being willing to drop one's beliefs at the first sign of trouble. In fact, it depends on advocacy because in the absence of advocates an idea may be prematurely discarded. We should not show disdain at Eddington and Dyson's skill at artfully presenting their science to the public. It is mistaken to believe that the truth needs no advocate. This need for advocacy applies not only to the public but also within science. Of course, advocacy is often partial and biased. But that is the price we pay for having it. In this respect science is like a court of law. Failing to find an advocate for the innocence of the accused will merely condemn them to conviction. Points of view that are not argued for will go unheard and unconsidered. It was a good thing that a leading theorist was, unusually, involved in the 1919 eclipse expedition because without Eddington, the theorists' insight—that Newton's theory was no longer tenable in its original form—would not have been represented. Without Eddington the importance of the test might not have been properly recognized.

It surely happens that experimenters have hopes for the directions their research might take. They think that if the resonance they are investigating is sharp enough, what a research tool it will prove to be. If the oscillation they are measuring turns out to be sufficiently stable, what a useful timing instrument it could make. Obviously, experimental science, like all science, has disappointments, but it also has moments when hopes and dreams come to fruition. Is this not a kind of bias? If such experimental bias exists, does it invalidate all scientific experimentation? I think not. Experimenters must simply try to be honest with themselves and not let their hopes lead them astray. The only issue setting Eddington apart is that his hopes were related to a theoretical tool rather than an experimental one. He hoped that general relativity would prove itself and open up the vistas that Einstein's innovation of metric theories promised. One of the many roles of a theory is as a simple tool for theoreticians. Just as an experimenter may hope that an experimental result will vindicate the use of a favorite tool, so a theorist may hope for the same thing. We need to recognize that the theoreticians' art is just as important as the experimenters' and just as likely to evolve. In essence, Eddington was in that uncomfortable position of being between paradigms. The old worldview had been overthrown. A new one was not yet firmly in place. The eclipse was exciting just to the extent that it might give a clue to the right path forward. It might suggest that the metric paradigm was viable, helping Eddington and other theorists out of their dilemma. This does not mean he would have manipulated results to make this happy event occur. As he himself wrote, the challenge of the unknown excited him, no matter what his experimental measurements threw at him. But it is hard to see an open ocean of discovery with no far shore in sight without some trepidation. How much more attractive is an ocean with a continent known to exist on the other side, as yet unexplored? Obviously, Eddington leaned toward the latter result.

Does it seem troubling that scientists believe in their theories and that this belief lets them work wonders? Does this reduce science to the status of another myth, something that vanishes when people cease to believe in it? The term *myth* has a pejorative aspect today and is more or less synonymous with falsehood. But it also refers to a way of explaining the world around us, and one of the attractive aspects of myth is the way that good myths are fecund. A myth builds on itself, generating new stories about its characters. Viewed in this way, a myth is a good model for science. One of the most important attributes of a scientific theory is its fecundity. If it fails to give rise to new questions, new concepts, and new research, then it is of little practical value. In this way relativity has been an extraordinarily fecund theory. It has given birth to ideas about the world that never existed before, such as gravitational waves, black holes, neutron stars, and a cosmology in which the geometry of the universe is not necessarily Euclidean. Some of these ideas were still hidden from view in 1919, but Eddington and Einstein knew enough to see the outlines of great discoveries ahead. They did not propose to install general relativity as the new theory of gravity without a test. They could have attempted that without the trouble of proposing tests or going to Africa at all. But whenever anyone undertakes a great endeavor, they harbor hopes for the possible outcome. Eddington and Einstein harbored great hopes for a new era of relativistic gravity. That their hopes came to pass does not, in itself, invalidate their efforts to test whether it really was the right path forward. Looking back a century later, we can certainly imagine that they would be proud of the successes of modern gravitational theory, all made possible by the observations of 1919.

Epilogue

WHERE ARE THEY NOW?

The two astrographic lenses continued their work on the *Carte du Ciel* project after their return from the eclipse. Restored to their original telescopes, they resumed the nightly round of mapping the sky. The project itself was never quite fully completed, and the numerous astronomers' and observatories' enormous dedication to it has sometimes been portrayed in a negative light. Those who focused on this project, it is believed, missed the revolution in astronomy that attended our discovery of the cosmos beyond the Milky Way and its expansion. Nevertheless, today the positions recorded in the *Astrographic Catalogue* and on the *Carte du Ciel* plates themselves play an important role in determining the proper motions of nearby stars when combined with the work of modern projects like NASA's Hipparcos mission.

When the Greenwich observatory moved out of London to Herstmonceux Castle in Sussex, the astrographic telescope that had measured star positions decades earlier under Dyson's direction moved with it, to study the proper motions of stars. When the Royal Greenwich Observatory (RGO) moved again to Cambridge, the telescope remained at Herstmonceux. Today, it occupies dome D in the Observatory Science Center there and is open to the public at times. When visitors use the telescope, they do not actually look through the astrographic lens, which was designed to focus light for photographic plates. Viewed with the eye, it produces fuzzy images. The astrographic camera is mounted on the side of another refractor that serves as its guide telescope. In use, an operator finds the correct star field with the guide scope and then uses the astrographic to take photographs. The equatorial mount of the telescope today is not the original but was built in the sixties by Grubb-Parsons, the successor to Grubb's old Dublin firm.

The Oxford astrographic was also mounted on the side of another refractor, in this case an earlier Grubb refractor built for the Oxford University observatory when it opened in the University Parks (this observatory is not to be confused with the older Radcliffe Observatory, also in Oxford). After the development of the aluminizing method of coating mirrors, which does not tarnish like the old silvering method, observatories switched to reflectors and away from the previously preferred refractors. This left many universities with large refractors that required big domes and took up a great deal of space on campus for little scientific reward. As a result, some of these old refractors have been offered to anyone who will take them away to a good home, and this is what happened to the Oxford Grubb refractor. In 1962 it went to the new Keele University in Staffordshire (Maddison 2011). It still has the brackets on the side for mounting the astrographic camera (Maddison 2011), but the astrographic itself had already been removed earlier the same year. The stellar proper motion project in which the RGO was engaged aimed to include most of the northern sky, including the Oxford zone of the *Carte du Ciel*. As it transpired, the Oxford and the Greenwich astrographic lenses, though both manufactured by Grubb, were made to very different designs. It was necessary to photograph the new star positions from the Oxford zone with the same lens, so the Oxford lens was loaned to Herstmonceux and fitted into the Greenwich astrographic camera when needed. This lens is currently on display at Herstmonceux, which now has both astrographic lenses, one from each of the expeditions of 1919.

The four-inch lens and its Grubb coelostat made its way back to Ireland and is today on display in Dublin at the Dunsink Observatory. Lenses and mirrors cannot speak and identify themselves, and Grubb made two very similar coelostats for the 1900 eclipse, one for the Royal Irish Academy and one for the Royal Dublin Society, who cosponsored the expedition. The Royal Irish Academy mirror was the one loaned to Father Cortie and used in Sobral. After 1919 Ireland achieved its independence, and the post of Astronomer Royal for Ireland ceased to exist. Since the Astronomer Royal had a dual responsibility as director of the Dunsink Observatory, the observatory remained closed for a long period. However, in 1947 the Irish government appointed a new director at Dunsink, the German astronomer Hermann Brück. As a young man, Brück's first position had been as Freundlich's research assistant at the Einstein Tower in the Potsdam Observatory (Brück 2000). Casting around for usable equipment in Dublin, Brück realized that the four-inch lens made by Grubb for the 1900 eclipse had been purchased for Dunsink. He asked for its return, and so the lens and the accompanying coelostat returned to Dublin. For a long time, the wrong mirror was assumed to be the one used in Sobral. But eventually, a logbook from the Grubb-Parsons firm came into the hands of Dunsink astronomer Tom Ray, and this gave the serial numbers of all Grubb's instruments, along with where they were sold or sent. Using this book, Ray was able to identify the correct coelostat, the one Crommelin had used and that had returned to Dunsink in 1947. He also found the four-inch lens languishing in a box in the basement. He identified it by measuring its focal length, which had been quoted precisely in the paper published by the 1900 Irish expedition. Both instruments have now been fully restored and are on display at Dunsink on the outskirts of Dublin.

Grubb's logbook ended up in Newcastle after Sir Charles Parsons purchased the company, which was in financial difficulty, from Grubb in 1925. The company was, from then on, known as Grubb-Parsons. This brought the firm's history full circle, since Parsons was the son of William Parsons, the famous Earl of Rosse, with whom Thomas Grubb had worked on telescopes in the nineteenth century. But

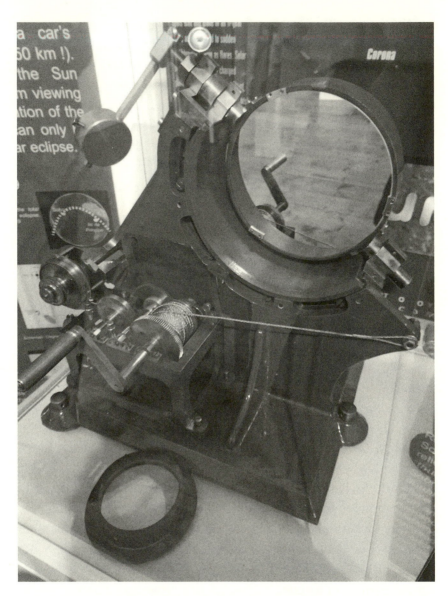

FIGURE 23. The four-inch lens and accompanying coelostat mirror made by Howard Grubb of Dublin for the Irish solar eclipse expedition of 1900, sponsored by the Royal Irish Academy and the Royal Dublin Society. The Royal Irish Academy and Dunsink Observatory in Dublin loaned these pieces to Father Aloysius Cortie for use in a number of eclipses, and they were taken to Sobral in 1919 as a backup instrument to the astrographic lens from Greenwich. The instrument was operated by Andrew Crommelin at Sobral, producing the high-quality images that provided the strongest evidence against Newton's theory of gravity and for general relativity. The lens and mirror are now on display at Dunsink Observatory in Dublin.
(Courtesy of Julia Kennefick.)

Parsons' principal motivation in taking the company on seems to have been patriotic rather than familial, in recognition of the importance of high-precision optics for military applications. As the inventor of the steam turbine, he had revolutionized naval warfare, making the modern battleship possible. The first such vessel, HMS *Dreadnought*, was powered by his turbines. More importantly, his invention provided an enormous boon to modern life by transforming the way electricity is generated. The glass used by firms like Grubb was, before the war, mostly made in Germany. As the war continued, the British government was forced to set up a special plant in Derby to produce the required glass. This plant would have gone bankrupt after the war if Parsons had not purchased it to maintain this important industrial capacity for Britain (Manville 1971). When Howard Grubb retired in 1925 and his firm, the main client of the Derby plant, might also have gone under, it was natural for Parsons to buy it as well and move it closer to his existing business in Newcastle, in the northeast of England.

Grubb himself moved back to Dublin for the later years of his life, so he does not seem to have been estranged from the new independent Ireland that the political heirs of his Fenian workforce created (Manville 1971). However, independent Ireland did lose out in the field of astronomy, which had been dominated by Anglo-Irish families like the Grubbs and Parsons. Eddington himself had experienced this now-vanished world of Irish gentlemen astronomers when, as a young man, he tutored the son of William Edward Wilson at one of the most important Irish observatories,[1] Daramona House, near Streete, County Westmeath, in the Irish midlands (Glass 2006, 200). By the 1920s these private observatories were long gone, following the demise of the old landed gentry,[2] and the new state could not afford to maintain observatories like Dunsink at the forefront of twentieth-century astronomy. The Crawford Observatory in Cork, equipped by Grubb, was built with a private donation in the late nineteenth century but went unused in the twentieth century for lack of an astronomer on the faculty at the university. In the latter part of the twentieth century, most Irish astronomers, even if educated in Ireland, went abroad to work in the field. Only in the twenty-first century has astronomy in Ireland begun to recover its former status. Cork now has an astronomer, in Paul Callanan, who has expertly restored the observatory; and visitors there will see Howard Grubb's instruments in their original setting, since he also designed the building. It comes complete with one of the astrographic lenses made by Grubb, similar to those used in 1919, and a coelostat mirror also made by Grubb.

The learned societies that sponsored the expeditions are still going strong. The Royal Society has moved to Carlton House Terrace, but the Royal Astronomical Society (RAS) is still based at Burlington House, where the famous joint meeting took place. They still publish their learned journals, and the RAS Gold Medal is still awarded every year. The last hiatus since 1920 was during World War II. The medal was awarded to Eddington in 1924, to Dyson in 1925, and to Einstein in 1926. The observatories have not fared so well. The Cambridge Observatory no longer exists as an independent entity but has merged into Cambridge University's Institute of Astronomy, which is still situated on the Madingley Road outside town. The original observatory buildings and telescopes still form part of the institute.

The Royal Observatory, Greenwich, was moved in the fifties to Herstmonceux Castle in Sussex when light pollution drove it out of London. While there, it was referred to as the Royal Greenwich Observatory, or RGO, and it continued to play a leading role in the life of British astronomy. In 1990 it was moved to Cambridge, where it occupied a site close to the Institute of Astronomy. Thus, the two observatories that had cooperated so famously in 1919 ended up as neighbors. However, in 1998 the RGO, after existing for over three centuries, was closed, its remaining functions taken over by the Royal Observatory in Edinburgh in an effort to save money by consolidating Britain's two state observatories.[3] British kings had always felt the need to duplicate functions between their two kingdoms, England and Scotland. Accordingly, there was a Scottish Astronomer Royal as well as an English one. Dyson had held the Scottish post before becoming Astronomer Royal for England. Today, though both posts continue as ceremonial appointments, there is only one national observatory, and the RGO has ceased to exist. The original observatory buildings in Greenwich, home to the prime meridian, are now a museum.

The Lick Observatory is still in use on Mount Hamilton in Northern California, although its future is said to be uncertain.[4] Its astronomers are now at the University of California at Santa Cruz, part of a general trend in which astronomers are now usually based at universities rather than at observatories. This is partly because of the increasing importance of their teaching roles and partly because the best modern telescopes are usually cited at relatively remote locations, such as mountaintops. In fact, Lick was the first of the mountaintop observatories. Mount Wilson followed, though it too is no longer financially supported or operated by its founding institution, the Carnegie Institute. In the late twentieth century, astronomers became frequent flyers as they regularly traveled to conduct their observing runs at distant telescopes. Nowadays, there is a trend toward conducting these runs by communicating remotely with the telescope operators via the Internet. In many cases the astronomers no longer leave the university campus. Many still like to travel to eclipses but often in an essentially amateur capacity, as I myself have done (I have gone one for two, being clouded out on one occasion). Although the RGO has closed, the *Observatory* magazine, founded at Greenwich in 1877, still exists. It continues to report on meetings of the RAS, as it did in 1919, and is still so widely read by amateur as well as professional astronomers that it has a more diverse readership than most scientific journals.

The *Anselm* continued to ply her trade back and forth between Liverpool and the Amazon River for the Booth Line until 1922, when the decline in the Brazilian rubber trade made this route less profitable (Haws 1998, 132). She was then sold to an Argentinian company and renamed the *Commodoro Rivadavia*. In 1931 she ran aground in the Straits of Magellan and was salvaged by the Chilean navy. Eventually, she returned to Argentina, having been purchased by the government there, and was renamed the *Rio Santa Cruz* in 1944, in keeping with other Argentinian government vessels named after rivers. Similarly, her original name reflected the Booth Line's custom of naming ships after saints of medieval England. She had been launched from the Workman and Clark yards in Belfast in 1905 and survived until 1959, when she was broken up for scrap at the end of a long working life. She even outlived most of the astronomers who sailed on her in 1919.

The human protagonists themselves all enjoyed long careers in science. Einstein and Eddington are often cited as prime examples of scientists who did their best work while young. In the words of Einstein's chief biographer, he might as well have gone fishing after 1925 (Pais 1994, 43). This is somewhat unfair, as it is only true when compared with the extraordinary success of their earlier work. Admittedly, both men concentrated, somewhat unprofitably, on unified field theory work in their later years. Eddington's work received even more criticism than Einstein's for its idiosyncrasy. Most of those involved with the eclipse measurements did not live to see general relativity's renaissance. Dyson did not live to see World War II begin, dying on board ship while returning from a visit to Australia. He was buried at sea (Wilson 1951). He would have shown no surprise that the eclipse expeditions are now universally remembered as being Eddington's affair. His biographer, who was also his daughter, quoted him as saying, "If I'm remembered in the future it will be because of my association with Eddington. People will say—'Dyson? Oh yes—he was Astronomer Royal, when Eddington was Chief Assistant'" (Wilson 1951, 157).

Like Dyson, Cottingham, Campbell, and Crommelin were all born in the mid-nineteenth century. Like him, they did not live to see the postwar period at all, and neither did Eddington, who, uniquely among the main actors in our story, did not live to see seventy years of age. He passed away in 1944 after an operation to treat cancer, with his sister by his side. The other younger men—Einstein, Trumpler, and Freundlich—lived on into the postwar period. Einstein died in 1955, too soon to attend a meeting held to commemorate the anniversaries of relativity theory in that year (fifty years of the special theory and forty years of the general). This meeting helped spark the renaissance in the general theory that entered full force in the sixties. The last of the eclipse men to leave the scene was one of the older generation, Davidson, who lived until 1970, when he died at the age of ninety-five. By then the term *black hole* had been coined, and the golden age of relativity theory was in full swing. During his long life, Davidson was made a Fellow of the Royal Society, a rare achievement for someone who began his working life in science as a computer.

Freundlich, who ended up being skeptical of relativity, lived well into the renaissance. This period revived interest not only in Einstein's theory but also in alternatives to it, and Freundlich found support for his ideas in some quarters, including from Einstein's old friend Max Born, though not from Einstein himself (Batten 1985). But he was long gone from the wonderful observatory he had built in Berlin by then, and his team was dispersed. After the Nazis came to power, Freundlich, who was partly of Jewish descent, had to leave Germany. He spent the years 1933 to 1936 in Istanbul, where he helped build up and equip the observatory there. He then took a position in Prague at the same German university where Einstein had worked when the two men began to tackle the light deflection problem together. He had built up a group there when the Nazis invaded in 1938 and forced him to flee again. This time he was fortunate enough to win a position at the University of St. Andrews in Scotland, where he ultimately held a newly endowed chair in astronomy. He settled happily in Scotland, using his Scottish middle name Finlay as if it were part of his surname. He developed a strong research group and built a new telescope. As late as 1954, he traveled to an eclipse in Sweden with the goal of finally settling the light deflection question. He was then nearly seventy years old but approached the task with an energy and enthusiasm that excited his students (Batten 1985). Again, he was foiled by poor weather, though others not far away had clear skies. He tried again the following year, returning to the East Indies region, where he met with success in 1929. The eclipse of 1955 was part of the same saros cycle as the 1919 eclipse and was, in fact, the longest total solar eclipse since the Middle Ages. Hopes must have been high, but one last time he was clouded out (von Klueber 1965). He passed away in 1966, having returned to Germany in his last years.

The phenomenon of gravitational light deflection, so difficult to detect in 1919, is now one of the most important tools of modern astronomy. Objects with far greater gravitational fields than the Sun, such as galaxies and black holes, can affect light from distant stars and galaxies so strongly as to focus the light from behind them. In this way, galaxies can be seen at far greater distances than would otherwise be possible. What is needed is a fortunately placed intermediate galaxy that operates as a gravitational lens to augment the light that reaches us from the more distant galaxy. This is of central importance to our study of galaxy evolution and cosmic structure. Gravitational lenses often produce such a large light deflection that not only is the star seen to be deflected on one side of the lens, but its light can also travel around the other side of the lens, so that we see the distant galaxy in two images on either side of the lensing galaxy. In the most perfect case, where the lensed galaxy lies directly behind the lensing galaxy, the lensed galaxy is visible as a ring of light, known as an Einstein ring, around the lens. This was all predicted by Einstein himself, although he only published the result in 1936 after being pestered to do so by an enthusiastic admirer (Renn and Sauer 2000). Einstein's objection to publishing had been that he thought astronomers would never be able to observe gravitational lensing!

If we view the two theories as the main protagonists of the eclipse story, then both are currently in rude health. Newton's law of gravity continues to be taught in every school, and the astronomers of 1919 will be relieved to hear that for many astrophysical applications, astronomers still reach for that familiar old equation to perform their calculations. When Joseph Taylor discovered the first binary pulsar in 1974, he told me that his wife was highly amused when he set off to the bookstore to buy textbooks on general relativity. He realized that he had discovered the first stellar system whose motion was governed by the strong gravitational fields predicted by Einstein's theory. Although he had taken a class in the theory as a student, he felt he was in need of a refresher course, since most practicing astronomers rarely use it. Even today most people who have not attended a graduate school in physics or astronomy would be unlikely to have taken a class in general relativity or to have studied it in any detail. Nevertheless, the theory has gone from strength to strength since 1919, passing every experimental test and giving rise to a new universe of astonishing phenomena, many of which, including black holes, gravitational waves, and dark energy, have passed into the vernacular and become the stuff of popular science programs on television. We have now reached a centenary of the theory's fame since it first hit the headlines in 1919. It is once again in the public view after the first detection of gravitational waves, and there seems no immediate likelihood that it will be overthrown. This is not for lack of trying. Alternative theories of gravity have been produced in plentiful supply since the 1950s, though none have looked likely to displace general relativity. Ironically, 1919 might be regarded as a low point for numbers of rival gravitational theories in existence! But few theory tests in the history of science can match the 1919 eclipse for drama, adventure, publicity, and consequence. Those involved made history by giving focus to a desire for peace and reconciliation after a terrible war. They also left their mark on our view of the universe around us. They showed that light has weight and that spacetime is curved, and these two paradoxical discoveries have changed physics forever.

ACKNOWLEDGMENTS

Portions of this book, especially from chapters 10, 11, and 12, appeared previously in Kennefick (2012). Although much of the material has been edited considerably for this book, it has been reproduced here by the kind permission of the original publisher, Birkhauser, now a part of Springer. A section of chapter 14 previously appeared in the *Collected Papers of Albert Einstein*, volume 15. Although edited somewhat for this book, it appears here by kind permission of the Einstein Papers Project and its director, Diana Buchwald, and the original publisher, Princeton University Press.

This book is dedicated to my wife, Julia Kennefick, who advised me that I should look into whether the 1919 eclipse plates could be remeasured and reanalyzed using modern equipment and techniques. Without her I would never have asked the crucial questions or heard the important answers.

The original suggestion that I go to the archives and closely research this topic came from my mentor Diana Buchwald, who, as director of the Einstein Papers Project, sent me to England in 2003. During that trip I was able to obtain the archival material that forms the basis of this book. Her encouragement and advice were critical at all stages of the project.

The project could not have been completed without the help of my friend Charlie Johnson of Flint, Michigan. Charlie has undoubtedly done more work in the 1919 eclipse archives than anyone living (for his only possible rival, see below). His indefatigable search for more relevant material and his constant willingness to share his discoveries with me has made this book possible. In addition, he has provided his invaluable insights based on his familiarity with all of the relevant documents in the case. He read this book several times in manuscript form and offered much helpful advice.

Adam Perkins of the Cambridge University Library, who then had charge of the Royal Greenwich Observatory (RGO) archive, was the one who told me about the 1978 reanalysis of the Sobral plates and who provided great assistance in going through the RGO papers relevant to the eclipse. His successor, Emma Saunders, offered much-appreciated help and advice in the final stages of writing the book.

Adam also very kindly gave me the name of Andrew Murray, whose idea it was to undertake the data reanalysis in 1978. Andrew very thoughtfully corresponded with me via e-mail and provided fascinating background to the plates and their analysis. It was the Cambridge astronomer Donald Lynden-Bell who actually put me in touch with Andrew. I should also thank Martin Rees, one of Eddington's successors as Plumian Professor of Astronomy, who as director of the Institute of Astronomy met with me in 2003 as part of a fruitless effort to track down the Principe plates.

Raymond Walter, then an undergraduate student at the University of Arkansas, combed through the data analysis sheets with me and helped me to puzzle out the steps taken in the crucial analysis that led to Dyson's decision to discount the Sobral data.

My friend and colleague at the Einstein Papers Project, Jozsef Illy, cowrote with me the original draft of the section of chapter 14 titled "Finding Our Place in the Galaxy." Although that draft has been considerably rewritten for this book, it still bears unmistakable signs of our original collaboration, and I am grateful for Joska's permission to use it here.

Tom Ray, of the Dublin Institute of Advanced Studies and Dunsink Observatory, very generously showed me the four-inch lens and accompanying coelostat mirror and gave me great insights into how the instruments were used both in 1919 and at earlier eclipses.

Paul Callanan of University College Cork (UCC), my alma mater, was my host throughout the writing of this book, showed me the Grubb instruments at UCC that are very similar to the lens and mirrors used in 1919, and devoted endless time to helping me understand the context of this instrumentation. I also thank the head of the physics department at UCC, John McInerney, and the other friends and colleagues in the department for all their kindness and hospitality as I worked on the book.

Ian Glass, an Irish-born astronomer working in South Africa, made himself available by e-mail to answer many thorny questions, which were as much help as his wonderful books and articles on the period.

Mark Hurn of the Cambridge Institute of Astronomy library answered many questions about papers there (and the absent Principe plates) and also provided me with an image of Halley's eclipse map. An old colleague, Gavin Dalton of Oxford University's astronomy department, helped me track down the whereabouts of the Oxford astrographic lens, along with Graham Dolan, Derek Jones of the Institute of Astronomy, and Sandra Voss, director of the Observatory Science Center at Herstmonceux. Graham was so kind as to provide me with several images used in this book and a great deal of information about the origins of these and other images.

Colin Partridge was a comrade in arms during the early days of this work, as we were both skeptical of the modern attack on Eddington and company and exchanged encouraging e-mails with each other. He kindly let me use a copy of his excellent master's thesis on the subject.

I had the opportunity to discuss, several times, important issues in the light deflection experiment with a leading expert on the subject today, astronomer Brad Schaeffer of Louisiana State University (LSU). I am also grateful to Toby Dittrich, who led the effort to raise funds for a modern redo of the experiment on a massive scale using citizen-scientist volunteers.

Jeffrey Crelinsten is the only person who can rival Charlie Johnson for time spent in the archives, and his book was indispensable to my research as I worked on this book. I am also in debt to the work of Matt Stanley, Alistair Sponsel, and Alex Soojung-Kim Pang for their groundbreaking research on various aspects of eclipse expeditions in the late nineteenth and early twentieth centuries. Klaus Hentschel,

Virginia Holmes, and Andy Warwick also wrote important books on the same period of scientific history, which were essential in writing my book.

Roni Grosz, director of the Albert Einstein Archive at the Hebrew University of Jerusalem, kindly helped me track down Einstein's letter to Hale; and his predecessor, Ze'ev Rozenkranz, my friend from the Einstein Papers Project, also thoughtfully advised me in many ways, including putting me in touch with historian Alfredo Tolmasquim. He, along with Luis Crispino, gave me invaluable advice about the eclipse in Brazil. Carlos Veiga and Katia Santos of the Observatório Nacional in Brazil generously provided me the photograph of the astronomers at Sobral. Gerard Whelan of the Royal Dublin Society was very helpful in identifying photographs of Howard Grubb.

At the University of Arkansas, I received very valuable assistance from our excellent librarians, especially Robin Roggio and Stephanie Pierce.

Finally, I had a number of very careful readers who made excellent suggestions that improved the manuscript, including my wife, Julia Kennefick; Charlie Johnson; Joe Martin; Dennis Lehmkuhl; Virginia Trimble; Ze'ev Rosenkranz; Jeroen van Dongen; and my editor, Al Bertrand, who was a very encouraging and helpful enabler of the book.

I leave the last word to my daughter, Dara. As a historian, I like her definition of science because learning about the past is included in it: "I think that science is important because you can learn a lot about the planet, universe, and past and future, and because you can make great discoveries and make the world a better place."

APPENDIX

The Sobral data analysis sheets are preserved at the RGO's archive in the Cambridge University library (Papers of Frank Dyson, RGO Archive 8, fol. 150). They are the work of Charles Davidson, Herbert Furner, and Frank Dyson. All the measurements of star positions were made by Davidson and Furner independently, and the average of their values was used in the calculations that followed (Dyson, Eddington, and Davidson 1920, 302). Key pages are in Dyson's handwriting. Particularly important is the calculation discussed in chapter 12 (on page 249), which concludes on page 210 with a mean light deflection of 1".98. This calculation begins on page 208 of the sheets (see figure 24).

At the top of page 208, we see a table very similar to the one that appears at the bottom of page 311 of the report (Dyson, Eddington, and Davidson 1920, table 2 below).

There are a few very minor differences, but otherwise it is the same. It consists of a set of equations, one each for the sixteen plates from the astrographic instrument, from which the data was taken. Thus, the calculations on page 208 clearly refer to the astrographic telescope, not the four-inch (with which fewer plates were taken). Comparing the table on page 210 (whose mean is the value of 1".98 mentioned above) with the one on page 208, we see there is every reason to believe that it also refers to the astrographic plates. There are sixteen quantities, numbered 1–12 and 15–18. This is the same numbering as on page 208 and page 311 of the report. These numbers refer to the eighteen plates taken with the astrographic telescope, but with plates 13 and 14 excluded because they were not used in the data analysis (probably because cloud obscured the stars on those two

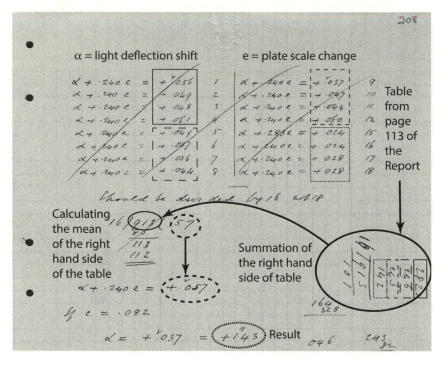

FIGURE 24. Page 208 of the Sobral data analysis sheets: At the top is a set of equations with only minor differences from the table at the bottom of page 311 of the published report (Dyson, Eddington, and Davidson 1920, table 2 below). The next step is to take the mean of the values on the right-hand side of these equations. But the value 0.913 is not the correct sum of the numbers in this table. It comes instead from inside the oval on the right, which is probably the sum of the values from the table multiplied by about 1.3, to give the total plus maximum error. The sum inside the oval seems to be obtained by summing the values in the table (multiplied by 1.3) in groups of four, as indicated. The mean of this sum (which is 0.57) is then evaluated and put in equation form. The equation is solved with a given value of e, the plate scale change given in the report. This gives the solution for the light deflection α at the limb of the Sun as 1".43 (marked *Result* here). This would be the largest value of the deflection consistent with the random errors and the measured scale change for the astrographic telescope. It is obviously inconsistent with the Einsteinian deflection of 1".75.
(Cambridge University Library, RGO Archive 8, fol. 150, 208. Reproduced by permission of the Science and Technology Facilities Council and the Syndics of Cambridge University Library.)

plates). By contrast, the four-inch lens took only eight plates because it was a slower lens that required longer exposure times.

So the deflection of 1".98 at the limb does not refer to a calculation based on four-inch data. It refers to astrographic data. Yet famously, the astrographic instrument produced a value for the deflection at the limb of the Sun that was 0".93, quite incompatible

TABLE 2.

Plate	e	α at limb	α + 0.24 e	α	α + 0.24 e
		From the report		From the data analysis sheets	
	p. 310	p. 310	p. 311	p. 210 (fig. 25)	p. 208 (fig. 24)
1	0.089	1".28	0.056	0.063	0.056
2	0.059	0".97	0.049	0.054	0.049
3	0.101	1".09	0.047	0.079	0.048
4	0.091	1".28	0.059	0.086	0.061
5	0.076	0".97	0.050	0.061	0.049
6	0.082	0".82	0.059	0.063	0.057
7	0.119	0".00	0.036	0.073	0.036
7	0.166	0".00			
8	0.064	0".82	0.046	0.049	0.044
9	0.129	0".31	0.035	0.055	0.037
10	0.045	1".01	0.048	0.047	0.047
11	0.061	1".24	0.045	0.058	0.044
12	0.102	1".91	0.059	0.085	0.060
12	0.114	0".74			
15	0.036	0".70	0.026	0.027	0.024
16	0.037	0".70	0.024	0.026	0.024
17	0.109	0".47	0.028	0.068	0.028
18	0.000	1".17	0.029	0.021	0.028

Data given in the report (Dyson, Eddington, and Davidson 1920) and the data analysis sheets (see figures 24 and 25) for the astrographic plates. The first column gives the number of the plate from which the data was taken. The second column gives the scale change *e* given in the report for that plate, and the third column gives the light deflection (α) at the limb of the Sun (ibid., 310). Column four gives the value in the right-hand side of the equation for $\alpha + 0.24 e$ from the report (ibid., 311). Column 5 gives the light deflection (α) for each plate as calculated by the alternative method given in the data analysis sheets (210, figure 25). Column 6 gives the values on the right-hand side of the equation for $\alpha + 0.24 e$ given in the data analysis sheets (208, figure 24), which are nearly identical to the value from the report given in column 4. All values are given in revolutions of the micrometer screw, except for values in column 3, which are in arc seconds.

with the value of 1".98 obtained from the four-inch instrument. What is going on here? By what method of analysis can the data from the astrographic be made to agree with the result from the four-inch? Obviously, if such a data-reduction method existed, it might explain why Dyson would then pronounce his willingness to accept the four-inch result, as he does at the bottom of page 210 (see chapter 12).

Let us go back to the top of page 208 to the table that is very similar to one that appears in the report. It represents a considerable reduction of the raw data from the astrographic, since each line in the double-column table refers to a given plate and gives an equation relating the average light deflection for stars on that plate α to the difference of plate scale *e* between the eclipse and the comparison plates. The right-hand side of each of the equations is a number obtained from a detailed reduction of star positions on the eclipse and comparison plates. It is different for each plate because of measurement errors unique to each plate. In the report the mean of these numbers from the sixteen equations is taken on page 312. This mean immediately gives the quoted result for the astrographic of α at the limb of the Sun as 0".93, given a scale change *e* of 0.r082 deduced from the same data. A value written 0.r082 refers to a turn of 0.082 revolutions of the micrometer screw on the plate-measuring machine. By a specific constant, it can be converted into a value in seconds of arc. For instance, the quoted value for α at the limb of the Sun of 0".93 corresponds to a shift in stellar positions on the plate of 0.r024. So we see that in this case the shift in positions of the stars due to a change in scale between the eclipse and the comparison plates was four times greater than the shift due to light deflection, according to the calculation given in the report.

Thus, we can see why someone would be worried that the scale change for the astrographic plates is uncomfortably large. But how could anyone start from this same table on page 208 of the data analysis sheets and end up with a light deflection value of 1".98 at the limb of the Sun two pages later?

Right under the table on page 208 is the memo "should be divided by 16 not 18," a necessary reminder since the last line in the table is labeled *18*, but only sixteen values are included because of the absence of plates 13 and 14. Then comes the division, obviously intended to be taking the mean of the values on the right-hand side of each line of the table. The value 918 is divided by 16, giving the answer 57. Below this comes the resulting mean equation, reading "α + 2.40

$e = +$ r.057." Then, we read, "If $e = .082$ $\alpha = +$ r.037 $= +$ 1".43." This is a surprise, since the report gets a value of 0."93 by taking the mean of essentially the same table. The explanation is that 0.918 is not the sum of the values in the table on page 208. Why write down a table of values and then take the mean of some different values? The only clue lies off in the right margin of the page, written at 90 degrees to the text. Here is a sum that adds up to 913, but it is a sum of four numbers, 282, 246, 243, and 142, which appear to have little to do with the table. One possibility is that, since the table consists of sixteen numbers, each of these four numbers might be the sum of four numbers from the table. That does not work because the sum of numbers 1–4 from the table is 214. Then, we have 186 from plates 5–8, 188 from plates 9–12, and 104 from plates 15–18. As noted in the report, the values from plates 15–18 are unusually low. The same is true for the fourth number above, 142. It is noticeably lower than the other three. Sure enough, the numbers 282, 246, 243, and 142 are all larger than the numbers 214, 186, 188, and 104 by close to a factor of 1.3. Maybe what is going on here is a calculation of the largest possible value consistent with the results from the astrographic plates. In that case 0."93 from the report would represent the result, while 1".43 would represent the top of the error bar. We could present this result as 0".93 ± 0".5, where 0".5 = 1".43 – 0".93.

How plausible is this interpretation? Let us look back to page 310 of the report where the value of α at the Sun's limb is given for each plate separately in arc seconds (see table 2, column 3). We notice first that there is considerable scatter. Indeed, plate 7 actually gives the null light deflection result. Obviously, since every other plate taken by both expeditions shows a positive shift, this result stands out—so much so that the plate was completely remeasured against a different comparison plate. This is why two lines in the table were devoted to it. The second effort gave an even larger scale change between plates but still a precisely zero result for deflection at the limb. Plate 12 also gave problems. Initially, it had a value for α at the limb of 1".91, completely out of step with the other astrographic plates. After remeasurement against a different comparison plate, it gave a value of 0."74! Thus, this plate alone yielded results that could be said to favor either "Newton" or Einstein. The table gives the mean of α from all eighteen plates (sixteen plates were actually used since 13 and 14 were discarded, but plates 7 and 15 were measured twice). If we now calculate the difference of the α of each plate from the mean and take the mean of those variances, then we find that the mean variance is 0".35. This is a rough estimate of the error to be expected from the plates. Since the result accepted by the expedition is close to 1 arc second (0".93), it follows that the error is something close to (indeed greater than) 30 percent. Therefore, it is a reasonable supposition that what is happening on page 208 is that instead of adding up all the values of α in the table to get 0.918, Dyson and Davidson have instead totaled up the means plus variance to get this maximum value. It is worth noting that the table at the top of page 208 has been struck through, as if someone decided not to use those values in taking the mean at that point. Unfortunately, they did not take the trouble to completely write out the calculation that they actually did (at least not on this page), and we must guess at the procedure.

If we accept that the value of 1".43 in the middle of page 208 represents the highest possible deflection associated with the astrographic plates, based on the actual mean plus the maximum error, then we must allow that we are still not near the Einsteinian value and nowhere near the mysterious value of 1".98 quoted on page 210. Let us now turn to the top of that page and read the text there:

"There is an apparent difference of scale +r0.82. If however, it may be assumed that there is no net difference of scale between the eclipse and comparison plates, except the part accounted for by refraction and aberration, the value of α found from the several plates would be," and then follows the table mentioned above, which gives a value for each plate of the sixteen mentioned, whose average allegedly gives 1".98 shift at the limb.

So what is happening here is the same calculation mentioned briefly in the report on page 312. It assumes that the change of scale was small, instead of the large value derived from the measurements. In the report that calculation gave a value of 1".52, which is clearly different from 1."98. But we note right away that 1".98/1".52 = 1.3, so once again we must suspect that this is a calculation of a value for α

There is an apparent difference of scale
+.082. If however, it may be assumed
that there is no real difference of scale
between the Eclipse & Comparison plates, except
the part accounted for by refraction and
aberration, the values of α found from
the several plates would be

<table>
<tr><td>1</td><td>+·063</td><td>5</td><td>+·061</td><td>9</td><td>+·055</td><td>15</td><td>+·027</td><td rowspan="4">Light
Deflection
values (α)
for each
astrographic
plate</td></tr>
<tr><td>2</td><td>+·054</td><td>6</td><td>+·063</td><td>10</td><td>+·047</td><td>16</td><td>+·026</td></tr>
<tr><td>3</td><td>+·079</td><td>7</td><td>+·073</td><td>11</td><td>+·058</td><td>17</td><td>+·068</td></tr>
<tr><td>4</td><td>+·086</td><td>8</td><td>−·049</td><td>12</td><td>+·038</td><td>18</td><td>+·021</td></tr>
</table>

The mean of these quantities is (+.051) Mean of α

or (1.98) at the limb.
Result

The results from the Astrographic telescope
support those from the "four-inch" telescope in
showing an outward deflection. For the reasons
given above they are not of the same value,
and the figures obtained from the 4-inch telescope
are to be accepted as the result of the expedition
to Sobral

FIGURE 25. Page 210 of the Sobral data analysis sheets: This is the page on which Dyson (it is his handwriting) decided to reject the data from the astrographic lens. At the top of the page, he states that he will calculate the light deflection by assuming the minimum possible plate scale change. The tabulated values below this statement are these light deflection values (for α) for each of the sixteen measured plates. They almost certainly differ from the values on page 208 by having this different minimum value for e substituted in, as is suggested by a simple analysis of the figures. Then, the mean for α is given (the mean of these tabulated values), which is labeled *Mean* here. This corresponds to a deflection of 1".98 at the limb of the Sun, labeled *Result* here. This would be the largest value of the light deflection consistent with a minimum scale change (i.e., a properly functioning telescope) and random errors all trending the same way. It is consistent with the Einsteinian deflection of 1".75 at the limb. With this result Dyson declares that he will accept the result of the four-inch data, which, coincidentally, is identical with this alternative result from the astrographic data.

(Cambridge University Library, RGO Archive 8, fol. 150, 210. Reproduced by permission of the Science and Technology Facilities Council and the Syndics of Cambridge University Library.)

plus the error, to give a maximum possible light deflection. This is an easy matter to check. Looking back to page 208 (figure 24), we see that the calculation there, based on the measured value of the scale change $e = 0.082$, went as follows:

$\alpha = 0.057 - 0.24*e = 0.057 - 0.24*0.082 = 0.037$, which corresponds to 1".43 at the limb.

Now, jumping ahead to page 210, the new calculation is described as rejecting the value of $e = 0.082$ and instead using the minimal theoretical scale change of $e = 0.02$. This would give us $\alpha = 0.057 - 0.24*e = 0.057 - 0.24*0.02 = 0.0522$, which agrees pretty well with the value of $\alpha = 0.051$ given on page 210. This is equivalent to a value of 1".98 at the limb of the Sun.

This means that the error bar of the 1".52 result does include the Einstein result but not the Newtonian result. In fact, the error bar just stretches as far as the mean value from the four-inch plates so that the two instruments' results overlap (on the basis of certain assumptions). This is true because the value on page 210 was obtained by a two-step process. One step was to recalculate α for each plate using not the measured e but the theoretical e. The other step (based on the calculation on page 208) was to use not the mean value but the mean plus variance in order to get a maximum possible value, given the small scale change. The mean of this table, 1".98, is then the maximum possible value that can be gotten from the Sobral astrographic, assuming that the plate scale did not change (except by the known theoretical amount) and assuming the maximum possible error. Since this value is consistent with the four-inch plates, we are not surprised to find that Dyson concluded he was confident in using the four-inch plates as the official results of the expedition.

Thus, the alternative data analysis method used by Dyson and company gives a result with an error bar of 1".52 ± 0".46. Comparing this with the result from the 1979 reanalysis, they gave 1".55 ± 0".34 for the same plates. These are remarkably similar figures, suggesting that Dyson and Davidson were correct in their analysis. Obviously, they were not in a position to assert that the 1."52 figure was the true one, but nevertheless, their logic is clear. If the lower 0".93 figure is correct, then the scale change was enormous—much bigger than it should have been. If the scale change was minimal, then the 1".52 figure is correct. In the former case, there is no justification, apart from a nationalistic love of Newton, for using the results because the scale change is unacceptably high. The instrument was dysfunctional. In a normal experiment, this run would have been discarded; and a new run, perhaps with a change of some items of equipment, made instead. Since this could not be done at the eclipse, the data was rejected. Of course, something might still be rescued for Newton if one performed a calculation designed to see if a result obtained with a small-scale change still favored him. But, no, it turns out that such an assumption favors Einstein instead. Accordingly, given that another instrument did favor Einstein, it makes the most sense to reject the poor-quality astrographic data and accept the other data. That is indeed what we find Dyson doing on page 210 of his notes.

Chapter 1. The Experiment That Weighed Light

1. Lunar eclipses are not as scientifically interesting and can be observed more easily, since the area of Earth over which a lunar eclipse is visible is very large.

2. Russell, letter to A. Vibert Douglas, October 10, 1953, Allie Vibert Douglas fonds, 2303.9-1-4, Queen's University Archives, Kingston, Ontario.

3. Although Dyson shares the same family name as the eminent physicist Freeman Dyson, they are not related, as Freeman Dyson has confirmed to me. But the younger man can recall Sir Frank Dyson speaking on the radio when Freeman was young and Freeman's family taking note that someone who shared their surname was such an eminent British scientist.

4. *Encyclopedia Britannica*, 11th ed. (1911), s.v. entry on "De la Rue, Warren."

Chapter 2. Eclipses

1. Letters from A. S. Eddington to Sarah Ann Eddington, fol. Trinity/EDDN/A2, Trinity College, Cambridge, https://janus.lib.cam.ac.uk/db/node.xsp?id=EAD%2FGBR%2F0016%2FEDDN.

2. The mine would have been deployed by a German minelaying submarine.

Chapter 3. Two Pacifists, Einstein and Eddington

1. We can think of this in either of two ways. One is that any given part of the disk circumference is shrunk by length contraction. Since the disk now has insufficient circumference to make a circle, it breaks apart under the strain. Alternatively, consider that in the rotating frame of reference, the circumference of a circle is greater than in a stationary frame of reference because pi has increased. The disk, which was constructed for a stationary frame of reference, has the wrong circumference for this new geometry and breaks apart.

2. Most geometers would prefer to say that the lines of longitude are not parallel, which means there are no parallel geodesics on a sphere. That is fine, but for our purposes they have, at the equator, just the property that Omar Khayyam demanded for parallel lines, and yet they are congruent.

3. The universe could also be negatively curved. Such a geometry is harder to imagine, though it was in fact the first kind of non-Euclidean geometry to be discovered. In this geometry, geodesics do not converge; they diverge. They move farther away in both directions. If a universe has too little mass, it will be negatively curved, which means it will be infinite in extent. We could be living in such a universe.

4. They have this name because if light rays had clocks, the rest of us would say they were stopped due to special relativistic time dilation. The closer to light speed one travels, the slower clocks run. At light speed, they appear to be stopped altogether. Thus, time does not pass for light rays, and the time it takes them to travel anywhere is zero. For this reason the paths on which they travel through spacetime are called *null geodesics*.

Chapter 4. Europe in Its Madness

1. There seems to be some debate as to whether the bombs actually damaged the observatory in 1916. Only in World War II did the Germans finally deal significant damage to its buildings, during the Blitz.

2. This was not the first time the Royal Observatory had been the target of a bomb. An anarchist named Martial Bourdin blew himself up just outside the building in 1894, in a bizarre incident that forms the basis for Joseph Conrad's novel *The Secret Agent*. A more dangerous incident that year took place just before Christmas, when a high wind blew the shutter off the astrographic dome. Part of the dome landed inside the observatory, narrowly missing the young Charles Davidson, as reported by the then newly appointed Chief Assistant Frank Dyson in the observatory's journal (http://cosmicdiary1894.blogspot.com/2009/12/saturday -december-22-1894.html).

3. Kapteyn to Eddington, January 24, 1915, Eddington Collection, 3237.1-1-2, Queen's University Archives, Kingston, Ontario.

4. This may be a reference to the zeppelin raids, since these German airships were often referred to as baby killers, in reference to the indiscriminate nature of aerial bombardment.

5. If this is a reference to German schools declaring a holiday in celebration of the sinking, then it is most certainly not a fact. That was a successful bit of Entente propaganda. I write these words, incidentally, while looking out at the harbor mouth in Cork, through which the few survivors of the *Lusitania* were brought after being rescued off the Irish coast. The official German response to the sinking was frank. They regarded the *Lusitania* as a legitimate target because she carried ammunition in her hold. Over a century later, the sinking of this passenger liner still arouses great passions. At the time, it was shocking beyond words. The crew of one British ship operating out of Cork Harbor, the *Baralong*, murdered more than a dozen surrendering U-boat mariners in cold blood in the weeks after the sinking. This atrocity was in turn used by Germany to justify the no-warning torpedoing of ships by U-boats as late as the Second World War (before 1917, U-boats were supposed to surface and demand that a ship surrender).

6. The *Vanguard* was a battleship that blew up without warning in 1917 while anchored in Scapa Flow, a bay in the Orkney Islands used as the base of the British Grand Fleet in World War I. Somehow, ammunition in her magazines exploded, setting off a chain reaction that rapidly sank the ship with the loss of almost all hands aboard.

7. Newall was the director of the Solar Physics Observatory in Cambridge. He did not work at Eddington's University Observatory.

8. During World War II, Copenhagen continued to serve as the Central Bureau, with Swedish help, but the German office, which had moved to Berlin from Kiel, did not survive the partition of Germany after the war. A new, purely West German bureau was set up in Heidelberg, so the fault line in the astronomy news system was reestablished along the line of the iron curtain (Vinter Hansen 1955).

9. This day, November 9, is the German 9/11 (or 9.XI, as Einstein wrote it). It is an even bigger day in German history than the other 9/11 is in American history. On this day the Berlin Wall fell in 1989. It should be a German national day, but it is also the date of the Nazi Beer Hall Putsch and the Kristallnacht, the Night of Broken Glass, on which the Nazis proclaimed the end of German decency in an enormous state-coordinated pogrom against Jews. Tainted by this association, the day has a conflicted legacy that prevents its annual celebration.

Chapter 5. Preparations in Time of War

1. I am grateful to the Royal Astronomical Society (RAS) for permission to quote from the minutes of the Joint Permanent Eclipse committee and its sub-committees. These minutes are to be found in RAS Papers 54, Minutes of the Solar Eclipse Committee and the Joint Permanent Eclipse Committee, Vol. 1, Joint Permanent Eclipse Committee Minutes, RAS Library, London.

2. The principle of the constancy of the speed of light in vacuum applies only to inertial observers and therefore is not one of the principles of the general theory. Imagine if you were at the horizon of a black hole and shone a light beam upward. It would be unable to move upward at all because light cannot escape from the event horizon. The speed of upward-moving light, in your (unrealistic) frame of reference, would be zero.

3. And in general relativity, the effect Laplace was trying to calculate is actually of even higher order, fifth order in v/c, and therefore even smaller (Kennefick 2007).

Chapter 6. The Opportunity of the Century?

1. Presumably Harold Spencer Jones, who, much later, succeeded Dyson as Astronomer Royal and became a knight in his turn.

2. Minutes of JPEC meetings are preserved in the archives of the Royal Astronomical Society, Burlington House, London.

3. Another issue Eddington raised in response to Hicks' presentation before the RAS was that the region of Lake Tanganyika had not been thoroughly

surveyed. With latitude and longitude of locations by the lake not perfectly known, an expedition might inadvertently choose an observing site some distance off the center line of totality, something that apparently happened to Eddington in Brazil in 1912. Of course, an island like Principe was bad from this point of view anyway, since the center line in the vicinity of Principe was in the ocean.

4. La Paz in Bolivia seems never to have been considered, presumably because there the eclipse occurred at sunrise, again low on the horizon, which was just as bad as being near sunset at Lake Tanganyika.

5. This observatory was a successor to an observatory in Madras that played a central role in the history of eclipse astronomy, because of its part in the eclipse of 1868 and the discovery of Helium (Nath 2013).

6. The Irish astronomer Annie Maunder, famous for her study of sunspots, had earlier also wondered if the Earth could have an effect on the Sun's surface.

Chapter 7. Tools of the Trade

1. He arrived only a fortnight after the infamous attempted bombing of the observatory by an anarchist.

2. Admittedly, the admiral's resources were not as great as they might have been, as the observatory was still recovering from the disrepair it had fallen into following the Prussian siege of Paris in 1870 and the Paris Commune of the following year.

Chapter 8. The Improvised Expedition

1. The *Vanguard* was one of only two British battleships (or dreadnoughts) to be lost during the Great War; the other was sunk by a mine.

Chapter 9. Outward Bound

1. These letters are preserved in the Eddington Papers at Trinity College, Cambridge, fol. Trinity/EDDN/A4, https://janus.lib.cam.ac.uk/db/node.xsp?id=EAD%2FGBR%2F0016%2FEDDN. Permission to quote from them is gratefully acknowledged from The Master and Fellows of Trinity College Cambridge.

2. The SS *Anselm*, which was sunk by a U-boat while serving as a troopship in 1940, was not the same ship. She was a successor to Eddington and company's boat, owned by the same steamship line. Both ships were named after a medieval archbishop of Canterbury.

3. Edward VII had died in 1910.

4. Eddington's biographer Vibert Douglas records that audience watching at rural cinemas was something of a pastime of Eddington's, who was an avid hill walker (Douglas 1957, 35).

5. "Fiance" is written but struck out in the letter.

6. The island of Principe was uninhabited when the Portuguese first occupied it. By natives, Eddington presumably meant islanders descended from slaves taken from the nearby coast of Africa.

Chapter 11. Not Only Because of Theory

1. Dyson's biographer, who was his daughter, has it that Dyson said, "If you do, Eddington will go off his head and commit suicide and you will have to come home alone" (Wilson 1951, 192).

2. This steamer later (in 1929) wrecked on the coast of Principe's neighboring island of São Tomé.

3. Jones (the Mr. Jones of the JPEC minutes) was discussing Eddington's trip to Malta to check a discrepancy in its measured longitude. This was one of his duties while serving as chief assistant at Greenwich.

4. Quoting from several different reviewers, from a version of the Amazon website, accessed June 27, 2007, http://www.amazon.com/gp/product/customer-reviews/0192805673/sr=8-1/qid=1182979869/ref=cm_cr_dp_all_helpful/104-5654845-3967159?ie=UTF8&n=283155&qid=1182979869&sr=8-1#customerReviews.

5. Apparently, the use of this term to describe the eclipse cycle may be a mistake by modern scholars trying to interpret ancient Babylonian texts in which the number saru would figure repeatedly.

6. The author is grateful to the Science and Technology Facilities Council and the Syndics of Cambridge University Library for permission to quote from the RGO Archive.

7. Eddington took two different brands of photographic plates with him to Principe, one of which he was unable to develop on the island, as he tells us on page 116 of his book *Space, Time and Gravitation* (Eddington 1987).

8. This famous anecdote was told to Chandra by Eddington himself and is related in Chandrasekhar 1987, 117.

9. Silberstein had clearly gathered that the conditions of 1919 were especially favorable and obviously worried that this was something that would not recur again in his lifetime.

Chapter 12. Lights All Askew in the Heavens

1. Stratton was made a member of both Britain's Distinguished Service Order and France's Legion d'Honneur.

2. Donald Lynden-Bell, private e-mail communication to the author, May 15, 2003.

3. At the joint meeting, Dyson pointed out, in response to Newall, that the corona does not appear symmetrical (including at the 1919 eclipse), but the star shifts measured were, as demanded by Einstein's theory.

4. For those who have visited Greenwich and have seen the prime meridian marker, Andrew Murray was the man who most recently defined the meridian at Greenwich around 1950. When the RGO moved to Herstmonceux, he also defined the meridian of the observatory there.

Chapter 13. Theories and Experiments

1. The story is told by Crelinsten (2006, 136–40) in his meticulously researched book on the early testing of general relativity.

2. See Cliff Will's book *Was Einstein Right?* for an authoritative account.

3. Presumably, Henry Norris Russell, who was awarded the medal the following year.

4. Article from *Washington Jewish Week*, accessed July 2018, http://washingtonjewishweek.com/15720/almost-100-years-ago-einstein-visited-d-c/featured-slider-post/.

5. The reason I know this is that the *Celtic* became a famous shipwreck in Cork Harbor when she ran aground under the lighthouse in 1928. The passengers were brought off safely and the ship broken up. Some of the passengers saved were enduring their second shipwreck of the voyage. They had started their journey from New York aboard the SS *Vestris*, which sank off Hampton Roads with much loss of life. Some of those rescued were brought aboard the *Celtic*, leading to newspaper speculation of a "Jonah" among them whose bad luck had wrecked two ships in one crossing!

6. Eddington to Lindemann, April 24, 1932, Lord Cherwell Collection D.57/4, Box D51-D69, Archives of Nuffield College, Oxford.

7. This short time of totality included an unusual eclipse in 1930 visible in California to which Lick sent a team even though totality lasted less than two seconds. It was an impressive test of twentieth-century astronomy's ability to predict eclipses with high accuracy, given that the track of totality was only one mile wide. This eclipse was not used to test Einstein's light deflection prediction (Pearson, Orchiston, and Malville 2011).

8. In his report Campbell refers to it as Ninety Mile Beach.

9. As we have seen, a group of Australian astronomers led by George Frederick Dodson did take up a station at Cordillo Downs sheep station in South Australia. Their plates were reduced by Davidson at Greenwich, so the Royal Observatory did end up playing a role in 1922 after all. Although the data from Cordillo Downs was of relatively poor quality, it was generally consistent with Einstein and inconsistent with the "Newtonian" result.

10. The Campbell family had done its bit for the war effort in the Great War. Elizabeth and Wallace's son Douglas Campbell was the first U.S.-trained air ace of the war ("Douglas Campbell, World War I Air Ace and Executive, 94," *New York Times*, October 17, 1990, 24).

11. This eclipse would have been viewable, for instance, from Baja California.

Chapter 14. The Unbearable Heaviness of Light

1. See *Cleveland Press*, May 25, 1921; letter from Miller to T. C. Mendelhaft, June 2, 1921, AIP Archive 70, 984.

2. George Ellery Hale to Joseph Larmor, December 21, 1921, Hale Papers, The Caltech Archives, Pasadena, CA.

3. The title of the most important scientific biography of Einstein, by Abraham Pais, *Subtle Is the Lord . . .* (Pais 1982), is borrowed from this quote.

4. The only total solar eclipse potentially visible to a large number of Americans between these dates was the eclipse of 1970, whose track of totality ran up along the Southeastern Seaboard but missed the great population centers of the Northeast. Unfortunately, clouds obscured the view along most of the track of totality within the United States (see *Weatherwise* 23, no. 100 [April 1970]).

5. *Kessel* (the German word for a cauldron) is a term that refers to the surrounding or encirclement of troops, made possible by the use of panzers in the famous blitzkrieg warfare of World War II.

6. $G_{\mu\nu}$, known as the Einstein tensor, describes the curvature of spacetime in general relativity and therefore the effects of the gravitational field. When Richard Feynman wanted to find his way to a conference of general relativists half a century later, he asked the taxi dispatcher which way a group of men who all walked around saying things like "Gmunu, Gmunu" had gone the previous day. He was taken to the right place (Kennefick 2007).

Epilogue

1. Wilson was the amateur astronomer who had gone with Grubb on the Irish expedition to observe the 1900 eclipse (Joly et al. 1901).

2. It was the struggle over this issue, known as the Land War, that had so excited Grubb's workforce in the 1880s.

3. Arguably, the Royal Observatory, Edinburgh, owed the rise in status that saw it "eclipse" its more famous sibling, the RGO, to the same Hermann Brück who had restarted astronomy at Dunsink. He left Ireland to take up the post of Astronomer Royal for Scotland and helped give the Edinburgh observatory an international reputation in instrumentation.

4. Article by Dennis Overbye in the *New York Times*, June 2, 2014.

BIBLIOGRAPHY

Adams, Walter S. 1925. "The Relativity Displacement of the Spectral Lines in the Companion of Sirius." *Proceedings of the National Academy of Sciences* 11:382–87.

Aitken, R. G. 1939. "Frank Watson Dyson, 1868–1939." *Publications of the Astronomical Society of the Pacific* 51:336–38.

Almassi, Ben. 2008. "Trust in Expert Testimony: Eddington's 1919 Eclipse Expedition and the British Response to General Relativity." *Studies in History and Philosophy of Modern Physics* 40:57–67.

Anderson, Alexander. 1919. "The Displacement of Light Rays Passing near the Sun." *Nature* 104:354.

Anonymous. 1932. "Adolph Friedrich Lindemann." *Monthly Notices of the Royal Astronomical Society* 92:256–58.

Ball, David. 2018. "Cottingham, Edwin Turner (1869–1940)." *Ringstead People*. http://ringstead.squarespace.com/ringstead-people/2010/10/14/cottingham -edwin-turner-1869-1940-modern-times.html.

Barrow-Green, June. 1999. "'A Corrective to the Spirit of Too Exclusively Pure Mathematics': Robert Smith (1689–1768) and His Prizes at Cambridge University." *Annals of Science* 56:271–316.

Batten, Alan H. 1985. "Erwin Finlay-Freundlich, 1885–1964." *Journal of the British Astronomical Association* 96:33–35.

Bauer, Louis Agricola. 1920. *Résumé of Observations Concerning the Solar Eclipse of May 29, 1919 and the Einstein Effect*. Archive 8, fol. 123. Frank Dyson Papers. Royal Greenwich Observatory Archives. Cambridge: Cambridge University Library.

Bertotti, Bruno, Brill, Dieter, and Krotkov, Robert. 1962. "Experiments on Gravitation." In *Gravitation: An Introduction to Current Research*, edited by Louis Witten, 1–48. Hoboken, NJ: John Wiley and Sons.

Born, Max. 2004. *Born-Einstein Letters, 1916–1955: Friendship, Politics and Physics in Uncertain Times*. London: Macmillan.

Brück, Hermann A. 2000. "Recollections of Life as a Student and a Young Astronomer in Germany in the 1920s." *Journal of Astronomical History and Heritage* 3:115–29.

Bruns, Donald G. 2018. "Gravitational Starlight Deflection Measurements during the 21 August 2017 Total Solar Eclipse." *Classical and Quantum Gravity* 35: 075009.

Brush, Steven G. 2014. "Why Was Relativity Accepted?" *Physics in Perspective* 1:184–214.

Burman, R. R., and Jeffery, P. M. 1990. "Wallal: The 1922 Eclipse Expedition." *Proceedings of the Astronomical Society of Australia* 8:312–13.

Callaprice, Alice. 2011. *The Ultimate Quotable Einstein*. Princeton, NJ: Princeton University Press.

Campbell, William Wallace. 1920. "Resignation of Dr. Curtis." *Publications of the Astronomical Society of the Pacific* 32:201–2.

Campbell, William Wallace. 1923. "The Total Eclipse of the Sun, September 21, 1922." *Publications of the Astronomical Society of the Pacific* 35:11–44.

Castle, Ian. 2015. *The First Blitz: Bombing London in the First World War*. Oxford: Osprey.

Chadwick, James. 1961. "Frederick John Marrian Stratton, 1881–1960." *Biographical Memoirs of Fellows of the Royal Society* 7:280–93.

Chandrasekhar, Subramanian. 1976. "Verifying the Theory of Relativity." *Notes and Records of the Royal Society of London* 30:249–60.

Chandrasekhar, Subramanian. 1987. *Truth and Beauty: Aesthetics and Motivations in Science*. Chicago: University of Chicago Press.

Chant, C. A., and Young, R. K. 1923. "Evidence of the Bending of the Rays of Light on Passing the Sun, Obtained by the Canadian Expedition to Observe the Australian Eclipse." *Publications of the Dominion Astrophysical Observatory* 2:275–85.

Christie, W.H.M., and Dyson, Frank Watson. 1896. "On the Determination of Positions of Stars for the Astrographic Catalogue at the Royal Observatory, Greenwich." *Monthly Notices of the Royal Astronomical Society* 56:114–35.

Coles, Peter. 2001. "Einstein, Eddington and the 1919 Eclipse." In *Historical Development of Modern Cosmology, ASP Conference Proceedings Vol. 252*, edited by Vicent J. Martínez, Virginia Trimble, and María Jesús Pons-Bordería, 21–42. San Francisco: Astronomical Society of the Pacific.

Collins, Harry M. 1974. "The TEA Set: Tacit Knowledge and Scientific Networks." *Science Studies* 4:165–86.

Collins, Harry M. 1985. *Changing Order: Replication and Induction in Scientific Practice*. Chicago: University of Chicago Press.

Collins, Harry M., and Pinch, Trevor. 1993. *The Golem: What Everyone Should Know about Science*. Cambridge: Cambridge University Press.

Cortie, Aloysius L. 1915. "Preliminary Report on the Total Solar Eclipse of 1914 August 21. (Observed by the Expedition of the Joint Permanent Eclipse Committee to Hernösand, Siveden.)" *Monthly Notices of the Royal Astronomical Society* 75:105–17.

CPAE (*Collected Papers of Albert Einstein*). Princeton, NJ: Princeton University Press. 15 vols. https://einsteinpapers.press.princeton.edu/. Selected individual volumes are listed in this bibliography under Einstein's name.

Crelinsten, Jeffrey. 2006. *Einstein's Jury: The Race to Test Relativity*. Princeton, NJ: Princeton University Press.

Crispino, Luis Carlos Bassalo, and de Lima, M. C. 2016. "Amazonia Introduced to General Relativity: The May 29, 1919, Solar Eclipse from a North-Brazilian Point of View." *Physics in Perspective* 18:379.

Crommelin, Andrew Claude De la Cherois. 1919a. "The Deflection of Light during a Solar Eclipse." *Nature* 104:372–73.

Crommelin, Andrew Claude De la Cherois. 1919b. "The Eclipse Expedition to Sobral." *Observatory* 42:368–71.

Crommelin, Andrew Claude De la Cherois. 1919c. "The Eclipse of the Sun on May 29th." *Nature* 102:444–46.

Crommelin, Andrew Claude De la Cherois. 1919d. "Results of the Total Solar Eclipse of May 29 and the Relativity Theory." *Nature* 104:280–81.

Curtis, Heber. 1919. "The Einstein Effect: Eclipse of June 8, 1918." In "Papers and Abstracts of Papers to be Presented at the Pasadena Meeting of the Society." *Publications of the Astronomical Society of the Pacific* 31:190. San Francisco: Astronomical Society of the Pacific.

Davidson, Charles R. 1922. "Observation of the Einstein Displacement in Eclipses of the Sun." *Observatory* 45:224–25.

Davidson, Charles R. 1923. "The Amount of the Displacement in Gelatine Films Shown by Precise Measurements of Stellar Photographs." *Transactions of the Optical Society* 24: 41–46.

Davidson, Charles R. 1940. "Andrew Claude de la Cherois Crommelin." *Monthly Notices of the Royal Astronomical Society* 100:234–36.

Denieffe, Joseph. 1906. *Recollections of the Irish Revolutionary Brotherhood.* New York: Gael. https://archive.org/details/personalnarrativ1906deni/page/n5.

de Sitter, Willem. 1916a. "On Einstein's Theory of Gravitation, and Its Astronomical Consequences." *Monthly Notices of the Royal Astronomical Society* 76:699–728.

de Sitter, Willem. 1916b. "On Einstein's Theory of Gravitation, and Its Astronomical Consequences, Second Paper." *Monthly Notices of the Royal Astronomical Society* 77:155–84.

DeWitt, Cécile. 1957. *The Role of Gravitation in Physics: Report from the 1957 Chapel Hill Conference.* United States Air Force Technical Report. Edition Open Sources (website). Edited by Dean Rickles. http://www.edition-open-sources.org/sources/5/toc.html.

Dodwell, George F., and Davidson, Charles R. 1924. "Determination of the Deflection of Light by the Sun's Gravitational Field from Observations Made at Cordillo Downs, South Australia during the Total Eclipse of 1922 September 21." *Monthly Notices of the Royal Astronomical Society* 84:150–62.

Douglas, A. Vibert. 1945. "Arthur Stanley Eddington." *Journal of the Royal Astronomical Society of Canada* 39:1.

Douglas, A. Vibert. 1956. "40 Minutes with Einstein." *Journal of the Royal Astronomical Society of Canada* 50:99–102.

Douglas, A. Vibert. 1957. *The Life of Arthur Stanley Eddington Nelson.* Nashville: Nelson.

Dyson, Frank Watson. 1915. "Report of the Astronomer Royal to the Board of Visitors of the Royal Observatory, Greenwich." *Greenwich Observations in Astronomy, Magnetism and Meteorology Made at the Royal Observatory, Series 2* 76:H1–H24.

Dyson, Frank Watson. 1917. "On the Opportunity Afforded by the Eclipse of 1919 May 29 of Verifying Einstein's Theory of Gravitation." *Monthly Notices of the Royal Astronomical Society* 77:445–47.

Dyson, Frank Watson. 1920. "Analysis of the Proper Motions of the Reference Stars in the Greenwich Astrographic Catalogue." *Monthly Notices of the Royal Astronomical Society* 80:633–36.

Dyson, Frank Watson. 1921a. *Astrographic Catalogue Greenwich Section: From Photographs Taken and Measured at the Royal Observatory, Greenwich Vol IV: Proper Motion and Photographic Magnitudes*. London: His Majesty's Stationary Office.

Dyson, Frank Watson. 1921b. "Relativity and the Eclipse Observations of May, 1919." *Nature* 106:786–87.

Dyson, Frank Watson, Eddington, Arthur Stanley, and Davidson, Charles R. 1920. "A Determination of the Deflection of Light by the Sun's Gravitational Field, from Observations Made at the Total Solar Eclipse of May 29, 1919." *Philosophical Transactions of the Royal Society, Series A* 220:291–330.

Dyson, Frank Watson, and Turner, Herbert Hall. 1917. "The Commencement of the Astronomical Day." *Observatory* 40:301–2.

Dyson, Frank Watson, and Woolley, R. 1937. *Eclipses of the Sun and Moon*. Gloucestershire, UK: Clarendon.

Earman, John, and Glymour, Clark. 1980. "The Gravitational Redshift as a Test of General Relativity: History and Analysis." *Studies in History and Philosophy of Science* 11 (September 1980): 175–214.

Earman, John, and Glymour, Clark. 1981. "Relativity and Eclipses: The British Eclipse Expeditions of 1919 and Their Predecessors." *Historical Studies in the Physical Sciences* 11:49–85.

Eddington, Arthur Stanley. 1913. "The Greenwich Eclipse Expedition to Brazil." *Observatory* 36:62–65.

Eddington, Arthur Stanley. 1915. "Some Problems of Astronomy XIX: Gravitation." *Observatory* 19:93–98.

Eddington, Arthur Stanley. 1919a. "The Deflection of Light during a Solar Eclipse." *Nature* 104:372.

Eddington, Arthur Stanley. 1919b. "The Total Eclipse of 1919 May 29 and the Influence of Gravitation on Light." *Observatory* 42:119–22.

Eddington, Arthur Stanley. 1924. "Relation between the Masses and Luminosities of the Stars." *Monthly Notices of the Royal Astronomical Society* 84: 308–32.

Eddington, Arthur Stanley. 1940. "Sir Frank Watson Dyson, 1868–1939." *Obituary Notices of Fellows of the Royal Society* 3:159–72.

Eddington, Arthur Stanley. 1941. "Edwin Turner Cottingham." *Monthly Notices of the Royal Astronomical Society* 101:131–32.

Eddington, Arthur Stanley. 1987. *Space, Time and Gravitation: An Outline of the General Relativity Theory*. Cambridge: Cambridge University Press.

Eddington, Arthur Stanley, and Cottingham, Edwin Turner. 1920. "Photographs Taken at Principe during the Total Eclipse of the Sun, May 29th." *Report of the Eighty-Seventh Meeting of the British Association for the Advancement of Science*. London: John Murray, 156–57.

Eddington, Arthur Stanley, and Davidson, Charles Rundle. 1913. "Total Eclipse of the Sun, 1912 October 10: Report on an Expedition to Passa Quatro, Minas Geraes, Brazil." *Monthly Notices of the Royal Astronomical Society* 73:386–90.

Einstein, Albert. 1907. "Relativitätsprinzip und die aus demselben gezogenen Folgerungen." *Jahrbuch der Radioaktivität* 4:411–62.

Einstein, Albert. 1911. "Ueber den Einfluss der Schwerkraft auf die Ausbreitung des Lichtes." *Annalen der Physik* 35:898–908.

Einstein, Albert. 1912. "Lichtgeschwindigkeit und Statik des Gravitationsfeldes." *Annalen der Physik* 38:355–69.

Einstein, Albert. 1915. "Erklärung der Perihelbewegung des Merkur aus der allgemeinen Relativitätstheorie." *Preussische Akademie der Wissenschaften, Sitzungsberichte*, part 2, 831–39.

Einstein, Albert. 1919. "Prüfung der allgemeinen Relativitätstheorie." *Naturwissenschaften* 7: 776.

Einstein, Albert. 1995. *The Collected Papers of Albert Einstein, Volume 4: The Swiss Years: Writings, 1912–1914*. Edited by Martin J. Klein, A. J. Kox, Jürgen Renn, and Robert Schulmann. Princeton, NJ: Princeton University Press.

Einstein, Albert. 2004. *The Collected Papers of Albert Einstein, Vol. 9: The Berlin Years: Correspondence, January 1919–April 1920*. Edited by Diana Kormos Buchwald, Robert Schulmann, Jozsef Illy, Daniel Kennefick, and Tilman Sauer. Princeton, NJ: Princeton University Press.

Einstein, Albert, and Fokker, Adriaan D. 1915. "Nordströmsche Gravitationstheorie vom Standpunkt des absoluten Differentialkalküls." *Annalen der Physik* 44:321–28.

Einstein, Albert, and Grossmann, Marcel. 1913. "Entwurf einer verallgemeinerten Relativitätstheorie und einer Theorie der Gravitation." *Zeitschrift für Mathematik und Physik* 62:225–61.

Everitt, C. W. Francis. 1980. "Experimental Tests of General Relativity: Past, Present and Future." In *Physics and Contemporary Needs*, Vol. 4, edited by Riazuddin, 529–55. Berlin: Springer.

Forbes, Eric Gray. 1963. "A History of the Solar Redshift Problem." *Annals of Science* 17:129–64.

Forsyth, A. R. 1935. "Old Tripos Days at Cambridge." *Mathematical Gazette* 19:162–79.

Fowler, Alfred. 1915. "The Total Solar Eclipse, 1914 August 20–21." *Monthly Notices of the Royal Astronomical Society* 75:315–17.

Fowler, Alfred. 1919. "Meeting of the Royal Astronomical Society, Friday, June 13, 1919." *Observatory* 42:261–63.

Freeth, Tony. 2009. "Decoding an Ancient Computer." *Scientific American* (December): 76–83.

Freundlich, Erwin Finlay. 1913. "Über einen Versuch, die von A. Einstein vermutete Ablenkung des Lichtes in Gravitationsfeldern zu prüfen." *Astronomische Nachrichten* 193:369–72.

Freundlich, Erwin Finlay. 1923. "Hollaendisch-Deutsche Sonnenfinsternis-Expedition nach Christmas Island." *Astronomische Nachrichten* 218:13–16.

Freundlich, Erwin Finlay, von Klueber, Harald, and von Brunn, Albert. 1931. "Ergebnisse der Potsdamer Expedition zur Beobachtung der Sonnenfinsternis von 1929, Mai 9, in Takengon (Nordsumatra). 5. Mitteilung. Über die Ablenkung des Lichtes im Schwerefeld der Sonne." *Zeitschrift für Astrophysik* 3:171–98.

Freundlich, Erwin Finlay, von Klueber, Harald, and von Brunn, Albert. 1932. "Bemerkung zu Herrn Trümplers Kritik." *Zeitschrift für Astrophysik* 4:221–23.

Fuller, R. Buckminster. 1975. *Everything I Know*. Video recording. Buckminster Fuller Institute Archives. https://archive.org/details/buckminsterfuller. Transcripts available at https://www.bfi.org/about-fuller/resources/everything-i-know.

Galison, Peter. 1987. *How Experiments End*. Chicago: University of Chicago Press.

Galison, Peter. 2008. "The Assassin of Relativity." In *Einstein for the 21st Century: His Legacy in Science, Art, and Modern Culture*, edited by Peter L. Galison, Gerald Holton, and Silvan S. Schweber, 185–204. Princeton, NJ: Princeton University Press.

Glass, Ian S. 1997. *Victorian Telescope Makers: The Lives and Letters of Thomas and Howard Grubb*. Bristol, UK: Institute of Physics. See also *Telescopes and Other Instruments by Thomas and Howard Grubb*. http://www.saao.ac.za/~isg/g.html.

Glass, Ian S. 2006. *Revolutionaries of the Cosmos: The Astro-Physicists*. Oxford: Oxford University Press.

Glass, Ian S. 2010. "The Grubb Contribution to Telescope Technology." In *Astronomy and Its Instruments before and after Galileo*, edited by L. Pigatto and V. Zanini. Paris: International Astronomical Union. http://www.saao.ac.za/~isg/page1.html.

Harvey, G. M. 1979. "Gravitational Deflection of Light: A Re-examination of the Observations of the Solar Eclipse of 1919." *Observatory* 99:195–98.

Hawking, Stephen. 1988. *A Brief History of Time*. London: Bantam Press, 1988.

Haws, Duncan. 1998. *Merchant Fleets: Lamport & Holt and Booth*. Hereford, UK: TCL Publications.

Henroteau, F. 1927. "The Greenwich Observatory." *Journal of the Royal Astronomical Society of Canada* 21:289–91.

Hentschel, Klaus. 1993. "The Conversion of St. John: A Case Study on the Interplay of Theory and Experiment." *Science in Context* 1:137–94.

Hentschel, Klaus. 1997. *The Einstein Tower: An Intertexture of Dynamic Construction, Relativity Theory, and Astronomy*. Palo Alto, CA: Stanford University Press.

Hinks, Arthur R. 1917. "Geographical Conditions for the Observation of the Total Solar Eclipse, 1919 May 28–29." *Monthly Notices of the Royal Astronomical Society* 78:79–82.

Holberg, Jay B. 2010. "Sirius B and the Measurement of the Gravitational Redshift." *Journal for the History of Astronomy* 41:41–64.

Holmes, Virginia Iris. 2017. *Einstein's Pacifism and World War I*. Syracuse, NY: Syracuse University Press.

Illy, Jozsef. 2006. *Albert Meets America: How Journalists Treated Genius during Einstein's 1921 Travels*. Baltimore: Johns Hopkins University Press.

Jeffery, P. M., Burman, R. R., and Budge, J. R. 1989. "Wallal: The Total Solar Eclipse of 1922 September 21." In *Proceedings of the Fifth Marcel Grossmann Meeting on*

General Relativity, edited by D. G. Blair and M. J. Buckingham, 1343–50. Singapore: World Scientific.

Joly, Charles Jasper, Wilson, William Edward, Grubb, Howard, and Rambaut, Arthur Alcock. 1901. "The Total Solar Eclipse of 1900: Report of the Joint Committee Appointed by the Councils of the Royal Dublin Society and Royal Irish Academy." *Transactions of the Royal Irish Academy Section A: Mathematical, Astronomical, & Physical Science (1902–1904)* 32:271–98.

Jones, Dennis. 1991. "E. T. Cottingham, FRAS." *Antiquarian Horology and the Proceedings of the Antiquarian Horological Society* 19:593–617.

Jones, D.H.P. 1988. "The Greenwich and Oxford Astrographic Telescopes, 1958–1987." In *Mapping the Sky: Past Heritage and Future Directions: Proceedings of the 133rd Symposium of the International Astronomical Union, Held in Paris, France, 1–5 June 1987*, edited by Suzanne Debarbat, 33–37. Dordrecht, The Netherlands: Kluwer Academic.

Jones, H. Spencer. 1919a. "Discussion on the Theory of Relativity." *Monthly Notices of the Royal Astronomical Society* 80:96–117.

Jones, H. Spencer. 1919b. "Joint Eclipse Meeting of the Royal Society and the Royal Astronomical Society Friday, 1919, November 6." *Observatory* 42:389–98.

Jones, H. Spencer. 1919c. "Meeting of the Royal Astronomical Society Friday, 1919 July 11." *Observatory* 42:297–306.

Jones, H. Spencer. 1919d. "Meeting of the Royal Astronomical Society Friday, 1919 June 13." *Observatory* 42:261–72.

Jones, H. Spencer. 1919e. "Meeting of the Royal Astronomical Society Friday, 1919 November 14." *Observatory* 42:421–31.

Jones, H. Spencer. 1939. "Sir Frank Watson Dyson." *Observatory* 62:179–87.

Jones, H. Spencer, and Whittaker, E. T. 1939. "Arthur Stanley Eddington." *Monthly Notices of the Royal Astronomical Society* 105:68–79.

Kennefick, Daniel. 2007. *Travelling at the Speed of Thought: Einstein and the Quest for Gravitational Waves*. Princeton, NJ: Princeton University Press.

Kennefick, Daniel. 2012. "Not Only Because of Theory: Dyson, Eddington and the Competing Myths of the 1919 Eclipse Expedition." In *Einstein and the Changing World Views of Physics*, edited by Christoph Lehner, Juergen Renn, and Matthias Schemmel, 201–32. Basel, Switzerland: Birkhäuser.

Lalli, Roberto. 2012. "The Reception of Miller's Ether-Drift Experiments in the USA: The History of a Controversy in Relativity Revolution." *Annals of Science* 69:153–214.

Laney, C. D. 1995. *History of the South African Astronomical Observatory*. South African Astronomical Observatory. https://www.saao.ac.za/about/history/.

Lindemann, A. F., and Lindemann, F. A. 1916. "Daylight Photography of Stars as a Means of Testing the Equivalence Postulate in the Theory of Relativity." *Monthly Notices of the Royal Astronomical Society* 77:140–51.

Longair, Malcolm. 2015. "Bending Space-Time: A Commentary on Dyson, Eddington and Davidson (1920) 'A Determination of the Deflection of Light by the Sun's Gravitational Field.'" *Philosophical Transactions of the Royal Society A: Mathematical, Physical and Engineering Sciences* 373:20140287.

Maddison, Ron. 2011. *The First Fifty Years of the Keele Observatory.* https://www.keele
.ac.uk/media/keeleuniversity/microsites/keeleobservatory/downloads/KOP2.pdf.

Manville, G. E. 1971. *Two Fathers & Two Sons.* Newcastle upon Tyne, UK: Grubb
Parsons. https://archive.org/details/TwoFathersTwoSons.

Martins, Roberto de Andrade. 2011. "Searching for the Ether Leopold Courvoisier's
Attempts to Measure the Absolute Velocity of the Solar System." *DIO: The International Journal of Scientific History* 17:3–30.

Matzner, Richard. 1975. "Testing Relativity in the Desert: The Texas-Mauritania
Eclipse Expedition." *Physics Teacher* 13:215–22.

Maunder, E. Walter. 1900. *The Royal Observatory Greenwich: A Glance at Its History
and Work.* London: Religious Tract Society. Includes the *Annual Report of the
Astronomer Royal in 1900,* edited by Eric Hutton. http://atschool.eduweb.co
.uk/bookman/library/ROG/INDEX.HTM).

McCausland, Ian. 1999. "Anomalies in the History of Relativity." *Journal of Scientific
Exploration* 13:271–90.

McCrea, W. H. 1982. "F.J.M. Stratton, DSO, TD, FRS, 1881–1960." *Quarterly Journal of the Royal Astronomical Society* 23:358–62.

Mehra, Jagdish, and Rechtenberg, Helmut. 1987. *The Historical Development of Quantum Theory,* Vol. 5, *Erwin Schroedinger and the Rise of the Wave Mechanics.*
In part 2, *The Creation of Wave Mechanics: Early Response and Applications,
1925–26.* Berlin: Springer.

Melotte, P. J. 1940. "Dr. A.C.D. Crommelin." *Observatory* 63:11–13.

Melotte, P. J. 1953a. *The Annual Report of the Astronomer Royal in 1900.* In Maunder,
Royal Observatory Greenwich, chap. 5.

Melotte, P. J. 1953b. "Herbert Henry Furner." *Monthly Notices of the Royal Astronomical Society* 113:305–6.

Michelson, A. A., Pease, F. G., and Pearson, F. 1929. "Repetition of the Michelson-
Morley Experiment." *Nature* 123:88.

Michelson, Albert A., and Gale, Henry G. 1925. "The Effect of the Earth's Rotation
on the Velocity of Light." *Astrophysical Journal* 61:140–45.

Mikhailov, A. A. 1959. "The Deflection of Light by the Gravitational Field of the
Sun." *Monthly Notices of the Royal Astronomical Society* 119:593–608.

Miller, Dayton C. 1922. "Ether-Drift Experiments at Mount Wilson Solar Observatory." *Physical Review* 19:407–8.

Miller, Dayton C. 1925. "Experiments at Mount Wilson." *Science* 61:617–21.

Miller, Dayton C. 1933. "The Ether-Drift Experiment and the Determination of the
Absolute Motion of the Earth." *Reviews of Modern Physics* 5:203–42.

Mills, A. A. 1985. "Heliostats, Siderostats, and Coelostats: A Review of Practical
Instruments for Astronomical Applications." *Journal of the British Astronomical
Association* 95:89–99.

Morley, Edward W., and Miller, Dayton C. 1905. "Report of an Experiment to
Detect the FitzGerald-Lorentz Effect." *Philosophical Magazine* 9:680–85.

Morley, Edward W., and Miller, Dayton C. 1907. "Final Report on Ether Drift
Experiments." *Science* 25:525.

Mota, Elsa, Crawford, Paulo, and Simões, Ana. 2009. "Einstein in Portugal: Eddington's Expedition to Principe and the Reactions of Portuguese Astronomers (1917–25)." *British Journal for the History of Science* 42:245–73.

Murray, C. Andrew, and Wayman, P. A. 1989. "Relativistic Light Deflections." *Observatory* 109:189–91.

Nath, Biman. 2013. *The Story of Helium and the Birth of Astrophysics*. New York: Springer-Verlag.

Nicol, Stuart. 2001. *McQueen's Legacy Vol. 2: Ships of the Royal Mail Line*. Stroud, UK: Tempus.

Norton, John. 1993. "Einstein and Nordström: Some Lesser Known Thought Experiments in Gravitation." In *The Attraction of Gravitation: New Studies in History of General Relativity*, edited by J. Earman, M. Janssen, and J. Norton, 3–29. Boston: Birkhauser.

Oort, Jan H. 1927. "Observational Evidence Confirming Lindblad's Hypothesis of a Rotation of the Galactic System." *Bulletin of the Astronomical Institutes of the Netherlands* 3:275–82.

Pais, Abraham. 1982. *Subtle Is the Lord: The Science and the Life of Albert Einstein*. Oxford: Oxford University Press.

Pais, Abraham. 1994. *Einstein Lived Here*. Oxford: Oxford University Press.

Pang, Alex Soojung-Kim. 1993. "The Social Event of the Season: Solar Eclipse Expeditions and Victorian Culture." *Isis* 84:252–77.

Pang, Alex Soojung-Kim. 1996. "Gender, Culture, and Astrophysical Fieldwork: Elizabeth Campbell and the Lick Observatory-Crocker Eclipse Expeditions." *Osiris* 11:17–43.

Pang, Alex Soojung-Kim. 2002. *Empire and the Sun: Victorian Solar Eclipse Expeditions*. Stanford, CA: Stanford University Press.

Partridge, Colin. 2006. "The Life and Times of a History of Science Paper: An Analysis of the Reach, Influence and Content of 'Relativity and Eclipses: The British Eclipse Expeditions of 1919 and their Predecessors' by John Earman and Clark Glymour." Master's thesis, Imperial College, London.

Pearson, John C., Orchiston, Wayne, and Malville, J. McKim. 2011. "Some Highlights of the Lick Observatory Solar Eclipse Expeditions." In *Highlighting the History of Astronomy in the Asia-Pacific Region*, edited by Wayne Orchiston, Tsuko Nakamura, and Richard G. Strom, 243–337. Berlin: Springer.

Peebles, Phillip James Edwin. 2017. "Robert Dicke and the Naissance of Experimental Gravity Physics, 1957–1967." *European Physical Journal H* 42:177–259.

Perrine, Charles Dillon. 1923. "Contribution to the History of Attempts to Test the Theory of Relativity by Means of Astronomical Observations." *Astronomische Nachrichten* 219: 281–84.

Perryman, Michael. 2010. *The Making of History's Greatest Star Map*. Heidelberg: Springer.

Perryman, Michael. 2012. "The History of Astrometry." *European Physical Journal H* 37:745–92.

Popper, Karl. 1963. "Science as Falsification." In *Conjectures and Refutations: The Growth of Scientific Knowledge*. New York: Basic Books. See also a second edition by Routledge, 2002.

Renn, Jürgen, and Sauer, Tilman. 2000. *Eclipses of the Stars—Mandl, Einstein, and the Early History of Gravitational Lensing*. Berlin: Max Planck Institute for the History of Science.

RGO (Royal Greenwich Observatory) Archive. Frank Dyson Papers. Cambridge: Cambridge University Library. https://janus.lib.cam.ac.uk/db/node.xsp?id=EAD%2FGBR%2F0180%2FRGO%208.

Rosenthal-Schneider, Ilse. 1980. *Reality and Scientific Truth*. Detroit: Wayne State University Press.

Russell, Henry Norris. 1945a. "Arthur Stanley Eddington, 1882–1944." *Astrophysical Journal* 101:133.

Russell, Henry Norris. 1945b. "Sir Arthur Stanley Eddington, O.M., F.R.S." *Observatory* 66:1–12.

Sciama, Dennis William. 1969. *The Physical Foundations of General Relativity*. New York: Doubleday.

Shankland, R. S, McCuskey, S. W., Leone, F. C., and Kuerti, G. 1955. "New Analysis of the Interferometric Observations of Dayton C. Miller." *Reviews of Modern Physics* 27:167–78.

Shapin, Steven. 1989. "The Invisible Technician." *American Scientist* 77:554–563.

Shears, Jeremy. 2014. "The British Astronomical Association and the Great War of 1914–1918." *Journal of the British Astronomical Association*, August 2014, 187–197. https://arxiv.org/abs/1309.5205.

Silberstein, Ludwik. 1920. "The Eclipse Results and the Contraction of Photographic Images." *Monthly Notices of the Royal Astronomical Society* 80:630–31.

Sponsel, Alistair. 2002. "Constructing a 'Revolution in Science': The Campaign to Promote a Favourable Reception for the 1919 Solar Eclipse Experiments." *British Journal for the History of Science* 35:439–67.

Stachel, John. 1986. "Eddington and Einstein." *Prism of Science: Boston Studies in the Philosophy of Science* 95:225–50.

Stanley, Mathew. 2003. "An Expedition to Heal the Wounds of War: The 1919 Eclipse and Eddington and Quaker Adventurer." *ISIS* 94:57–89.

Stanley, Matthew George. 2007. *Practical Mystic: Religion, Science and A. S. Eddington*. Chicago: University of Chicago Press.

Stevenson, Toner M. 2014. "Making Visible the Women Who Measured Stars in Australia: The Measurers and Computers Employed for the Astrographic Catalogue." *Publications of the Astronomical Society of Australia* 31:e018.

Strömberg, Gustaf. 1926. "Miller's Ether Drift Experiment and Stellar Motions." *Nature* 117:482–83.

Swenson Jr., Loyd S. 1972. *The Ethereal Aether: A History of the Michelson-Morley-Miller Aether Drift Experiments, 1880–1930*. Austin: University of Texas Press.

Texas Mauretanian Eclipse Team. 1976. "Gravitational Deflection of Light: Solar Eclipse of 30 June 1973 I: Description of Procedures and Final Results." *Astronomical Journal* 81:452–54.

Treschman, Keith John. 2014. "Early Astronomical Test of General Relativity: The Gravitational Deflection of Light." *Asian Journal of Physics* 23:145–70.

Trumpler, Robert J. 1932. "The Deflection of Light in the Sun's Gravitational Field." *Astronomical Society of the Pacific* 44:167–73.

Turner, Herbert Hall. 1912. *The Great Star Map: Being a Brief General Account of the International Project Known as the Astrographic Chart*. London: John Murray.

Turner, Herbert Hall. 1926. "The Reverend Aloysius James Cortie SJ." *Monthly Notices of the Royal Astronomical Society* 86:175–77.

van Biesbroeck, G. 1950. "The Einstein Shift at the Eclipse of May 20, 1947, in Brazil." *Astronomical Journal* 55:49–53.

van Biesbroeck, G. 1953. "The Relativity Shift at the 1952 February 25 Eclipse of the Sun." *Astronomical Journal* 58:87–88.

van Delft, Dirk. 2006. "Albert Einstein in Leiden." *Physics Today* 59:57–62.

Vinter Hansen, Julie. 1955. "The International Astronomical News Service." Special supplement no. 3, *Journal of Atmospheric and Terrestrial Physics*. In Vol. 1, *Vistas in Astronomy*, edited by Arthur Beer, 16–21. Oxford: Pergamon.

von Klueber, H. 1960. "The Determination of Einstein's Light Deflection in the Gravitational Field of the Sun." *Vistas in Astronomy* 3:47–77.

von Klueber, H. 1965. "Erwin Finlay-Freundlich." *Quarterly Journal of the Royal Astronomical Society* 6:82–84.

von Soldner, Johann Georg. 1801. "Ueber die Ablenkung eines Lichtstrals von seiner geradlinigen Bewegung, durch die Attraktion eines Weltkörpers, an welchem er nahe vorbei geht." *Berliner Astronomisches Jahrbuch*. Berlin: 161–72. https://en .wikisource.org/wiki/Translation:On_the_Deflection_of_a_Light_Ray_from _its_Rectilinear_Motion.

Waller, John. 2002. *Fabulous Science: Fact and Fiction in the History of Scientific Discovery*. Oxford: Oxford University Press. Published in the United States as *Einstein's Luck: The Truth behind Some of the Greatest Scientific Discoveries*. New York: Oxford University Press.

Warwick, Andrew. 2014. *Masters of Theory: Cambridge and the Rise of Mathematical Physics*. Chicago: University of Chicago Press.

Wayman, Patrick A. 1988. "The Grubb Astrographic Telescopes, 1887–1896." In *Mapping the Sky: Past Heritage and Future Directions: Proceedings of the 133rd Symposium of the International Astronomical Union, Held in Paris, France, 1–5 June 1987*, edited by Suzanne Debarbat, 139. Dordrecht, Netherlands: Kluwer Academic.

Weinstein, Galina. 2012. *The Einstein-Nordström Theory*. https://arXiv.org:1205 .5966.

Will, Clifford. 1993. *Was Einstein Right?* New York: Basic Books.

Wilson, Margaret. 1951. *The Ninth Astronomer Royal: The Life of Frank Watson Dyson*. Cambridge: Heffer and Sons.

Woolley, R. 1971. "Charles Rundle Davidson." *Biographical Memoirs of Fellows of the Royal Society* 17:193–94.

Ziegler, Charles A. 1989. "Technology and the Process of Scientific Discovery: The Case of Cosmic Rays." *Technology and Culture* 30:939–63.

INDEX

Page numbers in *italics* refer to figures and tables.

A NOTE ON THE TYPE

This book has been composed in Adobe Text and Gotham. Adobe Text, designed by Robert Slimbach for Adobe, bridges the gap between fifteenth- and sixteenth-century calligraphic and eighteenth-century Modern styles. Gotham, inspired by New York street signs, was designed by Tobias Frere-Jones for Hoefler & Co.